Ecological Restoration and the U.S. Nature and Environmental Writing Tradition

Laura Smith

Ecological Restoration and the U.S. Nature and Environmental Writing Tradition

A Rewilding of American Letters

palgrave
macmillan

Laura Smith
Department of Geography
University of Exeter
Exeter, UK

ISBN 978-3-030-86147-6 ISBN 978-3-030-86148-3 (eBook)
https://doi.org/10.1007/978-3-030-86148-3

Cover illustration: Maram_shutterstock.com

This Palgrave Macmillan imprint is published by the registered company Springer Nature Switzerland AG
The registered company address is: Gewerbestrasse 11, 6330 Cham, Switzerland

For my family

Preface

On November 8, 2016, I watched the through-the-night BBC reporting as Donald J. Trump defeated Hillary Rodham Clinton to emerge as the President-elect of the United States. Even now, I'm not sure why I felt compelled to watch election night in its entirety. Curiosity, absolutely. Hope. Unease, too. But I think mostly it was because I wanted to know—*needed* to know—as early as possible, what the election result meant for President Barack Obama's conservation legacy. For the fight for a proposed Bears Ears National Monument. For America's national monuments and parks system. For public lands. For climate change. For environmental policy and legislation. For a book proposal I'd nearly finished, and was days away from submitting.

By the morning of November 9, 2016, that book proposal was headed for the trash. Already, it no longer made sense. It couldn't—wouldn't—work. Five months and a major overhaul and rewrite later, I had a proposal again—and, by the end of the year, a book contract.

One thing's for sure—the Trump administration's repeated environmental rollbacks, and assaults on public lands, have certainly made writing this book an unforgettable (and frequently infuriating) experience. As this book project entered its final year, it had to further navigate the uncertainties and anxieties of the global COVID-19 pandemic. I had intended to spend the summer of 2020 in the United States for research. But international travel restrictions canceled any research plans, and the research quickly moved online.

Then, suddenly, it was November 3, 2020.

I again followed the televised coverage of the 2020 presidential election throughout the night. And for the same reason that I had done so four years earlier. To learn the fate of America's public lands. Of the dramatically downsized Bears Ears and Grand Staircase-Escalante national monuments. But it felt much more urgent second time around. With so much more at stake. I watched in the hope that maybe, *maybe*, I could end my book with the uplift and optimism for public lands protections I'd wanted to open with. Four days later, and Joseph R. Biden Jr. was declared President-elect of the United States, with Kamala Harris as Vice President-elect. My book had its hopeful ending.

The writing of this book has been bookended by two U.S. presidential elections. It is the product of the intervening four years. But the reading and thinking—and some of the conversations—reach back to mid-2007, in the midst of my doctoral research (funded with an Economic and Social Research Council Postgraduate Studentship (+3), award number PTA-030-2005-00680, 2005–2008).

Exeter, UK Laura Smith

Acknowledgments

The earliest idea for this book emerged over the course of the summer I spent with the Walden Woods Project and Thoreau Institute at Walden Woods in Lincoln, Massachusetts during my Ph.D. I would like to thank the staff at the WWP for hosting my doctoral research placement in 2007, and for continuing to support my research endeavors in the years since. I had the good fortune to meet with many others during my placement, and those conversations about (and hikes through) Walden Woods proved invaluable—my thanks to the Walden Pond State Reservation, the Walden Pond Board of Directors, the Massachusetts Department for Conservation and Recreation, the Thoreau Society, the Concord Free Public Library, the Massachusetts Audubon Society, Sasaki Associates, Inc., Friends of Thoreau Country, the Estabrook Woods Alliance, RESTORE The North Woods, as well as the Caddo Lake Institute, and the Sand County Foundation. Other research in the years immediately following my Ph.D. introduced me to the work of the Glen Canyon Institute and Restore Hetch Hetchy. For this book project a decade later, my thanks again go to the Walden Woods Project, RESTORE The North Woods, the Glen Canyon Institute, Restore Hetch Hetchy, and the Sand

County Foundation, together with Yosemite National Park, the Aldo Leopold Foundation, the University of Wisconsin-Madison Arboretum, Friends of the Everglades, the State Archives of Florida, Grand Staircase Escalante Partners, and Torrey House Press. A huge thanks also to the environmental writers and activists who shared memories, stories, and experiences of conservation campaigns and lobbying. Thank you all for your time and patience, for your enthusiasm for my research, and for meeting online and happily accommodating sometimes considerable time zone differences. Thank you for the awesome book post, too.

Colleagues and friends past and present in the Cultural and Historical Geographies Research Group at the University of Exeter have provided much support and good humor over the past three-and-a-half years I've been working on this book. The opportunity to share early (and outrageously long) chapter drafts at our research retreats twice a year made the process of writing this book much less daunting—thank you. I promise next time to stick to the word count. Maybe. Our daily 'Shut Up & Do' online space begun in lockdown has been motivating and cathartic, too. To Ian Cook, Matt Finn, Cordy Freeman, Jen Lea, Fran Ryfield, Phil Nicholson, Catherine Cartwright, Daisy Curtis, Harriet Earle-Brown, Gemma Lucas, and many others: thank you again. My thanks also to other Exeter Geography colleagues, especially Henry Buller and Tom Roland, for frequent office chats and indulgences on my book's progress. Any persisting errors and omissions are mine alone.

This book is for my family—my parents, Sheila and Martin, my sister, Emma, my brother-in-law, Aaron, and my niece, Iris, who arrived just in time for this last (and, I think, best) mention. Thank you for being part of this adventure. And for memorizing author names, places, book titles, and other trivia along the way.

Exeter, UK
March 2021

Laura Smith

Contents

Abbreviations

ALF	Aldo Leopold Foundation
AOU	American Ornithologists' Union
BEITC	Bears Ears Inter-Tribal Coalition
BLM	U.S. Bureau of Land Management
CCC	Civilian Conservation Corps
C&SF	Central and Southern Florida Project for Flood Control and Other Purposes
CCHS	Concord-Carlisle Regional High School
CERP	Comprehensive Everglades Restoration Plan
cfs	cubic feet per second
CHC	Concord Historical Commission
CPW	Concord Public Works
CR	Conservation Restriction
CRC	Colorado River Compact
CRSP	Colorado River Storage Project
DCR	Massachusetts Department of Conservation and Recreation
DPR	California Department of Parks and Recreation
DWR	California Department of Water Resources
EAA	Everglades Agricultural Area
EDD	Everglades Drainage District

EES	Ecological Extension Service (Mass Audubon)
EF!	Earth First!
EPA	Environmental Protection Agency
EPI	Elliotsville Plantation Inc.
EWA	Estabrook Woods Alliance
FFWC	Florida Federation of Women's Clubs
FMF	Fill Mead First
FOE	Friends of the Everglades
FOTC	Friends of Thoreau Country
GCI	Glen Canyon Institute
GSEP	Grand Staircase Escalante Partners
HCNR	House Committee on Natural Resources
HHWP	Hetch Hetchy Water and Power
IBA	Important Bird Area
ICF	International Crane Foundation
IIF	Internal Improvement Fund of the State of Florida
LMR	Leopold Memorial Reserve
maf	million acre-feet
MWNP	Maine Woods National Park and Preserve
NPS	National Park Service
OSRP	Open Space and Recreation Plan
PLI	Utah Public Lands Initiative
RESTORE	RESTORE The North Woods
RHH	Restore Hetch Hetchy
SAAMI	Sporting Arms and Ammunition Manufacturers' Institute
SCF	Sand County Foundation
SER	Society for Ecological Restoration
SFPUC	San Francisco Public Utilities Commission
SFWMD	South Florida Water Management District
SUWA	Southern Utah Wilderness Alliance
TAS	Tropical Audubon Society
TCCA	Thoreau Country Conservation Alliance
THP	Torrey House Press
UDB	Utah Diné Bikéyah
USACE	U.S. Army Corps of Engineers
USBR	U.S. Bureau of Reclamation
USDA	U.S. Department of Agriculture
USFS	U.S. Forest Service
USFWS	U.S. Fish and Wildlife Service

USGS	U.S. Geological Survey
UWMA	University of Wisconsin-Madison Arboretum
WCA	Water Conservation Area
WFW	Walden Forever Wild
WPBOD	Walden Pond Board of Directors
WPSR	Walden Pond State Reservation
WWP	Walden Woods Project and Thoreau Institute at Walden Woods
YNP	Yosemite National Park

List of Figures

1

A Storying—Restorying—Restoring of the Land: Rethinking Ecological Restoration Through Literature

1.1 A Brief History of Ecological Restoration

Monday, May 15, 2006. Concord, Massachusetts. The dedication ceremony of Thoreau's Path on Brister's Hill.

I had been invited by the Walden Woods Project (WWP) to attend the official opening of a one-mile interpretive loop trail through a northern part of Walden Woods, honoring the philosophy and writings of nineteenth-century Transcendentalist and Concord native, Henry David Thoreau. Thoreau's Path was also the finale to a multi-year ecological restoration program on Brister's Hill, that restored a forest stand denuded and abandoned by sand and gravel quarrying in the 1960s. A lot had happened in the year prior to that Monday morning. I had completed a Masters degree and enrolled on a Ph.D. program at Cardiff University. And I had gone from not knowing who Thoreau was, to a recommendation to read Thoreau's (2004) *Walden*, to finding walden.org, the website of the Walden Woods Project and the Thoreau Institute at Walden Woods, to securing the Walden Woods Project as a case study in my doctoral research on the changing nature(s) of ecological restoration.

L. Smith, *Ecological Restoration and the U.S. Nature and Environmental Writing Tradition*, https://doi.org/10.1007/978-3-030-86148-3_1

Which is how, less than eight months into my Ph.D., I found myself in the U.S. for the first time. Sat in the audience in a large marquee in the woods on Brister's Hill, rain gently bouncing on the roof, a couple hundred feet from Walden Pond, I listened to a humbling suite of speakers celebrate Thoreau and his environmental and social justice legacy. Kathi Anderson, the executive director of the Walden Woods Project, delivered the opening remarks. She was followed by Massachusetts Congressman Marty Meehan, Don Henley, the founder and president of the Walden Woods Project, Harvard University biologist Edward O. Wilson, Arun Gandhi, grandson of Mohandas K. Gandhi and founder and president of the M.K. Gandhi Institute for Nonviolence, civil rights leader and Georgia Congressman John Lewis, Virginia McIntyre, chair of the Concord Board of Selectmen, and Stephen Burrington, commissioner of the Massachusetts Department of Conservation and Recreation. The ceremony was brought to a close by Massachusetts Senator Edward M. Kennedy.

Other special guests in the audience included descendants of some of America's foremost environmental writers—the great-great-granddaughter of Ralph Waldo Emerson, Margaret Emerson Bancroft, the great-grandson of John Muir, Michael Muir, the grandson of Chief Luther Standing Bear, Oglala/Sicangu Lakota Calvin Standing Bear, and the adopted son and grandnephew of Rachel Carson, Roger Christie. The bibliophile in me couldn't quite believe it. Still doesn't.

After the tree plantings and speeches, I set out in the drizzle to explore Thoreau's Path on Brister's Hill alone, my copy of *Walden* in my daypack. There was something about being in Walden Woods that afternoon, and reading *Walden*, sheltering under the pines, that was inspiring, intimate. And the enthusiasm earlier that morning of the staff of the Walden Woods Project, and the many invited speakers and guests, was infectious. As introductions to new research go, that day was pretty spectacular. Thoreau and Walden Woods have been a central part of my scholarship and research ever since. For that week in May, I split my time between exploring Walden Pond, returning to Brister's Hill, and visiting the Walden Woods Project and Thoreau Institute campus ahead of a three-month research placement the following summer.

The summer of 2007 was one of walking, reading, and talking. That summer was also the perfect opportunity to continue my education in the American environmental writers who came after Thoreau, that I'd first been fleetingly introduced to the previous spring. When not exploring Walden's forest trails, I could most likely be found in one of four places—the Henley Library at the Thoreau Institute, the Concord Free Public Library, the Thoreau Society Shop at Walden Pond, or the Concord Bookshop. And with that, the seed of an idea for this book took root, although I neglected it for more than a decade, unsure of how to nurture it.

Ecological Restoration and the U.S. Nature and Environmental Writing Tradition: A Rewilding of American Letters presents a study of the intersections, collisions, and contingencies between American environmental literature and ecological restoration policy and practice. It charts the history and evolution of an ecological restoration sensibility in American environmental writing—it describes how environmental writing from across the nineteenth and twentieth centuries has come to take hold as a practice and ideology in mid-twentieth-century and early-twenty-first-century ecological restoration programs. It is an analysis of environmental writing and the canon's foundational legacy to restoration ecology. It is a search for ecological restoration policy in environmental writing. And it begins with Thoreau in Walden.

The remainder of this chapter briefly charts the genesis and evolution of ecological restoration and restoration ecology in the U.S., and introduces the principles of cultural landscape restoration. To explore the cross-over of an ecological restoration rhetoric with environmental literature—framed throughout the book as a 'restor(y)ing' of the land, a composite of storying and restoring—the chapter next introduces the philosophy and writings of five authors whose legacies have inspired ecological restoration programs and campaigns, and who populate in turn the central chapters of this book. The featured authors are Henry David Thoreau at Walden Pond, John Muir in the Hetch Hetchy Valley, Aldo Leopold beside the Wisconsin River, Marjory Stoneman Douglas in the Everglades, and Edward Abbey in Glen Canyon. While each author presents a strong message and doctrine on ecological restoration, when taken together in conversation, these five authors not only reveal the evolution of an ecological restoration sensibility in American nature and environmental writing, but also reveal a recasting and rethinking of environmental activism through literature. The close of the chapter offers an outline of the book's structure and format.

1.1.1 The Political Theater of America's Public Lands During the Trump Administration

Before this chapter continues, it is important to introduce the extraordinary political context that envelops this book. It is a political and legislative storying of public lands composed and endorsed over the four years of the Trump administration. Where climate science—and science more generally—has been disputed and ignored. Where environmental policy has been dismantled or reversed. Where national monument designations have been rescinded. Where vast tracts of public lands have been opened up to oil and gas extraction. The Trump administration's environmental rollbacks have not directly affected the restoration programs and campaigns central to this book, but echoes of these changes feature in some of the wider restoration stories included here.

The U.S. has witnessed a number of anti-environment presidents in its almost 250-year history, but conservationists and restorationists have had to confront an unprecedented assault on the environment from the Trump administration. Environmental literature and literary environmental activism have gained a new urgency as they both navigate a conservation battleground scarred—and redrawn—by the Trump administration's jettisoning of federal environmental protections, regulation, and legislation. And it is a conservation fight that is not yet over. This book joins that fight.

1.1.2 On Cultural Landscape Restoration, and a Literary Landscape Coda

This book coincides with an interesting cultural and political juncture for restorationists. In 2016, Edward O. Wilson's *Half-Earth: Our Planet's Fight for Life* was published, a powerful summons to protect and restore up to 50 per cent of the planet's land and oceans in the coming decades (Wilson 2016). But more significant, and further enriching the political visibility of an ecological restoration rhetoric, was the announcement by the United Nations General Assembly in New York on March 1, 2019, that 2021–2030 would be the U.N. Decade on Ecosystem Restoration (see UNEP 2019). 2019 also witnessed declarations by governments and institutions internationally of an environment and climate emergency. Extinction Rebellion marches were organized in cities around the world. Swedish climate activist Greta Thunberg continued her 'Skolstrejk för klimatet' (trans: School Strike for Climate). And April 22, 2020, marked the 50th anniversary of Earth Day.

Throughout this book, I examine the uniting of the American nature and environmental writing tradition and ecological restoration programs through ideas of cultural landscape restoration. George Perkins Marsh's *Man and Nature: Or, Physical Geography as Modified by Human Action*, originally published in 1864, is one of the earliest texts to introduce the ideas underpinning ecological restoration, such as advocating the 'restoration of disturbed harmonies' of nature (Marsh 2003: 3, 12, 35–36). The restoration of the Curtis Prairie at the University of Wisconsin-Madison Arboretum in the 1930s (featured in Chapter 4) is frequently identified as the world's first formal example of ecological restoration (Jordan 2003; Jordan and Lubick 2011). It was not until the establishment of the Society for Ecological Restoration (SER) in 1988 that ecological restoration emerged as a mainstream ecological term. SER defines ecological restoration as 'the process of assisting the recovery of an ecosystem that has been degraded, damaged, or destroyed' (Gann et al. 2019: S7). Restoration ecology is the science that underwrites and supports the practice of ecological restoration. As Zev Naveh (1998, also e.g. 1994, 2005, 2007) suggests, the goal of *cultural landscape* restoration is to restore cultural and historical values in/of landscapes. Cultural histories are entangled with, and retold and restored alongside, biological and ecological histories (for further discussion on the conservation and restoration of cultural landscapes, see also Higgs 2003; Jordan 2003; Agnoletti 2006; Jordan and Lubick 2011; Taylor et al. 2014).

But this book also collapses the taxonomy of rewilding into cultural landscape restoration, as a means of interrogating the often-central place of wilderness narratives in the environmental writings of Thoreau, Muir, Leopold, Douglas, and Abbey (on the history of the wilderness idea in America, see Oelschlaeger 1991; Buell 1995, 2005; Cronon 1995; Callicott and Nelson 1998; Lewis 2007; Nash 2014). As a concept, rewilding is hugely antagonistic (see the exchanges between Jørgensen 2015; Prior and Ward 2016; Cloyd 2016, also Monbiot 2014, Jepson and Blythe 2020), but I include it here precisely for its 'wild' etymology, and for the various ways 'wild' goes to, through, and out the other side of wildness and wilderness to disrupt and recast ideas of cultural landscape restoration.

Naveh's (1995) idea of cultural landscapes, of landscapes as physical, biological, ecological, sociological, anthropological, psychological, philosophical, and historical imbroglios, is instrumental in shaping the restoration rhetoric underpinning this book. This cultural landscape scaffolding is effectively captured in—and amplified by—wider conservation responses to the Anthropocene epoch (see Lorimer 2015; Minteer and Pyne 2015; Purdy 2015). In this and subsequent chapters, I argue that the literary landscapes scripted by environmental writers are also cultural landscapes, replete with autobiography, history, story, myth, and encounter, and can reveal emotive, affective, moral—and also speculative—ecologies (cf. Zapf 2008, 2016 on cultural ecology, also Weik von Mossner 2017 on affective ecologies). I have written elsewhere (Smith 2014) that cultural landscape restoration is also a proxy for ecological (and moral) redemption (cf. Rolston 1994; Gobster and Hull 2000; Jordan 2003; Clewell and Aronson 2006; France 2008; Basl 2010; Rohwer and Marris 2016). The evocative capacity of environmental literature to convey emotional and affective landscapes taps into a moral-emotional critique of restoration. I am interested in the moral, ethical, theological, and emotive language(s) (Van Wieren 2013) that are used to articulate ecological restoration issues and value.

Where environmental literature finds political purchase in cultural landscape restoration programs is in the composition of restoration reference models (Egan and Howell 2001; Clewell and Aronson 2013; Gann et al. 2019). The reference model is an assemblage of biological and ecological evidence (e.g. dendrochronology, palynology, geomorphology, hydrology, observational field evidence) and cultural evidence (e.g. archeology, ethnobiology, land surveys, maps, photographs, oral histories) that describes past environments and environmental conditions (after Egan and Howell 2001). Efforts to recast the reference model away from a purely scientific-technical guide for restorationists, narrowly focused on historic fidelity to a static, pre-disturbance condition, are realized in, for example, 'futuristic' (Choi 2004), 'dynamic' (Suding and Gross 2006; Hiers et al. 2012), or 'holistic' (Clewell and Aronson 2013) reference models, as well as 'novel' ecosystems (Hobbs et al. 2006; Jackson and Hobbs 2009; Higgs 2012; Hobbs et al. 2013; Miller and Bestelmeyer 2016). In this book, I argue that environmental writing should be added to the repertoire of cultural evidence that might inform and mold restoration reference models—as a cultural artifact reporting local, regional environmental history. I want to magnify the place of nature myths (Hall 2005)—those collective and shared beliefs, assumptions, and stories about the nature of nature—in dialogue and discussion on what to restore, in deciding, *which nature?* (after Hull and Robertson 2000). Environmental writing is recast—and politicized—as an ecological inventory, an almanac, a field guide (on restoration and environmental history, see Higgs 2003; Jordan 2003; Hall 2005, 2010; Jackson and Hobbs 2009; Jordan and Lubick 2011). *What would Thoreau/Muir/Leopold/Douglas/Abbey* (*delete as appropriate) do?* becomes a legitimizing—or, at least, a provocative—question. Two restoration goals developed by James Aronson et al. (1993) have proved instructive to this study—of 'restoration *sensu stricto*' (in a narrow or strict sense) and 'restoration *sensu lato*' (in a broad sense). The former adheres to historic fidelity, the latter embraces a more dynamic or holistic trajectory. As this book will demonstrate, in some cases, the literary landscape is used to endorse restoration *sensu stricto*, in others, to signal restoration *sensu lato*.

1.2 A Provocation on 'Restor(y)ing' the Land

Works of nature and environmental writing[1] are celebrated for their contributions to the conservation movement and environmentalism in the U.S. Yet this same genre is often overlooked or neglected in conversations on ecological restoration (although Barilla 2004, 2007 is an exception to this). But nature and environmental writing reveal a close scrutiny of the land and its environmental history that has much to offer a restoration ecology and especially cultural landscape restoration. There is the promise of the literary ecological imaginary to contribute in powerful ways to the understanding of landscapes that have been disrupted in various ways by human activity. A focus on environmental writing offers a response to Ben A. Minteer and Stephen J. Pyne's (2013) call for a reappraisal of restoration within and beyond American environmentalism's historic preservation/conservation schism (as explored in Chapter 3).

It begins with the idea of the literary landscape—and 'restoring storied pasts' (as Hall 2005). And that literary landscapes are cultural landscapes (on the making of American literary landscapes, see Turner 1989, also Fishkin 2015; Gatta 2018). That the landscape *on the page* matters—whether conceptually, ideologically, aesthetically—as much as the landscape *in place* in provoking emotional participation, attachment,

[1] Although this book uses both 'nature-' and 'environmental-' prefixes to describe the canon, it does so in order to illustrate its mutability and the recent shift in taxonomy. There has been a pivot from 'nature writing' toward 'environmental writing' or 'environmental literature' in the U.S., especially since the 1980s and 1990s (see also Buell 1995, 2001, 2005; Glotfelty and Fromm 1996; McKibben 2008; Prentiss and Wilkins 2017). In the U.K., this shift is made manifest in 'new nature writing' (see especially Cowley 2008; Smith 2017). If 'nature writing' discloses an intellectual curiosity in natural history, and the scientific study of natural phenomena, then 'environmental writing' looks beyond this, to also consider the emotional, moral, ethical, philosophical—and political—implications of society's (mis)uses of nature and landscapes. In name, 'environmental writing' no longer separates nonfiction nature writing from other important literary genres and categories of environmental communication (while the 'new nature writing' motif breaks with, and restarts, the history and trajectory of the genre). But perhaps more importantly, *environmental* writing offers a platform for understanding the intersection of society and nature, and how society affects nature. In doing so, it recognizes and amplifies feminist, BAME, Indigenous, and queer environmental voices. It recognizes gender, age, ethnicity, and class. It is finding different ways to engage with the environment and climate emergency. ~~Nature~~ environmental writing should be accepted as the default hereafter.

action (cf. Anderson 2014 on a page—place diptych). What is important is *storying place* and *placing stories* (Cronon 1992 celebrates 'a place for stories' in geographical inquiry, also Lopez 1989 esp. 'Landscape and Narrative' essay, Higgs 2003; Davis 2011; Cameron 2012; Daniels and Lorimer 2014; Weston 2016). The literary imaginary makes and remakes landscape—and landscape futures. Hubert Zapf's (e.g. 2008, 2016) framing of literature as a form of cultural ecology has particular resonance here. He asserts that literature fulfills a regenerative role within culture—that literary language has special properties that can effect ecocultural change. As Mark Tredinnick (2005) powerfully argues, the land inserts itself into the voices of environmental writers. The restoration of literary landscapes offers a provocation on what matters, on what is important, in cultural landscape restoration.

In order to unpack and interrogate the collisions between ecological restoration, story, and imagination, this book employs a storying—restorying—restoring triptych. I argue that through examining various *storyings* of landscapes by environmental writers, restorationists might find a roadmap to *restoring* (and rewilding) landscapes. Often hereafter, this is abridged, condensed to a restor(y)ing trope. This *restorying* motif plays with the entanglements of writing and rewriting stories about the land, and how restoration programs can reconceptualize, reorganize, reinterpret, recast, recycle, and retell these storyings of the land. This builds on the work of Terre Satterfield and Scott Slovic (2004), who, through interviews with nature writers, draw together ideas about how storytelling and narrative engages with moral, emotional, and aesthetic expressions of environmental value.

1.3 Introducing the Literary Landscapes of Thoreau, Muir, Leopold, Douglas, and Abbey

This book looks at five authors of seminal works of nature and environmental writing who also inspired ecological restoration programs and campaigns. Henry David Thoreau. John Muir. Aldo Leopold. Marjory Stoneman Douglas. Edward Abbey. A number of factors conspired to assemble this quintet of literary naturalists:

- *Historical longitude* / These authors were active from the mid-nineteenth century through to the close of the twentieth century, allowing for the tracing of evolving ideas on nature, wildness, wilderness, conservation, preservation, as well as restoration. From Thoreau's birth in 1817 to Douglas's death in 1998, there is no single year that one or more of these authors did not witness.
- *Geography* / Thoreau's environmental philosophy is most acutely molded by the forests and rivers of eastern Massachusetts and north-central Maine, Muir by the forests of Wisconsin and Canada, the forests and mountains of California, and the glaciers of Alaska, Leopold by the forests of Arizona and New Mexico, and the prairie farmlands of Wisconsin, Douglas by the swamps of the Florida Everglades, and Abbey by the forests of Pennsylvania, and the deserts of the American West, especially California, Utah, Arizona, and New Mexico. For restorationists, this multiplicity of landscapes invites reflection on the promise held in a literary restoration ethic.
- *An embryonic ecological restoration sensibility* / Even before restoration ecology became an established branch of applied ecology, many of these authors were adopting language, principles, and practices that would later come to be formalized under restoration science. I do not want to impose twenty-first-century thinking on restoration ecology, ecological restoration, and rewilding on the works of these authors, but rather acknowledge and amplify in these works some early, informal manifestations of restoration.
- *Environmental organizations* / Thoreau, Muir, Leopold, and Abbey have all inspired in different ways the creation of environmental nonprofits to defend the landscapes popularized in their writings—the Walden Woods Project, Restore Hetch Hetchy, the Aldo Leopold Foundation, the Glen Canyon Institute. Douglas set up her own environmental nonprofit to do so—Friends of the Everglades. Ecological restoration is a central theme across all five organizations (Muir also founded the Sierra Club in 1892, and Leopold was a cofounder of the Wilderness Society in 1935). Their literary legacies continue to be politicized in contemporary conservation and restoration battles.
- *Scalar politics* / Each of the restoration programs and campaigns profiled throughout this book operate at different (geographic,

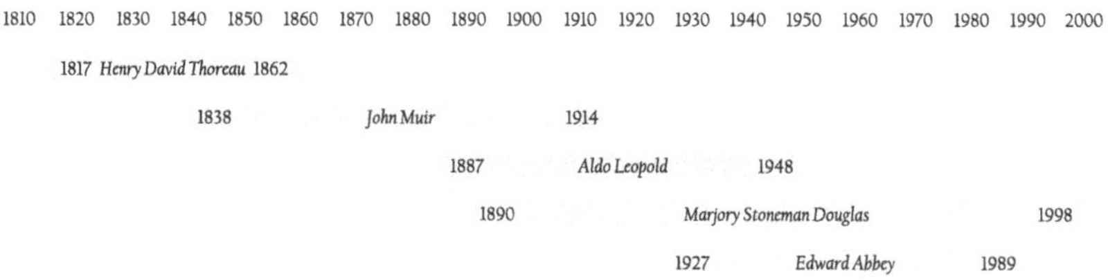

Fig. 1.1 Timelines of Thoreau, Muir, Leopold, Douglas, and Abbey

temporal) scales—some are small-scale, some are at a much grander, ecosystem scale. Some are completed, or in progress, while others remain speculative (Fig. 1.1).

I am interested in the ways literature and the literary imaginary are used to establish a restoration ethic. I am interested in the influence of these writers on the design and execution of *on the ground* restoration programs. Each of these literary landscapes is a laboratory for ideas and creative practices for restoration—for what matters in restoring the land—and, sometimes, they reveal a glimpse of what can happen when page and place meld in defense of the land.

This is a work of biography. A biography of literary naturalists and a biography of the places most closely associated with them. The two are indistinguishable, inseparable. But I am curious about why these literary names adhere to these places. Why Concord and Walden Pond so easily become part of 'Thoreau Country,' or why the Four Corners region of the American West becomes 'Abbey's Country,' for example—or why Leopold's 'Sand County' motif persists (as Leopold 1968), even though there is no Sand County in Wisconsin (but, many sand counties). And why it matters (and what it means) for restorationists working in those places. How places become a portmanteau for ecologies, histories, stories, ideas, activisms, and restorations. How the literary imaginary is political, and politicized, in defense of the land. So this book is, perhaps more than anything else, also an exposition of literary environmental activism.

The stories of these five authors are not the only stories that can be told about these places. There have been—and will continue to be—other names in all of these places. But *these* names, and *these* stories, have stuck because they agitate for conservation and restoration, for wildness, and wilderness protections on the page in ways that agitate for action in place. In some ways, this work is guilty of further centralizing and canonizing white, masculine intellectual authority. Many of these authors have already been thoroughly discussed in biographical and textual contexts elsewhere (and sometimes in combination), although not once as this particular ensemble. All five authors are Anglo-American, white, and middle-class. Four are men, only one is a woman. Douglas was the last of the book's quintet to pass away, in May 1998, almost a quarter-century ago. But other contemporary literary (and artistic) voices will also accompany, enter into conversation with, and challenge this central quintet—and many are the voices of women and postcolonial authors. And so, in aligning the history of ecological restoration with the history of nature and environmental writing, I want to expand the critical reading of the impact, significance, and relevance of some of the seminal nature and environmental writers of the nineteenth and twentieth centuries, and also reevaluate what is meant by ecological restoration—not only with regard to the longevity of the term, but also which disciplines might contribute to its articulation and development.

This is also a work spliced with other American histories. The history of the American conservation movement—with its victories, defeats, and lawsuits. The evolution of environmental legislation and regulation. The cultural geographies of national parks and national monuments. The idea of wilderness, the close of the frontier, and the reclamation of the American West. Military battles and campaigns. The Underground Railroad and the abolition of slavery. The dislocations—and relocations—of Native Nations. Nonviolent direct action and environmental protest. These are histories inhabited by a cast of ecological heroes and villains. Other environmental writers. Individual activists, and environmental nonprofits. The U.S.'s conservation presidents (and many of its anti-environment presidents). Federal agencies. Dam builders. Real estate developers. Lumber, cattle, and mining syndicates. And many more.

In tracing the genesis of a conservation and restoration sensibility across the philosophy and writings of Thoreau, Muir, Leopold, Douglas, and Abbey—and how this is translated into conservation and restoration actions by environmental organizations—this book challenges the under-representation of nature and environmental writing in ecological restoration rhetoric. And in turn, it advocates for the capacity of a literary imagination to revitalize and reinvigorate such restoration rhetoric. This book presents a literary intervention in restoration politics.

This book combines environmental history, literature, biography, philosophy, and politics in a commentary on considering (and developing) environmental literature's place in conversations on restoration ecology, ecological restoration, and rewilding. This book draws on published and unpublished writings, essays, journals, and correspondence from the featured authors, and the critical literary scholarship of others, together with archival materials, and interviews with key project managers, landscape architects, planners, ecologists, literary scholars, historians, and environmental writer-activists involved in the planning and implementation of ecological restoration programs and campaigns at Walden Pond, Hetch Hetchy, the Leopold Shack and farm property, the Everglades, and Glen Canyon.

1.4 The Poetics and Politics of Literary Environmental Activism

Literary activism is not a new addition to the campaign arsenal of American conservationists. Works of American nature and environmental writing have had a profound impact on federal, state, and local political agendas, and on ecological and political reform. Rachel Carson's (1962) *Silent Spring* and the banning of synthetic pesticide DDT is perhaps the most frequently cited. Environmental writing creates a constituency (or, ecological awareness, as Slovic 1992) for conservation and restoration through the confluence of ecological, biocentric, aesthetic, moral, spiritual, and ethical concerns (on the relationship between U.S. nature writing and environmental politics, see Payne 1996; Philippon 2005). But what is new is the formalization and legitimization of this literary activism for restoration within environmental organizations. Of environmental organizations honoring individual writers—and their restoration sensibilities—in place (as the Walden Woods Project, and the Aldo Leopold Foundation), or resuming or taking inspiration from dormant restoration campaigns begun on the page (Restore Hetch Hetchy, and the Glen Canyon Institute), or continuing to celebrate a literary and social activism restoration legacy (Friends of the Everglades). And with it comes the endorsement that literary naturalists and environmental writer-activists both past and present still have a powerful voice in contemporary conservation battles—and still have much to teach restorationists.

Daniel J. Philippon (2005), in *Conserving Words: How American Nature Writers Shaped the Environmental Movement*, has already explored the contributions of U.S. nature writers to the formation and development of environmental organizations. But his interest lies in the relationship between nature writers and the environmental movement, and these writers' 'ecology of influence,' rather than in the roadmap from literature to restoration ethic to conservation battle. While I owe a debt to Philippon's (2005) approach, and his close scrutiny of the layers of metaphor, narrative, and values implicit in writing that led to environmental groups (and, also, to Brulle 2000), this book does something different. I linger instead on the translation of literary ideas and proclamations into *on the ground* restoration programs.

I agree with Sean Prentiss and Joe Wilkins (2017) that all nature and environmental writing is political. And that all nature and environmental writing might be regarded as activist-writing. For nature and environmental writing is never just about nature and the environment. It is also about politics, globalization, urbanization, culture, technology, industrial agriculture, conflict and war, dispossession, exploitation, inequality, environmental injustice, pollution, contamination and toxicity, loss of wilderness, sustainability, climate change, health and wellbeing, spirituality, colonialism, postcolonialism, gender, race and ethnicity, queer politics, and more.

Where environmental writing excels as a tool in a conservation campaign arsenal is in its uniting of science and poetry. Literary environmental activism emerges as a creative yet gently subversive political performance, where texts are penned, composed by writers themselves—or adopted by conservationists, restorationists, and others—to speak back to an ethics of (ecological) care and responsibility, and to respond to environmental injustices (see Ammons 2010 for an excellent review of the progressive activist tradition in American literature, cf. Lindholdt 1996; Lauter 2001). One famous, albeit dramatic, example, is Abbey's (1975) protest novel *The Monkey Wrench Gang*, which inspired the formation of radical environmental group Earth First! (EF!) in 1980, and introduced the term 'monkeywrenching' to describe the practice of eco-sabotage (often abridged to 'ecotage').

While many of the texts encountered throughout this book were activist literary texts at the time they were written (whether subtly or radically, revolutionarily so), they have been politicized anew, to speak to new (and returning) environmental issues, and as provocations on ecological (and political) reform. And so environmental literature provides—and becomes—a site for restorationists to grapple with and work through questions of *why* and *how* nature should be restored.

1.5 How This Book is Organized

Ecological Restoration and the U.S. Nature and Environmental Writing Tradition is organized into seven chapters. The central chapters (Chapters 2 through 6) draw on five case studies of ecological restoration to explore how the literary imagination is molding ecological (and political) reform. Each chapter title adopts the featured author's description of that place.

This book has many different audiences. I hope this book will appeal not only to scholars in ecological restoration, cultural geography, and ecocriticism and the environmental humanities, but also to general readers interested in environmental history, and environmental policy. The chapters are presented historically, organized by date of birth of the featured author. But they can be read in any order (indeed, I wrote them nonlinearly), particularly if interest lies instead in wider themes of the American conservation movement, environmentalism, U.S. national parks history, or civil disobedience and environmental protest.

Chapter 2 begins with Henry David Thoreau's (1817–1862) two-year sojourn in a cabin at Walden Pond, from 1845 through 1847, and unpacks and examines his writings on nature, wildness, and wilderness at Walden. When Walden Woods is threatened by recreation and development pressures a little over a century later, Thoreau's environmental philosophy and writings become part of the arsenal of campaigners fighting to preserve and restore the woods. This campaign in the 1950s heralds the first of several confrontations in Walden Woods, and this chapter critiques the different restoration reference models and strategies that have played out over the last 70 years at Walden—and in particular, the work of the Walden Woods Project in restoring Thoreau's Walden. There is a second Thoreau narrative that runs through this chapter, and it follows Thoreau's trips to Maine in 1846, 1853, and 1857, and how his writings on the Maine woods are echoed in the proposal for a national park in Maine's North Woods. Here, the work of RESTORE The North Woods is used to tell the story of the decades-long campaign—and the place of Thoreau's legacy therein—together with the story of the designation of the Katahdin Woods and Waters National Monument in 2016.

John Muir's (1838–1914) California writings provide the context for Chapter 3, especially his writings on the Sierras and Yosemite. In this chapter, these writings are set against a number of events that reveal the genesis of Muir's battle to save the Hetch Hetchy Valley in Yosemite—from the signing of the Yosemite Grant by President Abraham Lincoln in 1864, to Yosemite's designation as America's third national park in 1890, to Muir's founding of the Sierra Club in 1892, and his Yosemite camping trip with President Theodore Roosevelt in 1903, to Roosevelt's signing of the 1906 Antiquities Act, and the 1906 San Francisco earthquake that also signaled the death knell for Hetch Hetchy with the passing of the Raker Act in 1913. The Hetch Hetchy Valley was submerged behind the O'Shaughnessy Dam and the rising waters of the Hetch Hetchy Reservoir in the 1910s and 1920s. Yet Muir's philosophy and writings continue to inspire campaigns fighting to restore Hetch Hetchy. This chapter examines the ongoing call for restoration through the work of Restore Hetch Hetchy over the last two decades, and RHH's legal campaign against the City and County of San Francisco.

Chapter 4 begins by tracing the development of Aldo Leopold's (1887–1948) conservation philosophy through his work with the U.S. Forest Service (USFS) in the American Southwest, and especially as a professor at the University of Wisconsin-Madison. Two experiments in ecological restoration confirm Leopold's position as one of the early pioneers of restoration—at the Curtis Prairie at the UW Arboretum, and at the Leopold family farm. Leopold's conservation and stewardship legacy, and his family's restoration efforts at the farm—'the Shack'—are continued through the work of the Aldo Leopold Foundation. This chapter also looks at Leopold's wilderness legacy, and what this means for ecological restoration—from his work with the USFS to establish the first wilderness area in the U.S., to his place as a cofounder of the Wilderness Society, and the achievements of the 1964 Wilderness Act.

Chapter 5 relocates to the Florida Everglades. It was not until 1947, which saw both the publication of Marjory Stoneman Douglas's (1890–1998) *The Everglades: River of Grass* (Douglas 2017), and the designation of Everglades National Park, that conservation and restoration conversations began to take hold in the Everglades. Two decades later, Douglas established Friends of the Everglades as a platform for discussion—and, importantly, action—on Everglades restoration. This chapter examines how Douglas emerged as a leading voice campaigning for restoration in the Everglades and traces the key legal battles and lawsuits filed (or supported) by Friends of the Everglades. The chapter also explores the legislation on, and strategies for, ecological restoration that have played out across the Everglades—culminating in the Comprehensive Everglades Restoration Plan. But the Everglades is not the only muse for Florida's environmental writer-activists—hurricanes also feature prominently in a reactive literary environmental advocacy, with Gulf Coast poets working to restore connections to correspondences with place.

When Glen Canyon Dam was authorized in 1956 as part of the Colorado River Storage Project Act, literature and the literary imagination emerged as a powerful platform for conservationists to voice concern over the loss of Glen Canyon, but to also call for its restoration—with Muir's battle for Hetch Hetchy retold in the Glen Canyon campaign. Edward Abbey (1927–1989) is perhaps Glen Canyon's most vociferous protagonist, and Chapter 6 examines his fiction and nonfiction writings on the Glen and its anticipated restoration. Abbey's works are also considered alongside other writers on Glen Canyon restoration, especially Katie Lee, Ellen Meloy, and Terry Tempest Williams. This chapter goes on to examine the work of the Glen Canyon Institute as one institutional, organizational response to Abbey's call-to-arms. This chapter also explores the near-failure of the dam in the summer of 1983, and how the Bureau of Reclamation (USBR, responsible for the operation of

Glen Canyon Dam), is responding to the sustained calls for restoration, as well as the place of Glen Canyon Dam and Lake Powell in wider political conversations on dam decommissioning in the U.S. This chapter also looks beyond Glen Canyon, to examine how literature has played a part in other conservation campaigns in southern Utah—chiefly, in the designation of Grand Staircase-Escalante National Monument in 1996, and Bears Ears National Monument in 2016.

The final chapter, Chapter 7, draws together the restoration ecologies of Henry David Thoreau, John Muir, Aldo Leopold, Marjory Stoneman Douglas, and Edward Abbey, and argues that, in conversation, these authors present a reading of ecological restoration and rewilding that reaffirms a commitment and intimacy with the land. The ecological restoration sensibilities of Thoreau, Muir, Leopold, Douglas, and Abbey—and the responses of the Walden Woods Project, Restore Hetch Hetchy, the Aldo Leopold Foundation, Friends of the Everglades, and the Glen Canyon Institute to restoration—demonstrate how environmental writing has become a tool for activism and advocacy. Yet this chapter also articulates the challenges and tensions facing this canon of work as a commentary on the environmental condition, following the recent rise of alt-facts and shock politics in the U.S.

Works Cited

Abbey, Edward. 2004 [1975]. *The Monkey Wrench Gang*. London: Penguin Books Ltd.

Agnoletti, Mauro, ed. 2006. *The Conservation of Cultural Landscapes*. Wallingford, Oxon: CABI Publishing.

Ammons, Elizabeth. 2010. *Brave New Words: How Literature Will Save the Planet*. Iowa City, IA: University of Iowa Press.

Anderson, Jon. 2014. *Page and Place: Ongoing Compositions of Plot*. Amsterdam and New York, NY: Rodopi.

Aronson, James, and Jelte van Andel. 2006. Challenges for ecological theory. In *Restoration Ecology: The New Frontier*, eds. Jelte van Andel and James Aronson, 223–233. Oxford: Blackwell Science Ltd.

Aronson, J., C. Floret, E. LeFloc'h, C. Ovalle, and R. Pontanier. 1993. Restoration and Rehabilitation of Degraded Ecosystems in Arid and Semi-Arid Lands. I. A View from the South. *Restoration Ecology* 1 (1): 8–17.

Barilla, James Jerome. 2004. *The Nature of Homelands: Narratives of Restoration and Return*. Unpublished Ph.D. thesis. Davis, CA: University of California, Davis. ProQuest/UMI 3148427.

Barilla, James Jerome. 2007. A Mosaic of Landscapes: Ecological Restoration and the Work of Leopold, Coetzee, and Silko. In *Coming into Contact: Explorations in Ecocritical Theory and Practice*, eds. Annie Merrill Ingram, Ian Marshall, Daniel J. Philippon, and Adam W. Sweeting, 128–140. Athens, GA and London: University of Georgia Press.

Basl, John. 2010. Restitutive Restoration: New Motivations for Ecological Restoration. *Environmental Ethics* 32 (2): 135–147.

Brulle, Robert J. 2000. *Agency, Democracy, and Nature: The U.S. Environmental Movement from a Critical Theory Perspective*. Cambridge, MA: MIT Press.

Buell, Lawrence. 1995. *The Environmental Imagination: Thoreau, Nature Writing, and the Formation of American Culture*. Cambridge, MA: Belknap Press of Harvard University Press.

Buell, Lawrence. 2001. *Writing for an Endangered World: Literature, Culture, and Environment in the U.S. and Beyond*. Cambridge, MA: Belknap Press of Harvard University Press.

Buell, Lawrence. 2005. *The Future of Environmental Criticism: Environmental Crisis and Literary Imagination*. Malden, MA: Blackwell.

Callicott, J. Baird, and Michael P. Nelson, eds. 1998. *The Great New Wilderness Debate*. Athens, GA: University of Georgia Press.

Cameron, Emilie. 2012. New Geographies of Story and Storytelling. *Progress in Human Geography* 36 (5): 573–592.

Carson, Rachel. 1962. *Silent Spring*. Boston, MA: Houghton Mifflin.

Choi, Young D. 2004. Theories for Ecological Restoration in Changing Environment: Toward 'Futuristic' Restoration. *Ecological Research* 19 (1): 75–81.

Clewell, Andre F., and James Aronson. 2006. Motivations for the Restoration of Ecosystems. *Conservation Biology* 20 (2): 420–428.

Clewell, Andre F., and James Aronson. 2013. *Ecological Restoration: Principles, Values, and Structure of an Emerging Profession*, 2nd ed. Washington, DC: Island Press.

Cloyd, Aaron A. 2016. Reimaging Rewilding: A Response to Jørgensen, Prior, and Ward. *Geoforum* 76: 59–62.

Cowley, Jason, ed. 2008. *Granta 102: The New Nature Writing*. London: Granta Publications.
Cronon, William. 1992. A Place for Stories: Nature, History, and Narrative. *Journal of American History* 78 (4): 1347–1376.
Cronon, William. 1995. *Uncommon Ground: Rethinking the Human Place in Nature*. New York, NY: W. W. Norton.
Daniels, Stephen, and Hayden Lorimer. 2014. Until the end of days: Narrating landscape and environment. *cultural geographies* 19 (1): 3–9.
Davis, Diana K. 2011. Reading landscapes and telling stories: Geography, the humanities, and environmental history. In *Envisioning Landscapes, Making Worlds: Geography and the Humanities*, eds. Stephen Daniels, Dydia DeLyser, J. Nicholas Entrikin, and Douglas Richardson, 170–176. Abingdon, Oxon: Routledge.
Douglas, Marjory Stoneman. 2017 [1947]. *The Everglades: River of Grass*, 70th Anniversary Edition. Sarasota, FL: Pineapple Press.
Egan, Dave, and Evelyn A. Howell, eds. 2001. *The Historical Ecology Handbook: A Restorationist's Guide to Reference Ecosystems*. Washington, DC: Island Press.
Fishkin, Shelley Fisher. 2015. *Writing America: Literary Landmarks from Walden Pond to Wounded Knee, A Reader's Companion*. New Brunswick, NJ: Rutgers University Press.
France, Robert L., ed. 2008. *Healing Natures, Repairing Relationships: New Perspectives on Restoring Ecological Spaces and Consciousness*. Sheffield, VT: Green Frigate Books.
Gann, George D., Tein McDonald, Bethanie Walder, James Aronson, Cara R. Nelson, Justin Jonson, James G. Hallett, Cristina Eisenberg, Manuel R. Guariguata, Junguo Liu, Fangyuan Hua, Cristian Echeverría, Emily Gonzales, Nancy Shaw, Kris Decleer, and Kingsley W. Dixon. 2019. International Principles and Standards for the Practice of Ecological Restoration. Second edition. *Restoration Ecology* 27 (S1): S1–S46.
Gatta, John. 2018. *Spirits of Place in American Literary Culture*. New York, NY: Oxford University Press.
Glotfelty, Cheryll, and Harold Fromm, eds. 1996. *The Ecocriticism Reader: Landmarks in Literary Ecology*. Athens, GA: University of Georgia Press.
Gobster, Paul H., and R. Bruce Hull, eds. 2000. *Restoring Nature: Perspectives from the Social Sciences and Humanities*. Washington, DC: Island Press.
Hall, Marcus. 2005. *Earth Repair: A Transatlantic History of Environmental Restoration*. Charlottesville, VA: University of Virginia Press.
Hall, Marcus, ed. 2010. *Restoration and History: The Search for a Usable Environmental Past*. New York, NY: Routledge.

Hiers, John K., Robert J. Mitchell, Analie Barnett, Jeffrey R. Walters, Michelle Mack, Bren Williams, and Rob Sutter. 2012. The Dynamic Reference Concept: Measuring Restoration Success in a Rapidly Changing No-Analogue Future. *Ecological Restoration* 30 (1): 27–36.

Higgs, Eric. 2003. *Nature by Design: People, Natural Process, and Ecological Restoration*. Cambridge, MA: MIT Press.

Higgs, Eric. 2012. History, Novelty, and Virtue in Ecological Restoration. In *Ethical Adaptation to Climate Change: Human Virtues of the Future*, eds. Allen Thompson and Jeremy Bendik-Keymer, 81–101. Cambridge, MA: MIT Press.

Hobbs, Richard J., Salvatore Arico, James Aronson, Jill S. Baron, Peter Bridgewater, Viki A. Cramer, Paul R. Epstein, John J. Ewel, Carlos A. Klink, Ariel E. Lugo, David Norton, Dennis Ojima, David M. Richardson, Eric W. Sanderson, Fernando Valladares, Montserrat Vilà, Regino Zamora, and Martin Zobel. 2006. Novel Ecosystems: Theoretical and Management Aspects of the New Ecological World Order. *Global Ecology and Biogeography* 15: 1–7.

Hobbs, Richard J., Eric S. Higgs, and Carol M. Hall, eds. 2013. *Novel Ecosystems: Intervening in the New Ecological World Order*. Chichester: Wiley.

Hull, R. Bruce, and David P. Robertson. 2000. Which Nature? In *Restoring Nature: Perspectives from the Social Sciences and Humanities*, eds. Paul H. Gobster and R. Bruce Hull, 299–307. Washington, DC: Island Press.

Jackson, Stephen T., and Richard J. Hobbs. 2009. Ecological Restoration in the Light of Ecological History. *Science* 325 (5940): 567–569.

Jepson, Paul, and Cain Blythe. 2020. *Rewilding: The Radical New Science of Ecological Recovery*. London: Icon Books Ltd.

Jordan, William R. III. 2003. *The Sunflower Forest: Ecological Restoration and the New Communion with Nature*. Berkeley and Los Angeles, CA: University of California Press.

Jordan, William R. III., and George M. Lubick. 2011. *Making Nature Whole: A History of Ecological Restoration*. Washington, DC: Island Press.

Jørgensen, Dolly. 2015. Rethinking Rewilding. *Geoforum* 65: 482–488.

Lauter, Paul. 2001. *From Walden Pond to Jurassic Park: Activism, Culture, and American Studies*. Durham, NC: Duke University Press.

Leopold, Aldo. 1968 [1949]. *A Sand County Almanac, and Sketches Here and There*. Oxford: Oxford University Press.

Lewis, Michael, ed. 2007. *American Wilderness: A New History*. Oxford: Oxford University Press.

Lindholdt, Paul. 1996. Literary Activism and the Bioregional Agenda. *ISLE: Interdisciplinary Studies in Literature and Environment* 3 (2): 121–137.
Lopez, Barry. 1989. *Crossing Open Ground*. New York, NY: Vintage Books.
Lorimer, Jamie. 2015. *Wildlife in the Anthropocene: Conservation after Nature*. Minneapolis, MN: University of Minnesota Press.
Marsh, George Perkins. 2003 [1864]. *Man and Nature: Or, Physical Geography as Modified by Human Action*. Seattle, WA: Weyerhaeuser Environmental Books/University of Washington Press.
McKibben, Bill, ed. 2008. *American Earth: Environmental Writing Since Thoreau*. New York, NY: Library of America.
Miller, James R., and Brandon T. Bestelmeyer. 2016. What's Wrong with Novel Ecosystems, Really? *Restoration Ecology* 24 (5): 577–582.
Minteer, Ben A., and Stephen J. Pyne. 2013. Restoring the Narrative of American Environmentalism. *Restoration Ecology* 21 (1): 6–11.
Minteer, Ben A., and Stephen J. Pyne, eds. 2015. *After Preservation: Saving American Nature in the Age of Humans*. Chicago, IL: University of Chicago Press.
Monbiot, George. 2014. *Feral: Rewilding the Land, Sea, and Human Life*. London: Penguin Books.
Nash, Roderick Frazier. 2014. *Wilderness and the American Mind*. 5th ed. New Haven, CT: Yale University Press.
Naveh, Zev. 1994. From Biodiversity to Ecodiversity: A Landscape-Ecology Approach to Conservation and Restoration. *Restoration Ecology* 2 (3): 180–189.
Naveh, Zev. 1995. Interactions of Landscapes and Cultures. *Landscape and Urban Planning* 32 (1): 43–54.
Naveh, Zev. 1998. Ecological and Cultural Landscape Restoration and the Cultural Evolution Towards a Post-Industrial Symbiosis Between Human Society and Nature. *Restoration Ecology* 6 (2): 135–143.
Naveh, Zev. 2005. Epilogue: Toward a Transdisciplinary Science of Ecological and Cultural Landscape Restoration. *Restoration Ecology* 13 (1): 228–234.
Naveh, Zev, ed. 2007. *Transdisciplinary Challenges in Landscape Ecology and Restoration Ecology—An Anthology*. Dordrecht, The Netherlands: Springer.
Oelschlaeger, Max. 1991. *The Idea of Wilderness: From Prehistory to the Age of Ecology*. New Haven, CT: Yale University Press.
Payne, Daniel G. 1996. *Voices in the Wilderness: American Nature Writing and Environmental Politics*. Hanover, NH: University Press of New England.

Philippon, Daniel J. 2005. *Conserving Words: How American Nature Writers Shaped the Environmental Movement*. Athens, GA: University of Georgia Press.

Prentiss, Sean, and Joe Wilkins. 2017. *Environmental and Nature Writing: A Writer's Guide and Anthology*. London: Bloomsbury Academic.

Prior, Jonathan, and Laura Smith. 2019. The Normativity of Ecological Restoration Reference Models: An Analysis of Carrifran Wildwood, Scotland, and Walden Woods, United States. *Ethics, Policy & Environment* 22 (2): 214–233.

Prior, Jonathan, and Kim J. Ward. 2016. Rethinking Rewilding: A response to Jørgensen. *Geoforum* 69: 132–135.

Purdy, Jedediah. 2015. *After Nature: A Politics for the Anthropocene*. Cambridge, MA: Harvard University Press.

Rohwer, Yasha, and Emma Marris. 2016. Renaming Restoration: Conceptualizing and Justifying the Activity as a Restoration of Lost Moral Value Rather Than a Return to a Previous State. *Restoration Ecology* 24 (5): 674–679.

Rolston, Holmes. III. 1994. Does Nature Need to Be Redeemed? *Zygon: Journal of Religion & Science* 29 (2): 205–229.

Satterfield, Terre, and Scott Slovic, eds. 2004. *What's Nature Worth? Narrative Expressions of Environmental Values*. Salt Lake City, UT: University of Utah Press.

Slovic, Scott. 1992. *Seeking Awareness in American Nature Writing: Henry Thoreau, Annie Dillard, Edward Abbey, Wendell Berry and Barry Lopez*. Salt Lake City, UT: University of Utah Press.

Smith, Jos. 2017. *The New Nature Writing: Rethinking the Literature of Place*. London: Bloomsbury Academic.

Smith, Laura. 2014. On the 'Emotionality' of Environmental Restoration: Narratives of Guilt, Restitution, Redemption, and Hope. *Ethics, Policy & Environment* 17 (3): 286–307.

Suding, Katharine N., and Katherine L. Gross. 2006. The Dynamic Nature of Ecological Systems: Multiple States and Restoration Trajectories. In *Foundations of Restoration Ecology*, eds. Donald A. Falk, Margaret A. Palmer, and Joy B. Zedler, 190–209. Washington, DC: Island Press.

Taylor, Ken, Archer St. Clair, and Nora J. Mitchell, eds. 2014. *Conserving Cultural Landscapes: Challenges and New Directions*. New York, NY: Routledge.

Thoreau, Henry David. 2004 [1854]. *Walden*. 150th Anniversary Edition, ed. J. Lyndon Shanley. Princeton, NJ and Oxford: Princeton University Press.

Tredinnick, Mark. 2005. *The Land's Wild Music: Encounters with Barry Lopez, Peter Matthiessen, Terry Tempest Williams & James Galvin*. San Antonio, TX: Trinity University Press.

Turner, Frederick. 1989. *Spirit of Place: The Making of an American Literary Landscape*. Washington, DC: Island Press.

United Nations Environment Programme. 2019. New UN Decade on Ecosystem Restoration offers Unparalleled Opportunity for Job Creation, Food Security and Addressing Climate Change. *United Nations Environment Programme*. https://unenvironment.org/news-and-stories/press-release/new-un-decade-ecosystem-restoration-offers-unparalleled-opportunity.

Van Wieren, Gretel. 2013. *Restored to Earth: Christianity, Environmental Ethics, and Ecological Restoration*. Washington, DC: Georgetown University Press.

Weik von Mossner, Alexa. 2017. *Affective Ecologies: Empathy, Emotion, and Environmental Narrative*. Columbus, OH: Ohio State University Press.

Weston, Daniel. 2016. Plots: The Narratives of Place in Contemporary Nature Writing. In *Affective Landscapes in Literature, Art, and Everyday Life: Memory, Place and the Senses*, eds. Christine Berberich, Neil Campbell, and Robert Hudson, 173–188. Farnham: Ashgate.

Wilson, Edward O. 2016. *Half-Earth: Our Planet's Fight for Life*. New York, NY: Liveright Publishing Corporation/W. W. Norton.

Zapf, Hubert. 2008. Literary Ecology and the Ethics of Texts. *New Literary History* 39 (4): 847–868.

Zapf, Hubert. 2016. *Literature as Cultural Ecology: Sustainable Texts*. London: Bloomsbury Academic.

2

'With Walden in Its Midst': Henry David Thoreau, Walden Pond, and the Walden Woods Project

2.1 Thoreau's Sojourn at Walden Pond, 1845–1847

When *Walden; or, Life in the Woods* was published on August 9, 1854, by Ticknor and Fields in Boston, Massachusetts, Concord Transcendentalist philosopher and naturalist Henry David Thoreau (1817–1862) noted simply in his journal,

> Aug. 9. *Wednesday.*—To Boston.
> "Walden" published. Elder-berries. Waxwork yellowing.
>
> —Thoreau 1906 *VI*: 429

His August 9 journal entry was succinct, laconic. The publication of Thoreau's second book (and the last to be published during his lifetime) appears in his journal inserted amidst the day's travel notes and botanizing. A modest success, it would take five years to sell the 2,000 copies printed on the first print run—an improvement on Thoreau's first book, which sold only a few hundred copies, with the publisher returning the rest to Thoreau who had self-financed the venture. But

L. Smith, *Ecological Restoration and the U.S. Nature and Environmental Writing Tradition*, https://doi.org/10.1007/978-3-030-86148-3_2

these two words—'"Walden" published'—belie an eco-literary behemoth that would go on to shape not only an American nature and environmental writing tradition, but also the birth of the environmental movement in the U.S. But also discernible in Thoreau's (2004a) *Walden* is the early foundation for what might be couched a Thoreauvian ecological restoration sensibility.

John Thoreau Jr., Thoreau's older brother, died suddenly in 1842 at the age of 26, and Thoreau's move three years later to Walden Pond,[1] a mile south of Concord, provided a solitude where he could not only mourn his late brother, but also write about their 1839 boating and hiking excursion from Concord, Mass., to Concord, New Hampshire, following the Concord and Merrimack Rivers (published in 1849 as *A Week on the Concord and Merrimack Rivers*, his first book, see Thoreau 2001a). In the early spring of 1845, 27-year-old Henry Thoreau started to build a one-room cabin on the northern shore of Walden Pond, on a 14-acre woodlot (the Wyman Lot) recently purchased by friend and mentor Ralph Waldo Emerson. Thoreau moved to Walden Pond on July 4—Independence Day—1845, where he would spend the next 26 months. And it is this sojourn that would later be immortalized in Thoreau's canonical work, *Walden* (Thoreau 2004a).[2]

Thoreau's (2004a) *Walden* stands as a primer for restorationists working in Walden Woods. Yet, at the same time, *Walden* was only the beginning. This chapter opens at the start of Thoreau's two-year sojourn in Walden Woods in the summer of 1845 and uses this episode to navigate a pathway through the origins of some of the pivotal influences on Thoreau's environmental philosophy, and the development of a Thoreauvian ecological restoration ethos. Discussion begins at Thoreau's cabin and bean field, and his beloved Walden Pond, before moving out to consider Walden's place in Thoreau's scientific inquiry, especially in his

[1] This was not Thoreau's first sojourn in Walden Woods. He had already spent six weeks in the summer of 1837 with his Harvard classmate Charles Stearns Wheeler in a cabin built by the latter on the shores of Flint's (or Sandy) Pond in Lincoln. When Thoreau's request to build a cabin at Flint's Pond in the early-1840s was declined by the owner, a local farmer who gave his name to the pond, he looked instead to Walden Pond (Thoreau would later lambast Flint's Pond in *Walden* [Thoreau 2004a: 195–196], see also Brooks 1976).

[2] *Walden* went through seven revisions and doubled in length in the nine years before it was published in 1854 (see Shanley 1957; Sattelmeyer 1990; Buell 1995).

theories of forest succession and seed distribution. A second, concurrent theme to run through this chapter is Thoreau's three excursions into Maine's North Woods, the first of which intersects with his Walden sojourn. Thoreau's experiences amidst the forests, lakes, and mountains of north-central Maine refocuses his attention on the wildlands of Concord, but also inspires some of the earliest declarations for wilderness preservation legislation. The chapter then moves to cast Thoreau as an early pioneer of ecological restoration, by way of his agitation at the logging of Walden, and his demand for town forest preserves, before examining his experiment in rewilding—a replanting of pine saplings on his former Walden bean field site in the spring of 1859.

The chapter then jumps to the early-1920s, and the first formalization and institutionalization of an ecological restoration rhetoric in Walden Woods, as a Deed of Gift to the Commonwealth of Massachusetts secures the protection of 80 acres along Walden's shore, and establishes the Walden Pond State Reservation. But this designation is not enough to secure the protection of Thoreau's Walden, and from the late-1950s onward, Walden restorationists would variously deflect, and on occasion be defeated by, issues including a beach expansion, proposed commercial development projects, shoreline erosion and bank collapse, options for the closed and capped municipal landfill, and new high school playing fields. An evolving, retaliatory restoration trajectory, inspired by Thoreau's ideas, is traced through the work of the Walden Woods Project and its predecessors.

From here, the chapter resumes—and concludes—its brief detour to Maine's North Woods, with the confluence of Thoreau's wilderness and restoration proclamations in the 2016 designation of the Katahdin Woods and Waters National Monument in northern Maine, and the accompanying work of RESTORE The North Woods and other organizations.

2.1.1 The Cabin and the Bean Field

Thoreau would often write of his Walden sojourn as an 'experiment'—an endeavor in self-reliance, to live simply and deliberately (Thoreau 2004a, also 1906 *IX*, 2000). At the epicenter of his experiment was his one-room cabin on the shore of Walden Pond, set back in a young stand of pitch pine and hickory, and surrounded by fruit bushes (Fig. 2.1).

But it is another central tenet of Thoreau's Walden experiment that would later come to have much import for a Thoreauvian ecological restoration rhetoric—his 2.5-acre bean field, located east of the cabin. Thoreau began tending his bean field more than a month before his move to Walden Pond. Such is the significance of the bean field to Thoreau's experiment at Walden that 'The Bean-Field' is the title and subject of *Walden*'s seventh chapter, and opens not only with a description of the plot, but also offers an insight into Thoreau's relationship with his plot (Thoreau 2004a). In an extended passage, Thoreau describes how he planted seven miles of bean rows on the 2.5-acre plot, and how, in tending to the bean field, he developed a commitment, investment, and intimacy with the beans that saw him desire to learn of and from the beans—to *know* beans—and the surrounding forest. Thoreau's intimacy with his bean field plot signals both the *intellectual* and *physical rigor* that he expends in understanding this tract of the Walden landscape.

Fig. 2.1 The site of Thoreau's cabin in Walden Woods, Concord, Massachusetts. Left: The view from the cabin site near Thoreau's Cove across Walden Pond toward Little Cove. Top right: The original site of Thoreau's cabin. Bottom right: The cairn, and 'I went to the woods...' Photographs by the author, June 2007

The bean field would enter into the early annals of ecological restoration later in the 1850s.

2.1.2 Walden Pond, and Thoreau's 1846 Survey

Not long after moving to Walden, Thoreau recalls in an August 1845 journal passage his first childhood visit to the pond when he was five years old (Thoreau 1906 *I*, also 2004a). Walden Pond and Walden Woods would become one of the most—if not *the* most—closely observed and studied landscapes in Thoreau's natural history writings. And when Thoreau (2004a) comments in the opening pages of *Walden* that he writes about himself so much because there is no one else with whom he is so closely acquainted, I have often thought that 'Walden Pond' could replace 'Thoreau' in this statement, and serve as an affidavit to Thoreau's ecological legacy. W. Barksdale Maynard's (2004) *Walden Pond: A History* is the first detailed study to consider the pond amidst Concord's intellectual history (Fig. 2.2).

Thoreau's curiosity (and fondness) of Walden Pond (and an uncertainty among his contemporaries as to the history of the pond)[3] led him to embark on a survey of the 61-acre pond during his first winter at Walden (described in Thoreau 2004a). Thoreau often worked as a land surveyor during his sojourn at Walden, and his journals and essays comprehensively record the various surveys and reports he completed on behalf of his Concord neighbors of woodlots, fields, swamps, and rivers. Thoreau would continue this practice after leaving Walden, spending much of December 1857 surveying Walden Woods (Thoreau 1906 *X*). The ventures of Thoreau-as-surveyor also extend beyond Walden, to Maine's North Woods, and to Cape Cod.

What began as an endeavor to record the depth of Walden Pond evolved into a mapping of the pond's basin. With more than one hundred measurements of the pond, Thoreau produced one of the

[3] Walden Pond (along with nearby Fairhaven Bay on the Concord River) is a kettle hole, a kettle pond, formed by the melting of Glacial Lake Sudbury and the Wisconsin glacier (for a glaciological history of the formation of Walden Pond, see e.g. Schofield 1989; Schofield and Baron 1993; McGregor 1997; NPS 2002; Maynard 2004; Thorson 2014).

Fig. 2.2 Walden Pond. Photograph by the author, August 2007

earliest maps of Walden Pond—'Walden Pond, A reduced Plan, 1846' (in Thoreau 2004a). Thoreau's pond survey would energize and become an anchor not just for the earliest drafts of *Walden*, but also for Thoreau himself, in Concord (Thoreau's surveying legacy is the focus of a meticulous study by Chura 2010, see also Botkin 2001; Van Noy 2003; Donahue 2004; Thorson 2017; Miller 2018).

2.1.3 Thoreau's Botanizing, and a Concord and Walden Phenology

Thoreau spent much of his life in Concord (15 miles west of Boston), and the town, together with the adjacent woods of Walden, would become his principal canvas for studies in botany and natural history. Thoreau's days were often bookended morning and evening by writing and journaling, with afternoons spent outdoors, exploring forest paths and trails (on Thoreau's Concord landscapes, see especially Gleason 1917, 1975; McGregor 1997; Foster 1999; Schneider 2000; Finley 2017). Thoreau first writes of 'the Walden wood' in his journal on April 26, 1838, in the poem 'The Bluebirds' (Thoreau 1906 *I*, cf. Blanding 1988). A letter from Thoreau's friend Daniel Ricketson written just three days after the publication of *Walden*, on August 12, 1854, confirms the conflation of Thoreau and Walden even during his lifetime, for it ends, 'Dear Mr. Walden' (Thoreau 1958: 335). Thoreau was an early pioneer of what would later become known as bioregionalism.

By the late-1840s, and following his Walden episode, Thoreau's interest in science had shifted from theory toward practice. Thoreau would often embark on 'botanical rambles' (Thoreau 1906 *XI*, 2001b esp. 'Autumnal Tints') through Walden—he was a local specimen collector for Harvard zoologist and geologist Louis Agassiz in May and June 1847, toward the end of his sojourn, and also amassed a large personal herbarium of more than 900 specimens (Angelo 1983a). Thoreau attended (and gave) lectures at the Concord Lyceum and elsewhere across New England, and also aligned himself with a group of Harvard vitalists (Robinson 2004; Arsić 2016; Raden 2017) to support and corroborate his own scientific inquiry. In December 1850, he was elected to the Boston Society of Natural History. Thoreau's own plant collections are held in the Henry David Thoreau Herbarium at Harvard University, his alma mater. Thoreau read extensively on the history of science and botany throughout the 1850s (on nineteenth-century philosophy and science and Thoreau, see the groundbreaking study by Walls 1995, and later 2000, 2016, 2017, 2018, also Baym 1965; Angelo 1983a; Richardson 1986, 1992; Schofield and Baron 1993; McGregor 1997; Botkin 2001; Thorson 2014; Arsić 2016; Finley 2017).

It was out of these rambles that Thoreau's early study of the systems and processes of forest succession and seed distribution emerged. Trees quickly became his principal focus. Commentaries and meditations on forest trees, native plants, and the cultivation of orchards emerged as anchors in his writing (on the history of Thoreau's scientific inquiry of forest trees, see also Walls 1995; McGregor 1997; Foster 1999; Botkin 2001; Higgins 2017). Through his prolonged immersion in, and studies around, Walden Woods of the changing forest history and forest growth,

Thoreau was attuned to and recognized patterns and consequences of succession, adaptation, pollination, and the historical role of different plant species in determining forest types.

It was on Brister's Hill, in the northeastern part of Walden Woods, that Thoreau witnessed most dramatically the consequences of felling, and the rupture to ecological processes—presenting him with a living laboratory to study succession, but also leading to his early plea for conservation. Thoreau presented his theory in lectures, and in the 1860 polemical essay 'The Succession of Forest Trees,' his only essay on botany to be published during his lifetime (two further botanical essays were published less than six months after his death—'Autumnal Tints' and 'Wild Apples'—and all three are included in Thoreau 2001b). His scrutiny of forest dynamics, and especially renewal, in 'The Succession of Forest Trees' is a hook for restorationists. While some botanical taxonomists and historians have maligned or discredited Thoreau's writings for their scientific inaccuracies, inconsistencies, and omissions (as outlined in Angelo 1983a, b), they often overlook his contribution to plant-animal interactions, plant population biology, and coevolution (see Nabhan 1993).

It would be another century-and-a-half almost before another of Thoreau's studies in this field would be published for the first time. In 1993, Bradley P. Dean assembled and edited *Faith in a Seed: The Dispersion of Seeds and Other Late Natural History Writings* (Thoreau 1993). *Faith in a Seed* borrows its title from Thoreau's declaration in the closing passages of 'The Succession of Forest Trees' (in Thoreau 2001b) and contains Thoreau's late essay 'The Dispersion of Seeds,' which expands and reaffirms his earlier theorizing to set out a treatise on seed dispersal. In 'The Dispersion of Seeds,' Thoreau (1993) presents his ambitious study of seed dispersal across Concord's fields and woodlots as a richly descriptive inventory of Concord tree seeds. His study is particularly innovative and pioneering, for it disputes popular nineteenth-century 'spontaneous generation' theories advanced by Louis Agassiz and George B. Emerson (Emerson's cousin), among others (on Thoreau's challenge to spontaneous generation theories, see Thoreau 1906 *XIV*, also Richardson 1993). For restorationists, the import of *Faith in a Seed* (Thoreau 1993) lies in Thoreau's recognition that forests *regenerate*, and in his signaling to the place of felling and clearing in forest restoration. (Thoreau had already laid the groundwork on nature's regenerative power earlier in 'The Succession of Forest Trees' and 'Autumnal Tints,' see Thoreau 2001b.) For Thomas Blanding (1988), 'The Dispersion of Seeds' celebrates nature's restorative agency.

Thoreau began (somewhat sporadically at first) keeping a journal in October 1837 at the suggestion of Ralph Waldo Emerson, but it was not until the early-1850s that Thoreau would regularly write, and assign dates to, journal entries. By the time of Thoreau's death on May 6, 1862, his journals totaled more than two million words. Thoreau's journals (collected as Thoreau 1906 *I–XIV*) became a platform for his accounts of excursions (on foot or by boat), his intense almost-daily natural history observations, recordings and reflections, stories of and meetings with acquaintances, and thoughts on scientific, historical, and philosophical volumes he had read. Thoreau's phenological studies are important to the history of ecology and would come to inform early developments in the field of evolutionary ecology over a century later. Thoreau emerges as a protoecologist ('oecologia' or 'ecology' was defined by German zoologist Ernst Haeckel in 1866, four years after Thoreau's death). Just as botanical taxonomists and historians maligned Thoreau's journals for scientific inaccuracies and omissions, so too have some literary critics, for a dearth of poetics in favor of detailed observation, scientific inquiry, and technical writing (highlighted in Case 2013).

One of Thoreau's last projects saw him begin to compile his own natural history compendium for Concord, and it is these writings—more than any other in his canon—that would help determine the ecological restoration reference models developed for Walden Woods. In a lengthy journal entry on September 7, 1851, Thoreau concedes that this phenological project is his life's work (Thoreau 1906 *II*). The beginnings of this ambitious project are trialed in Thoreau's earlier works—most notably in *Walden* (Thoreau 2004a, in the chapters 'The Ponds,' 'Brute Neighbors,' 'The Pond in Winter,' and 'Spring'), and in *The Maine Woods* (Thoreau 2004b, which closes with an extensive Appendix from Thoreau's 1857 excursion, cataloging observed trees, flowers and shrubs, plants, birds, quadrupeds, and more). In 'The Ecology of Walden Woods,' Edmund A. Schofield (1993) observes how the alliance between *Walden* the book and Walden the place resulted in the weaving of an extensive ecological inventory through a literary work (see also Angelo 1983b). But Thoreau's journals would emerge as the keystone to this project.

Beginning in the late-1850s, Thoreau started to overhaul and restructure his journals, compiling lists and charts of Concord's natural phenomena rearranged in order of appearance each year. There were lists tracking and detailing the first winter freeze and spring thaw (ice-out) dates of Walden Pond, the leafing out of trees, shrubs, and flowers in the spring, the ripening of fruits, and migratory bird arrivals and departures through the seasons, as well as sightings of flowers, insects, reptiles, fishes, and mammals (this project is further explored in Baym 1965; Angelo 1983b; Richardson 1993; Buell 1995; Walls 1995, 2017; McGregor 1997; Botkin 2001; Robinson 2004; Case 2013; Primack 2014; Higgins 2017). Thoreau halted work on his journal late in 1861, making his final entry on November 3, as his health deteriorated (Thoreau 1906 *XIV*). But his phenological studies continued, with a particular intensity from October 1861 through January 1862. By this time, Thoreau was no longer able to take long walks and relied on the observations of others.

Thoreau sometimes described his project as a Kalendar, or Book of Concord—anticipating reconfiguring and synthesizing several years of field observations into a single archetypal year (Thoreau 1906; cf. Richardson 1993; Walls 1995; Dean 2000), with a structure and style similar to *A Week on the Concord and Merrimack Rivers* (where his 13-day excursion is retold across seven days) or *Walden* (where his two-year experiment is condensed into one). But at the time of his death from tuberculosis in May 1862 at the age of 44, his elegiac master Kalendar project remained incomplete.

In the 1880s and 1890s, Thoreau's friend and correspondent H. G. O. Blake (to whom his journals passed upon the death of Thoreau's sister, Sophia, in 1876) edited four volumes drawn from the journals—*Early Spring in Massachusetts*, *Summer*, *Winter*, and *Autumn*. Blake's efforts may have been an attempt to honor Thoreau's last, unfinished project (as Richardson 1993).[4] It would not be until the turn of

[4] Thoreau's *Journals* would not be published until 1906, as 14 volumes in the 20-volume collection, *The Writings of Henry David Thoreau.*

the twenty-first century that further insight into Thoreau's ambitious Kalendar project would be revealed: in Bradley P. Dean's edited volume, *Wild Fruits: Thoreau's Rediscovered Last Manuscript* (Thoreau 2000). A compendium of the fruits of New England, *Wild Fruits* continues and extends Thoreau's scrutiny and rigor in local natural history studies. Thoreau's essays on 'Autumnal Tints,' 'Wild Apples,' and 'Huckleberries' (in Thoreau 2001b) together form the backbone of this larger *Wild Fruits* inventory project (cf. 'Thoreau's Kalendar: a digital archive of the phenological manuscripts of Henry David Thoreau,' at: thoreauskalendar.org). In March 2020, the Walden Woods Project completed its Thoreau Animal Index Blitz, a collaborative project with the Thoreau community to inventory the animals mentioned throughout Thoreau's journals—similar to Ray Angelo's (1983b) botanical index of Thoreau's journals.[5]

For restorationists, the scope and detail of Thoreau's cataloging of the natural history of Concord, Lincoln, and Walden provide an extensive botanical, ecological record for the middle decades of the nineteenth century, and a reference model for restoration programs. In *Walden Warming: Change Comes to Thoreau's Woods*, Boston University conservation biologist Richard B. Primack (2014) finds a new use for Thoreau's phenological notes. For an 11-year period beginning in 2003, Primack and his lab team repeated Thoreau's endeavor in recording the first blooming dates of wildflowers in Concord's woods. With Thoreau's 1851–1858 notes as a scientific baseline, Primack's modern records track the effects (and complications) of a warming climate on the flora and fauna of Concord, especially plant flowering times and bird migratory patterns. Primack's application of Thoreau's phenological notes is encouraging for the work of Walden restorationists—confirming, and renewing and reinvigorating, the place of Thoreau-the-field-scientist in contemporary conservation rhetoric. Thoreau's Kalendar, although unfinished, is at once retrospective and forward-looking. Across his works, Thoreau presents an historical genealogy of Walden's forest trees and shrubs.

[5] WWP online interview, July 23, 2020. WWP online interview, July 30, 2020. WWP email correspondence, July 31, 2020.

2.2 The Wild Recast: The Wilderness of Maine's North Woods

Half-way through his two-year sojourn at the cabin, in August 1846, Thoreau left Walden Pond for two weeks for his first excursion into Maine's North Woods to climb 5,276-ft. Mount Katahdin (today, the northernmost trailhead of the 2,190-mile Appalachian Trail) and explore the surrounding lakes and streams. Inspired by—*and* marking a departure from—fellow Concord Transcendentalists Amos Bronson Alcott, mentor Ralph Waldo Emerson, Nathaniel Hawthorne, and Margaret Fuller, Thoreau sought spiritual contact with wild nature—beyond the page, *in place*. Thoreau would make two further excursions to the Maine Woods—in September 1853 and July 1857—and his accounts of these three excursions would be published together posthumously in 1864 as the essays 'Ktaadn,' 'Chesuncook,' and 'The Allegash and East Branch,' in *The Maine Woods* (Thoreau 2004b).

Thoreau's Maine writings—much like his Walden writings—continue to see him positioned in the prose as narrator and traveler-explorer throughout, but they also mark the beginning of a new phase in Thoreau's later writings. Echoing Thoreau's burgeoning interest in scientific inquiry and rigor in the early-1850s, inspired by the works of Johann Wolfgang von Goethe, Alexander von Humboldt, Louis Agassiz, and Charles Darwin among others, his later Maine writings present a scrupulous, more detailed account of his botanizing, and focus more explicitly on natural history and the environmental condition.

But what is striking about Thoreau's accounts of his excursions into Maine's North Woods is how the wooded landscapes there forced him to confront, and begin to challenge and question, his own embryonic understanding of wildness and wilderness[6] developed at Walden Pond (on Thoreau's place in the history of the American wilderness idea, see e.g. Buell 1995; Cronon 1996; Dean 2007; Nash 2014). And yet, for all Thoreau's writings on the wilderness he found in the woods of Maine, he

[6] In distinguishing between wildness and wilderness, Thoreau regards the former as the (spiritual) *experience* in nature, and the latter as the *physical place* in nature—wildness could be experienced in the wilderness—although this distinction, again, is not without inconsistencies and switchbacks across his canon (see Schofield 2004).

would still value the wildness, wilderness of the partially cultivated lands of Concord above all else (as Thoreau 2004b).

It is during his first excursion that he first notes the disparity between the wilderness of Walden and Maine, when he acknowledges Walden Woods's history as a working landscape and how it has been shaped and altered by the actions and priorities of woodsmen and husbandmen (Thoreau 2004b). This disparity is revisited on his second excursion, when he again looks to Walden's place in village life, this time scrutinizing (almost scolding) its function as a village woodlot, with a history told through land deeds and boundary lines (Thoreau 2004b, also 1906 *X*).

Thoreau remains transfixed by his experience on Katahdin long after his return to Concord. Yet he rejects the harsh wilderness of Katahdin's summit in favor of a wilderness that encourages spiritual and creative contact with nature. Put simply, Walden 'was never Katahdin.'[7] Thoreau spent much of June 1853 auditing the wild tracts of Concord, as another way of building up—and preserving—an acquaintance with the town's pastures, woodlots, and orchards (Thoreau 1906 *V*).

For Thoreau, Concord's wilderness was inspiriting—a 'relief' (Thoreau 2001a) or 'tonic' (Thoreau 2001b e.g. 'Walking,' 2004a) to village life—and nowhere better represented this than in the swamps and bogs of Walden. The taxonomy of 'wild' is used by Walden restorationists[8] to describe the bogs, the Andromeda Ponds, Fairhaven Hill and the Cliffs, Heywood's Meadow, and Emerson's Cliff (also Blanding 1988). Concord's wild swamps were the antithesis of gardens and clearings. Such is Thoreau's praise and appreciation of swamps as the epitome of wildness and wilderness (e.g. Thoreau 1906 *IV–VI*, *IX–XI*, *XIV*, 2001b e.g. 'Walking'), that Rod Giblett (1996) declares Thoreau the patron saint of swamps (see also Garber 1977; Botkin 2001 on Thoreau and swamps). The swamp, rather than the mountain, becomes Thoreau's default allegory for sanctified wilderness—a wilderness bountiful, rather than barren. Thoreau's account of becoming lost during a portage in

---

[7] Sasaki Associates, Inc. interview, July 23, 2007.

[8] WWP interview, June 20, 2007. DCR interview, July 5, 2007. WPSR interview, July 13, 2007. Environmental historian telephone interview, July 24, 2007.

the Umbazooksus Swamp with Edward Hoar in Maine in July 1857 embodies this wilderness framing (Thoreau 2004b, also Botkin 2001).

And yet, Thoreau's Walden was not the old-growth forest wilderness recognized by early colonists—it was an agricultural landscape of woodlots and meadows, inhabited by freed slaves, railroad workers, and others, and fractured by the railroad causeway. As Thoreau opines in an October 18, 1860 journal entry, 'We find ourselves in a world that is already planted, but is also still being planted as at first' (Thoreau 1906 *XIV*: 146, 1993: 101), echoing Darwin, and codifying Walden as a cultural landscape. Indeed, Thoreau's discovery of arrowheads while hoeing his bean field causes him to speculate on the past cultivation of Walden Woods by Native American communities (see 'The Bean-Field' and 'Former Inhabitants, and Winter Visitors' chapters in Thoreau 2004a, also 1906, 2001a, b e.g. 'Walking,' 2004b).

* * *

A century before the signing of the Wilderness Act by President Lyndon B. Johnson on September 3, 1964, Thoreau anticipates the need for wilderness protections when he begins to advocate for 'national preserves' and to articulate a national parks idea in the eastern U.S. *The Maine Woods* was published the same year President Abraham Lincoln signed the Yosemite Grant (Chapter 3), and eight years before Yellowstone, the first U.S. national park, was designated (and 52 years before the National Park Service Organic Act, which established the National Park Service (NPS), was signed into law):

> The kings of England formerly had their forests "to hold the king's game," for sport or food, sometimes destroying villages to create or extend them; and I think that they were impelled by a true instinct. Why should not we, who have renounced the king's authority, have our national preserves, where no villages need be destroyed, in which the bear and panther, and some even of the hunter race, may still exist, and not be "civilized off the face of the earth,"—our forests, not to hold the king's game merely, but to hold and preserve the king himself also, the lord of creation,—not for idle sport or food, but for inspiration and our own true recreation? or

> shall we, like villains, grub them all up, poaching on our own national domains?
>
> —Thoreau 2004b: 156

This premise, together with a passage written in his journal on January 3, 1861, where he calls for the protection of 'rare' or 'delectable' landscapes, landscapes of 'rare beauty' (Thoreau 1906 *XIV*), establishes and confirms Thoreau's foresight on wilderness concerns (also 'Walking' in Thoreau 2001b). Thoreau's essay 'Walking' (Thoreau 2001b) and *The Maine Woods* (Thoreau 2004b) are two of the earliest works to advocate for wilderness preservation legislation.

2.3 Thoreau-as-Restorationist

In a journal entry on March 5, 1853, Thoreau describes himself as 'a mystic, a transcendentalist, and a natural philosopher to boot' (Thoreau 1906 *V*: 4). But 'a restorationist' could (and, perhaps, should) be included in this list. Henry David Thoreau is celebrated for his contributions on civil disobedience and social activism, political reform, and environmentalism—but often, his contribution to environmental concerns is framed in terms of his natural history commentary and botanizing, and stops short of ecological restoration. Yet Thoreau's ecological and natural history works reveal an early ecological restoration sensibility. The genesis of this restoration sensibility can be traced to two intersecting concerns of Thoreau's at Walden: logging in Walden Woods, and a need for 'primitive forests.'

In the nineteenth century, Walden Woods (and Walden Pond) was predominantly a working landscape. Thoreau records in *Walden* (Thoreau 2004a) instances of wood-chopping for fuel and timber, ice-cutting, peat-cutting, and fruit harvesting. Across the individual revisions of his *Walden* manuscript over a nine-year period, the felling of Walden gains increasing prominence in his narrative (see Shanley 1957; Sattelmeyer 1990; Buell 1995). The felling of Walden's woodlots is repeatedly and more thoroughly documented in Thoreau's journals throughout the 1850s (Thoreau 1906, see also Blanding 1988, Schofield

1989, 2004; Foster 1999; Sattelmeyer 2000; Botkin 2001; Donahue 2004; Thorson 2017). Thoreau's is an early voice cautioning against the unchecked, unregulated logging at Walden (and elsewhere), and questioning the consequences and repercussions for plant and animal communities in the woods.

During Thoreau's first excursion to Maine's North Woods in the summer of 1846, where logging occurred at a more industrial scale, servicing the mills around Bangor, his anxieties are only escalated. In 'Ktaadn,' the first essay in *The Maine Woods* (Thoreau 2004b), he singles out the New England Friction Match Company for its acquisition of white pine stands. Across his three Maine excursions, Thoreau is increasingly surprised—and dismayed—to discover the scarcity of white pine stands in the woods and abhors the selective commercial valuing of the woods. By his third trip, Thoreau is resigned to the scarcity of the white pines—abandoning his search for an undisturbed stand of white pine in the woods (Thoreau 2004b). The Pine Tree State was thinning.

The felling of Walden in the early-1850s for cordwood and lumber elicited some of Thoreau's most melancholy and morose commentaries and reflections on the exhausted and disappearing woods. Walden was moribund, and Thoreau was there to bear witness. Journal entries on December 30 and 31, 1851, present a somber vignette of the sawing of a 'noble pine tree' from Thoreau's vantage point on Fairhaven Hill (Thoreau 1906 *III*), and this timbre continues in journal entries into mid-January 1852, when Thoreau again pushes back against the felling (Thoreau 1906 *III*). Thoreau's sorrow at the felling culminates in the early spring of 1855, when he offers a pseudo-eulogy to Walden on March 6: 'Our woods are now so reduced that the chopping of this winter has been a cutting to the quick. At least we walkers feel it as such. There is hardly a wood-lot of any consequence left but the chopper's axe has been heard in it this season' (Thoreau 1906 *VII*: 231). In the same entry, he goes on to record the clear-cutting of Emerson's Cliff, White Pond, and other tracts throughout the woods. But it is not just trees that are lost to logging—there is also a loss of history, myth, story: 'The Anglo American can indeed cut down and grub up all this waving forest and make a stump speech and vote for Buchanan on its ruins, but he cannot converse with the spirit of the tree he fells—he cannot read the poetry and mythology which retire as he advances. He ignorantly erases mythological tablets in order to print his handbills and town meeting warrants on them' (Thoreau 2004b: 229).

Thoreau estimated that Concord's primitive and second growth forest had all but disappeared by October 1860 (Thoreau 1906 *XIV*, also Whitford 1950; Whitney and Davis 1986). The following month, Thoreau completed an audit of remaining forest stands in Concord (Thoreau 1906 *XIV*). Thoreau offers a remedy—albeit in jest—to the felling: growing his beard, to cover the exposed, denuded, and vulnerable ground (Thoreau 2004b).

Even after the completion of the Fitchburg Railroad infrastructure through the woods in 1844, the year before Thoreau moved to the pond, the railroad (later, the Boston and Maine Railroad, Fitchburg Division Line) still had a part to play in destroying the pine trees of Walden. Sparks and cinders from passing locomotives were notorious for setting many fires in Walden Woods. Despite his protests over the felling of trees for the causeway, for railroad ties, and to fuel the locomotives, and roiling against the forest fires caused by sparks from the locomotives, the railroad became part of Walden lore, part of Thoreau's Walden idyll. Thoreau would often also walk along the causeway to get to town.

Thoreau is not free from blame in the reduction of forest cover at Walden and admits in *Walden* (Thoreau 2004a) to having 'profaned' the woods. Thoreau's 'profaning' of Walden might include his felling of white pine to construct his cabin, his clearing of a plot for his bean field, or even his surveying (and dividing) of Walden's woodlots, but most probably the scenario that Thoreau alludes to occurred the year before he began his sojourn at Walden Pond, when a rogue campfire near Fairhaven Bay ignited a large swath of the surrounding forest. On a boating excursion, Thoreau and companion Edward Hoar had stopped for supper on the shore, away from the tree line, when a spark from their fire ignited the dry grass and quickly spread uphill into the woods. The April 30, 1844 fire burned over 300 acres and caused $2,000 of damage. Thoreau did not write of the accidental blaze in his journal for more than six years—until May 31, 1850 (Thoreau 1906 *II*, 2004a, also Blanding 1988; Harding 2015). John Pipkin's (2010) novel *Woodsburner* fictionalizes this April 1844 fire in the woods.

In the late-1850s, after bearing witness to this felling not only in and around Walden, but also in Maine's North Woods, and on Cape Cod, Thoreau offers a rebuttal to Walden's wood-chopping past, and avows,

> Each town should have a park, or rather a primitive forest, of five hundred or a thousand acres, where a stick should never be cut for fuel, a common possession forever, for instruction and recreation […] inalienable forever. Let us keep the New World *new*, preserve all the advantages of living in the country. There is meadow and pasture and wood-lot for the town's poor. Why not a forest and huckleberry-field for the town's rich? All Walden Wood might have been preserved for our park forever, with Walden in its midst, and the Easterbrooks Country, an unoccupied area of some four square miles, might have been our huckleberry-field.
>
> —Thoreau 1906 *XII*: 387

This earnest preservation sentiment is repeated almost verbatim in the 'Huckleberries' essay (in Thoreau 2001b) and to conclude *Wild Fruits* (Thoreau 2000). But it is in these latter two works that Thoreau also confirms and amplifies his position as an early pioneer and advocate of conservation and preservation practices, cast perhaps as a protoenvironmentalist (see also his prioritizing of preservation over destruction in Thoreau 2004b). For in addition to calling for public parks, he also proposes the establishment of town committees, and the appointment of forest wardens or rangers, to oversee their protection (Thoreau 1906 *XI*, *XIV*, 1993, 2000, 2001b, also Blanding 1988). (Earlier in 1852, writing in his journal on April 12, he as good as nominates himself as Walden's warden or guardian (Thoreau 1906 *III*).) He also alludes to introducing legislation that would allow towns to regulate which woodlots should be cut each winter. Furthermore, Thoreau not only advocates for township-level protections, but also suggests a state-level land preservation program (Thoreau 2000, 2001b). (But the *Journals* would not be published until 1906, and 'Huckleberries' and *Wild Fruits* later still—with Thoreau's counsel thus dampened to a slow burn.)

While his provocation on the need for 'primitive forests' and trepidation surrounding the felling of Walden Woods underpins Thoreau's restoration sensibility, there are a number of instances elsewhere in

Thoreau's writings where a restoration rhetoric is more ephemeral—as when Thoreau champions a 'revival' of trees (Thoreau 2004c), or forest 'renovation' (Thoreau 1993). William R. Jordan III (1993) has done some early work on the connections between ecological restoration and Thoreauvian thought, distinguishing three restoration themes across Thoreau's canon: the reinhabitation of nature, renewal, and the performative relationship with nature. But I want to further argue that a Thoreauvian restoration sensibility can be distilled to two motifs: restoration as an act of *healing* and as an act of *redemption.* When Thoreau writes in *A Week on the Concord and Merrimack Rivers* of, 'This poor globe, how it must itch in many places! Will no god be kind enough to spread a salve of birches over its sores?' (Thoreau 2001a: 126), the 'salve of birches' might be read as a forerunner of what would become a framing of ecological restoration as an act of healing (see especially Maser 1988; Higgs 1997, 2003; France 2008).

It is in his celebration of the wild swamps and bogs of Concord that Thoreau further unpacks a restoration rhetoric, and here recasts restoration as an act of ecological redemption to further endorse his claims of wilderness (see especially Garber 1977 on Thoreau's 'redemptive imagination,' of redemption through imagination, also Jordan 1993; Buell 1995; Marx 2000; Robinson 2004). Thoreau explores the idea and promise of 'redeeming' nature in Walden's swamps. In perhaps his most explicit use of the idea, he declares in his journal that, 'redeeming a swamp [...] comes pretty near to making a world' (Thoreau 1906 *IX*: 311). For Thoreau, the wildness found in the swamp becomes a tonic, a restorative ('A Walk to Wachusett,' 'Walking' in Thoreau 2001b). But confusion also cloaks Thoreau's use of 'redeem,' for while it could be read as preserving, conserving, restoring nature, Thoreau also uses it to describe those patches of marshes or swamps reclaimed—tamed—by farmers and others (Thoreau 1906 *III*, *XIV*, 1993, 2000, 'Walking' in 2001b, 2004a, c). The redemptive imagination, the redemptive consciousness (after Garber 1977) both enriches and reclaims wild nature. Walden is redeemed by the imagination.

Thoreau also demonstrates an early appreciation of the systems and processes of natural regeneration—observing forest trees 'lustily springing up' on cut-over lands (e.g. 'Huckleberries' in Thoreau 2001b,

2004a)—and this appreciation is central to his treatise on seed dispersal (Thoreau 1993, 'The Succession of Forest Trees' in 2001b, also Blanding 1988; Nabhan 1993; Foster 1999). Because of Thoreau's exposure to Walden Woods over many decades, he observes and tracks the sequential processes and dynamics of natural regeneration that take place in neglected, abandoned meadows and forests (Thoreau 1993)—of birth, death, decay, and rebirth. His treatise also identifies six categories of New England woodlands: the primitive wood, second growth, primitive woodland, cleared woodland, new woods, and artificial[9] woods (Thoreau 1906 *XIV*, 1993), charting succession. He takes this one step further, to ponder a Concord sans its townsfolk, where trees and shrubs reclaim and reoccupy the land (Thoreau 1906 *XIV*). Thoreau also notes the promise of natural regeneration held in the stewardship, or low disturbance, of woodlots (Thoreau 1906 *XIV*, 1993, 2000, and cf. Thoreau's account on December 11, 1856, of friend George Minot as an early adopter of provident, sustainable forest management, in Thoreau 1906 *IX*). For Thoreau, regeneration is most often impeded by human interference and intervention—by a lack of knowledge of forest processes among farmers and landowners (Thoreau 1906 *XIV*).

A restoration rhetoric in Thoreau's works also reveals the capacity for both ecological and social restoration, and the synergy between the two, as when he argues, 'If, then, we would indeed restore mankind by truly Indian, botanic, magnetic, or natural means, let us first be as simple and well as Nature ourselves, dispel the clouds which hang over our own brows, and take up a little life into our pores' (Thoreau 2004a: 78–79, cf. 'the regeneration of mankind,' 1906 *IX*). A social restoration rhetoric is further echoed in Thoreau's search for 'a natural life,' demanding that one become both naturalized *and* spiritualized in, by nature (Thoreau 2001a, cf. Robinson 2004).

Yet Thoreau does more than simply offer commentary on the need to restore nature. On three consecutive days in mid-April 1859, nearly a decade-and-a-half after concluding his sojourn, Thoreau returns to

---

[9] Thoreau's framing of 'artificial woods' here (with the phrase appearing again in Thoreau 2004b, c) is interesting in teasing out Thoreau's restoration ethos, not least because a nature—artifice diptych has long raged at the forefront of philosophical, ethical, and moral critiques of ecological restoration (most notably in Elliot 1982; Katz 1992).

his former bean field in Walden Woods to plant 400 white pine saplings transplanted from the woodlots of Brister Freeman and George Heywood, and from elsewhere on Emerson's Wyman Lot. This small-scale replanting project reveals an early example of ecological restoration and an explicit example of a Thoreauvian restoration sensibility. As Thoreau records in his journal,

> *April* 19. [...] P. M.—Began to set white pines in R. W. E.'s Wyman lot.
> *April* 20. [...] Setting pines all day.
> *April* 21. Setting pines all day. This makes two and a half days, with two men and a horse and cart to help me. We have set some four hundred trees at fifteen feet apart diamondwise, covering some two acres. I set every one with my own hand, while another digs the holes where I indicate, and occasionally helps the other dig up the trees. We prefer bushy pines only one foot high which grow in open or pasture land, yellow-looking trees which are used to the sun, instead of the spindling dark-green ones from the shade of the woods. Our trees will not average much more than two feet in height, and we take a thick sod with them fifteen to eighteen inches in diameter. There are a great many more of these: plants to be had along the edges and in the midst of any white pine wood than one would suppose.
>
> —Thoreau 1906 *XII*: 152

In rewilding the bean field, Thoreau recasts the role of 'gardener' as perhaps 'steward' (drawing parallels with Minot's stewardship of his woodlot, recounted in Thoreau 1906 *IX*) and begins a renewed commitment and intimacy with not only the plot, but also the wider woods. For Jordan (1993), Thoreau is teetering on the edge of an ecological restoration rhetoric in his management of the bean field and had he only developed his ideas further, might have become the first restorationist in the U.S.

A year later, writing in 'The Succession of Forest Trees,' Thoreau alludes perhaps to his replanting, to argue in favor of a greater alliance with—and respect, reverence for—nature: 'So, when we experiment in planting forests, we find ourselves at last doing as Nature does. Would it not be well to consult with Nature in the outset? for she is the most extensive and experienced planter of us all, not excepting the Dukes of

Athol' (in Thoreau 2001b: 438, also 1993: 134). In one respect, Thoreau even foretold his own replanting endeavor nearly a decade before, when in a September 4, 1851 journal passage he writes, 'We are a young people and have not learned by experience the consequence of cutting off the forest. One day they will be planted, methinks, and nature reinstated to some extent' (Thoreau 1906 *II*: 461–462, *XIV*). His own studies of succession and seed distribution could be read as the very 'learned experience' he writes of.

Thoreau is not the only Concord resident to recognize the importance of replanting Walden's forest, for he also notes in his April 21, 1859 journal entry that Emerson—on whose plot Thoreau's cabin and bean field plot were located—intended a similar practice, having recently purchased a quarter-pound of white pine seed (Thoreau 1906 *XII*). It would be Thoreau who would plant the white pine seed on Emerson's second Walden woodlot, on the pond's south shore.

Regrettably, Thoreau's replanted white pine stand would turn out to be ephemeral. Less than a half-century later, it was largely destroyed in first an 1872 and then an 1896 blaze caused by sparks from a locomotive (Blanding 1988; Maynard 2004; Dean 2005; Donahue 2013) and again during the 1938 hurricane. But white pine has reestablished on the bean field plot. The location of Thoreau's former bean field would not be confirmed until 2005 (see Dean 2005) (Fig. 2.3).

* * *

Thoreau's restoration sensibility is not confined to woods and forests. Fifty years before John Muir's unsuccessful campaign to prevent the construction of a dam and reservoir project on the Tuolumne River in Yosemite's Hetch Hetchy Valley in California (Chapter 3), and over a century before Edward Abbey began calling for the restoration of Glen Canyon in southern Utah by draining the waters of Lake Powell behind Glen Canyon Dam on the Colorado River (Chapter 6), Thoreau was speculating within the pages of *A Week on the Concord and Merrimack Rivers* on the possibility of dam removal on the Concord River and restoring the riparian landscape. Thoreau made an early case that removal would benefit both fish populations and farmers. And it is here, too,

Fig. 2.3 The site of Thoreau's former bean field. Courtesy of the Walden Woods Project and Thoreau Institute at Walden Woods

that Thoreau expresses the sentiment that would come to epitomize the fictional quartet of eco-saboteurs in Edward Abbey's comic and satirical novel, *The Monkey Wrench Gang* (Abbey 2004): 'and who knows what may avail a crow-bar against that Billerica dam?' (Thoreau 2001a: 20, on Thoreau and dams, see also 2004b, and for an extended account of Thoreau's surveying work in the late-1850s for the River Meadow Association in support of its case to secure the removal of the factory dam in Billerica, see Thorson 2017).

2.4 After the 1922 Deed of Gift

Before Thoreau's sojourn, Walden Pond had been a muse for Concord Transcendentalists including Alcott, Emerson, Hawthorne, and others.[10] But with the publication of Thoreau's *Walden* in 1854 (and his death in 1862), the pond and woods—the 'philosophers' sacred grove'—emerged as one of the earliest sites of literary pilgrimage in the U.S. (Gleason 1917; Blanding 1988; Buell 1995; Maynard 2004). Toward the end of the nineteenth century, Walden's place as a recreation destination was cemented—both figuratively and literally—with the establishment in 1866 of an excursion park on the shore of Ice Fort Cove by the Fitchburg Railroad, which, although expanded in the 1880s and 1890s, later burned down in 1902 and was never rebuilt. And the popularity of the automobile in the opening decades of the twentieth century brought yet more visitors to the pond, with the Town of Concord offering swimming lessons from 1913, followed four years later by the construction of new bathhouses.

Concerned for the fate of Walden Pond, and its legacy amidst the philosophy and writings of the Concord Transcendentalists, members of the Emerson, Forbes, and Heywood families (including two of Emerson's children, Edith Emerson Forbes and Edward Emerson) granted almost 80 acres of woodlots surrounding the pond to the Commonwealth of Massachusetts in mid-1922 (with responsibility later shifting to Middlesex County).[11] This Deed of Gift reaffirmed—and secured—Walden's place in Concord's literary history, and established the Walden Pond State Reservation (WPSR).

The Deed of Gift explicitly required the continued preservation of 'the Walden of Emerson and Thoreau,' while supporting recreation such as 'bathing, boating, fishing, and picnicking' (qtd. in Maynard 2004:

[10] The Emerson-Thoreau Amble, opened in 2015, is a 1.7-mile trail from Heywood Meadow near Emerson's home at 18 Cambridge Turnpike to Thoreau's cabin at Walden Pond and follows the approximate route the two men would walk between Concord and Walden.

[11] Edith Forbes Emerson had earlier purchased a woodlot on Fairyland Pond, in November 1883, to protect it from felling.

229), consistent with Thoreau's experiences.[12] In referencing Thoreau and Emerson, the donors note the literary, historical, cultural, as well as geographical significance of the landscape. It is the Walden found *on the page* that is celebrated. Less than five years later, the Fitchburg Railroad conveyed the remaining land that abutted the shore of the pond to the Commonwealth. The proclamations set forth in the 1922 Deed of Gift would become an important determinant of the reference models and restoration programs developed for Walden Woods and Walden Pond later in the twentieth century.

2.5 Walden Restoration: Politicizing 'the Walden of Emerson and Thoreau'

Thoreau was not ignorant of the threats facing Walden Woods. In *Walden* (Thoreau 2004a), he identifies a trifecta of woodchoppers, ice-cutters, and the Fitchburg Railroad as the key antagonists profaning Walden. Located on Concord's southern periphery, Walden Woods in the mid-nineteenth century held little value for the town beyond its pecuniary function as a woodlot and had consequently attracted the town's outcasts (see Blanding 1988; Lemire 2009). Historically, Walden was regarded as little more than a 'pitch-pine plain' (Thoreau 1993). And it was these acidic, sandy, droughty, sterile soils, and steep slopes, together with limited 'water privileges' beyond Walden Pond and Brister's Spring (Thoreau 2004a, also Schofield 1989; Collins et al. 2000), that ultimately protected the woods from development in the nineteenth century. But this would not long be the case. Soon the woods would be under siege.

It is in the posthumously-published essay 'Walking,' Thoreau's treatise on wilderness, that restorationists might find their greatest endorsement for politicizing Thoreau's philosophy and writings in defense of Walden Woods. The essay opens with the declaration that, 'I wish to speak a

[12] Revere Beach, the first public beach to be designated in the U.S. in the late-1890s, close to downtown Boston, was cited in the Deed of Gift as the antithesis of what the families desired for Walden.

word for Nature, for absolute freedom and wildness, as contrasted with a freedom and culture merely civil,—to regard man as an inhabitant, or a part and parcel of Nature, rather than a member of society' (in Thoreau 2001b: 225). It is this statement that carries through into, and has come to define, the ecological restoration reference models and programs that have played out over the last 70 years at Walden. E. L. Doctorow (1991), in his remarks at the April 1990 press conference for the newly established Walden Woods Project, articulates the need to safeguard both the book *and* the place, pointing to the power of literature in composing place, and how Walden the place now stands as a symbol of hope and possibility (cf. Blanding 1988). A power that rests with what is *suggested* by the word 'Walden.'

Thoreau's resolve to 'speak a word for Nature' would be retold across Walden Woods from the middle of the twentieth century onward, as the woods and pond variously fell prey to human folly—to a beach expansion, proposed commercial development projects, shoreline erosion and bank collapse, options for the closed and capped municipal landfill, and new high school playing fields. And each time, Thoreau's environmental stewardship philosophy—and ecological restoration sensibility—would be negotiated and renegotiated by conservationists in defense of Thoreau's Walden.

Walden the book (Thoreau 2004a) is a profoundly jeremiadic, political text, and it has in no small part helped to politicize Walden the place (see Wheeler 2005; Petrulionis and Walls 2007). For Jordan (1993), ecological restoration at Walden is but an extension of Thoreau's own Walden 'experiment.' Ecological restoration is, then, a legacy of Thoreau's (2004a) *Walden*. Nature here is political and politicized, with ecological restoration a politically powerful rhetoric upholding 'the Walden of Emerson and Thoreau' (qtd. in Maynard 2004). There is an historical, literary, cultural, political contingency of ecological restoration in Walden. And yet a paradox persists in restor(y)ing Thoreau's Walden: 'the Walden of Emerson and Thoreau' spans decades, and Walden Woods in the 1840s and 1850s was a working landscape, that by the late-1850s had been extensively felled for fuel and timber. By contrast, Walden in the twenty-first century is no longer a working landscape, and is certainly more fully wooded. No restoration program is going to aggressively advocate for felling. Robert Sattelmeyer (2000) makes an interesting observation when he remarks that contemporary reforestation—restoration—has resulted in a Walden Woods that is more like the Walden depicted in Thoreau's (2004a) *Walden* than the Walden of the 1840s. Here, it is the imagined Walden, the literary Walden, that lingers in an ecological restoration rhetoric. And yet, the denser vegetation in the woods masks a decline in species diversity. And degradation persists, as erosion and water quality worsen (Sattelmeyer 2000).

Across the various restoration programs in the woods, 'the Walden of Emerson and Thoreau' is a mutable, malleable reference model. And so *WWTD? What would Thoreau do?* becomes a legitimizing question (as Thoreau Farm Trust 2017). What is sought is Thoreau's constructed, curated, imagined, restor(y)ed—and botanized—Walden. Thoreau's Walden *idyll*.

2.5.1 Where It All Began: 1957, the Thoreau Society's Save Walden Committee, and the Campaign to Restore Red Cross Beach

The first assault on 'the Walden of Emerson and Thoreau, its shores and woodlands' (qtd. in Maynard 2004: 229) came just three-and-a-half decades after the Deed of Gift was granted. In the spring of 1957, the Middlesex County Commissioners authorized, at the request of the American Red Cross and other organizations, the expansion of a beach area on the northeast shore of Walden Pond, allowing greater numbers of children to enroll in swimming and water safety classes (part of a 20-year improvement plan for WPSR). Walden Pond is the deepest natural body of water in the Commonwealth of Massachusetts and flourished as a popular outdoor recreation site. In creating the expanded beach area in mid-June to mid-July 1957, part of a steep bank was regraded and the excavated material moved into the pond, more than a hundred trees were felled, and a gravel road was cut between the (expanded) parking lot and beach. On July 4, 1957, exactly 112 years after Thoreau began his sojourn, *The Concord Journal* printed a trio of photographs, each titled 'Walden improved,' celebrating the work of the Middlesex County

Commissioners.[13] Two week's earlier, in a June 20 editorial in the same newspaper, Ruth Wheeler had sharply remarked that an 'artificial and synthetic' Walden held greater lure for visitors than the 'natural' one (qtd. in Wheeler 2005). And all this activity took place on an acre-and-a-half of land earlier owned by C. Fay Heywood, included in the Deed of Gift.

The Red Cross Beach expansion was not the first change to Walden's landscape after the Deed of Gift. But it was the first to significantly alter—and destroy—a large tract of the shoreline and woods protected under the conveyances. The Great New England Hurricane of 1938 (the 'Long Island Express') had already leveled a large swath of the woods, and multiple stands had succumbed to chestnut blight and an infestation of gypsy moths. Between 1945 and 1950, a dirt roadway, a flight of concrete steps and a concrete ramp, an administrative building, two concrete bathhouses (replacing two smaller wooden bathhouses and outhouses), a pump house, a retaining wall, and a wooden raft extending into the pond were constructed on or near the most accessible beach on Walden Pond's eastern shore. By the late-1950s, there were also three parking lots along the east shore. Yet even with these additions, the shoreline had remained largely unaffected.

The July 1957 annual meeting of the Thoreau Society, held to coincide with Thoreau's birthday on July 12, also coincided with the beach expansion works. Although not primarily a conservation organization, the Thoreau Society established the Save Walden Committee that summer, led by vice president Gladys Hosmer, and began a legal campaign that fall to challenge the violation, repeal by the County Commissioners of the conveyances set down in the Deed of Gift (recounted across issues 61 through 65 [1957 through 1958] of the *Thoreau Society Bulletin*,[14] see also Maynard 2004; Wheeler 2005 on the battle for Red Cross Beach). Thoreau's (and Emerson's) writings became a trigger warning for the desecration facing Walden and were seized upon—and politicized—by the Save Walden Committee to defend the woods.

---

[13] 'Walden improved.' *The Concord Journal*. (July 4, 1957): 1, 2, 4 [Thoreau Society Collections—Archives of the Thoreau Society, at the Thoreau Institute at Walden Woods].

[14] Thoreau Society Collections—Archives of the Thoreau Society, at the Thoreau Institute at Walden Woods.

An early success of the Save Walden Committee in September 1957—obtaining a temporary injunction to halt construction of a hard road and bathhouse—was short-lived, with the Middlesex County Superior Court later reversing the order and dismissing the petition, forcing the Save Walden Committee to take its appeal to the Massachusetts Supreme Judicial Court. It was in the Save Walden Committee's campaign correspondence of the late-1950s that Walden *the place* was first politicized as a 'living monument,' and a 'historical shrine,' a 'literary shrine,' a 'national shrine,' a 'shrine for conservationists and naturalists'[15]—with this language carrying through into many of the legal briefs. Media coverage of Save Walden's campaign was widespread across New England and beyond.

The legal briefs prepared on behalf of the Save Walden Committee[16] record concerns regarding the scale of the beach works—the artificial contouring of the bank, increased water runoff, a lack of knowledge of indigenous trees and shrubs, and an absence of plans for replanting. For the plaintiffs, the changes were inconsistent with preserving 'the Walden of Emerson and Thoreau' (qtd. in Maynard 2004). Remarkably, Thoreau's philosophy and writings, and his actions in Walden Woods, were used by *both* parties to argue their cases. The plaintiffs made direct reference to Thoreau's elegiac descriptions of Walden Pond and Walden Woods (e.g. Thoreau 2004a: 185, 186),[17] to illustrate what

[15] Thoreau Society Collections—Archives of the Thoreau Society, at the Thoreau Institute at Walden Woods, and the Edwin Way Teale papers relating to Thoreau and to Concord, Mass., 1918–1980, at the Concord Free Public Library.

[16] First for John E. Nickols et al. v. William G. Andrew et al. 1958. Commonwealth of Massachusetts Superior Court. Middlesex SS. No. 20454, and later, John E. Nickols et al. v. William G. Andrew et al. 1960. Commonwealth of Massachusetts Supreme Judicial Court. Suffolk County. No. 5924 (A companion case, also filed by the Save Walden Committee, sought to transfer management of WPSR to the state's Department of Natural Resources.) [reports and proceedings held in the Thoreau Society Collections—Archives of the Thoreau Society, at the Thoreau Institute at Walden Woods].

[17] Cited in Hale and Dorr. 'Petitioners' Memorandum of Law in Support of Final Decree,' 3–4, 11, prepared for John E. Nickols et al. v. William G. Andrew et al. 1958. Commonwealth of Massachusetts Superior Court. Middlesex SS. No. 20454, also Fisher, Frederick G. Jr. 'Brief of the Petitioners in Both Cases,' 27, prepared for John E. Nickols et al. v. William G. Andrew et al. 1960. Commonwealth of Massachusetts Supreme Judicial Court. Suffolk County. No. 5924 [Thoreau Society Collections—Archives of the Thoreau Society, at the Thoreau Institute at Walden Woods].

was lost to the expansion of Red Cross Beach, while the defendants looked to Thoreau's own felling of white pines for his cabin, with the County Commissioners also asking what 'Thoreau's Walden' meant, over a century after his sojourn.[18]

The Save Walden Committee argued that the Middlesex County Commissioners should be ordered to restore Walden Pond—that the contours should be regraded, native trees and shrubs replanted, the 'Indian trail' restored, the roadway removed, and plans for a bathhouse abandoned. Save Walden also petitioned to have the erosion elsewhere along the shoreline included in the restoration program, and recommended the completion of a study into use and activity at Walden Pond prior to the start of any restoration program. Moreover, the committee requested that the County Commissioners be ordered to refrain from any future act inconsistent with 'the Walden of Emerson and Thoreau.'[19] Together these actions would, over time, the committee argued, remedy the damage and restore Walden's 'sylvan charm.' The term 'restoration' appears repeatedly throughout a later legal brief,[20] and this is one of the earliest instances of an ecological restoration rhetoric becoming institutionalized and politicized in Walden Woods.

On May 3, 1960, Justice R. Ammi Cutter, sitting on the Massachusetts Supreme Judicial Court, ruled in favor of the Save Walden Committee of the Thoreau Society, recognizing Walden as a literary shrine (the first in the U.S.), and requiring the Middlesex County Commissioners to restore the woodland and shoreline of Red Cross

---

[18] Commissioners of Middlesex County. 1958. 'Walden Pond' [Thoreau Society Collections—Archives of the Thoreau Society, at the Thoreau Institute at Walden Woods].

[19] Fisher, Frederick G. Jr. 'Brief of the Petitioners in Both Cases,' prepared for John E. Nickols et al. v. William G. Andrew et al. 1960. Commonwealth of Massachusetts Supreme Judicial Court. Suffolk County. No. 5924 (cf. 'Request by the Petitioners and Plaintiffs for Findings of Fact by the Master and Auditor,' prepared for John E. Nickols et al. v. William G. Andrew et al. 1958. Commonwealth of Massachusetts Superior Court. Middlesex SS. No. 20454) [Thoreau Society Collections—Archives of the Thoreau Society, at the Thoreau Institute at Walden Woods].

[20] Fisher, Frederick G. Jr. 'Brief of the Petitioners in Both Cases,' 8–9, 10, 12, esp. 29–31, prepared for John E. Nickols et al. v. William G. Andrew et al. 1960. Commonwealth of Massachusetts Supreme Judicial Court. Suffolk County. No. 5924 [Thoreau Society Collections—Archives of the Thoreau Society, at the Thoreau Institute at Walden Woods].

Beach.[21] Judicial notice was taken of Thoreau's *Walden* and *Journals* (with a footnote in the final decree extensively citing *Walden*), as well as Emerson's poetry. This decision would prove pivotal in the conservation and restoration of Walden. In 1962, a century after Thoreau's death, Walden Pond was added to the National Historic Landmark register (the first landmark to be listed under a 'Literature' theme). However, the damaged slope would remain a gravel pit for over two decades, until the management of Walden was transferred to the state in July 1974 (to the Department of Natural Resources [DNR], which became the Department of Environmental Management [DEM], and later the Department of Conservation and Recreation [DCR]), and restoration work began in 1980.

2.5.2 Shoreline Stabilization at Walden Pond

The increased popularity of Walden Pond in the postwar era introduced new threats to the Walden landscape. Traffic congestion, water pollution, and erosion (see Maynard 2004; Wheeler 2005) became major concerns. Erosion was recorded in the 1960 Supreme Judicial Court ruling.

By the end of Middlesex County's tenure of WPSR in the mid-1970s, there was severe erosion around the perimeter of Walden Pond. In adherence with the Deed of Gift, the County Commissioners had run WPSR as a recreation site, but had imposed very little environmental management or regulation at the state park. In the 1960s, the County Commissioners had introduced a pseudo-restoration in an effort to control erosion of the steep shoreline—by installing railroad ties and timber cribbing, stacked and angled up the slopes, like bleachers. But no corresponding effort was made to control visitor numbers. So over the years, erosion resumed, and the cribbing began to disintegrate and collapse. Richard A. Gardiner and Associates' (1974) *Walden Pond Restoration Study, Final Report*, recommended an agenda for restoration that not only prioritized the revegetation and stabilization of Walden's shoreline and trails, but also commented on infrastructure, pointing

[21] *Nickols v. Commissioners of Middlesex County*341 Mass. 13. 1960. 166 N.E.2d 911.

to improving parking facilities (and pedestrian access to the pond), restricting unauthorized, uncontrolled parking on soft verges along Route 126, capping the number of visitors allowed in the reservation, and improving the beach facilities. The release of the study coincided with the transfer of WPSR to DNR. When DNR took over the management of WPSR, the state park became part of the Massachusetts State Forests and Parks system (on DNR/DEM's management at WPSR, see, e.g., Couture 1993).

In 1980, a grassroots environmental campaign took hold at Walden Pond. Led by botanist and forester Mary P. Sherwood, Walden Forever Wild (WFW, previously the Walden Pond Society) lobbied to see the Walden Pond State Reservation redesignated as a sanctuary or refuge, to protect the Thoreauvian spirituality and solitude of Walden.[22] WFW also sought stricter environmental regulations and management at Walden, and petitioned to prohibit swimming at the pond, counter to the terms of the Deed of Gift (arguing it would reduce water contamination). In many respects, WFW was a continuation of the Save Walden Committee (although independent of the Thoreau Society) and sought to draw attention to the (ecological restoration) proclamations laid down in the 1960 ruling. Although WFW achieved success as a citizen action group, it was ultimately unsuccessful in its bid to have Walden Pond designated as a literary shrine. (Conservationists, often led by Sherwood, would also rally to have Walden come under the purview of the NPS throughout the 1980s and 1990s.) Of all the groups and organizations seeking to variously 'restore' Thoreau's Walden, Walden Forever Wild is the only one to institutionalize, politicize the idea of Walden-as-shrine, of Walden as a *sacred* site.

It was at the start of the 1980s that DEM began to restore the shoreline and slope at the center of the 1957 Red Cross Beach lawsuit. Between 1980 and 1984, park rangers introduced new glacial deposits

[22] Beginning in late-1983, and intermittently across the next decade, WFW filed a bill in the Massachusetts state legislature—variously titled the Walden Protective Act, the Walden Pond State Sanctuary Act—to designate Walden Pond a state sanctuary (Couture 1993; Maynard 2004) [also *Walden Forever Wild Committee Newsletter*, *Walden's Future Forever Wild Newsletter*, *The Voice of Walden: Walden Forever Wild Newsletter* in the Walden Forever Wild Research Collections, and Thoreau Society Collections—the Walter Harding Collection, at the Thoreau Institute at Walden Woods].

to replace those lost during the 1957 excavation on Walden's northern slopes, restored the original contours, and transplanted tree and shrub seedlings. (Sherwood was a key figure in this restoration effort, alongside local amateur archeologist Roland Wells Robbins (who had discovered Thoreau's cabin site in mid-November 1945), landscape architect J. Walter Brain, and others, and later in the 1980s proposed a similar replanting program for Walden's south shore, to combat erosion.[23]) Returning to the idea of (historical) fidelity in ecological restoration, the restoration of Red Cross Beach most closely aligns with James Aronson et al.'s (1993) restoration *sensu stricto* ('narrow sense'), of a restoration defined by strict historical fidelity goals. The restoration program at Red Cross Beach, perhaps more so than any other restoration program that followed elsewhere in Walden Woods, is the one that most closely adheres to—and demands—an historic reference. Red Cross Beach, for Walden restoration purists, 'is ecological restoration'[24] because of the replacement of the glacial deposits (Fig. 2.4). Subsequent restoration programs in Walden, in contrast, adopt or favor restoration *sensu lato* ('broad sense'), a 'functional' restoration (Aronson et al. 1993).

But DEM's restoration program did not stop with Red Cross Beach. In the mid-1980s, DEM revisited the restoration needs of Walden Pond, with a restoration program built around two issues: (i) the physical restoration, and (ii) limiting the number of visitors at WPSR at any one time.[25] Gardiner and Associates (1974) had earlier recommended a carrying capacity of 1,000 visitors at WPSR, and a parking lot already built in the 1970s for 350 vehicles complemented the cap. The parking lot could regulate visitor numbers, but still visitors found other places to park—whether at the nearby high school campus, or on the fringe of the woods along Route 2 and Route 126 (wooden posts were subsequently installed on Route 126 as a countermeasure). DEM worked with the Town of Concord, and the Concord-Carlisle Regional High School

[23] Shoreline stabilization and restoration efforts are recounted across *Walden Forever Wild Committee Newsletter*, *Walden's Future Forever Wild Newsletter*, *The Voice of Walden: Walden Forever Wild Newsletter* [Walden Forever Wild Research Collections, and Thoreau Society Collections—the Walter Harding Collection, at the Thoreau Institute at Walden Woods].

[24] FOTC interview, July 27, 2007.

[25] Mass Audubon interview, July 3, 2007.

Fig. 2.4 The restored shoreline and wooded slopes at Red Cross Beach, the center of the 1957 Save Walden Committee campaign. Photographs by the author, May 2006

(CCHS), to eliminate the high school as a parking venue (part of the CCHS campus would later become the site of another Walden conservation battle). At the start of the twenty-first century, another study investigated relocating Route 126 away from the eastern shore of the pond (Epsilon Associates 2001).

One early project to stabilize the banks along the shore included piling fieldstones from the water's edge up to the perimeter trail and then planting among the rocks. Although durable, the agreement among restorationists was that it was 'really not the right look'[26] for Walden. Next, fieldstones were placed just at the water line (to mitigate the wave action along the shore) and then in-filled above with topsoil and native planting—the fieldstones a small alliance with Thoreau's observations of the pond's 'stony shore' (Thoreau 2000, 2004a). This first shoreline restoration project also involved removing some of the timber cribbing introduced in the 1960s and exposing the sandy slopes (Jordan 1989).[27] Within a few years, the plantings had carpeted the slopes. This work started at Thoreau's Cove on the northwestern shore, passed below the railroad embankment, and continued partially around the south shore.

[26] Mass Audubon interview, July 3, 2007.

[27] WPBOD interview, July 9, 2007.

Fig. 2.5 Shoreline and slope stabilization along the Pond Path. Left: Shoreline stabilization. Center: Trail, and planting on slopes above Red Cross Beach. Right: Fieldstones as site stabilization along the trail. Photographs by the author, June–August 2007

During this initial shoreline restoration, restorationists found an inventory by Thoreau of plants in the immediate vicinity of the pond—and this list was used to select the different plant species introduced onto the pond slopes, including blueberry, summersweet, alder, and red maple (also Loewer 1996; Schofield and Bush-Brown n.d.).[28] Along with the fieldstones, it was at this time that light fencing was introduced along the perimeter trail to control pedestrian access to the pond (Fig. 2.5).

By the mid-1990s, the shoreline of Walden Pond was once again the focus of a comprehensive restoration program—with DEM and WPSR carrying out major trail improvements (including the removal of the incomplete boardwalk) and the associated restoration of the bank. But the focus had shifted away from hard engineering and toward soft engineering solutions (Weinreb 1985; DCR 2013). A two-phase restoration schedule around the pond was centered first on repairing the stretch of the Pond Path between Red Cross Beach and Thoreau's Cove (Fall 1996 through Summer 1997), before repairing the segment of the trail

[28] Mass Audubon interview, July 3, 2007.

between the bathhouse and Long Cove (Fall 1997 through Summer 1998).[29] To stabilize the shoreline, DEM pioneered a soil bioengineering model that included coir blankets, brush layering, bundled coir fascines, and live staking of, for example, willow and dogwood, as well as blueberry sod (cf. Maynard 2004),[30] with no need for fieldstones at the base. Planted amidst the bioengineered vegetation are more than 50,000 native trees and shrubs (a mix of 45 species),[31] grown from seeds and cuttings collected throughout the reservation (cf. Maynard 2004).

In the early years of DEM's tenure at WPSR, there was a shift away from a principal focus on delivering a recreation resource, toward a pairing (read also: compromise, balancing act,[32] as Couture 1993) of recreation with Walden's history—in an attempt to more closely honor Thoreau's Walden, and the conveyances set down in the Deed of Gift. And so other restoration programs carried out in the early-1980s included the mass demolition of one of the bathhouses, the concrete apron in front of the bathhouse, the dock, and the retaining wall along the eastern shore of Walden Pond (replaced with stones and native granite), and the remodeling of the remaining bathhouse (most notably the introduction of cedar shingle cladding), as well as the realigning and resurfacing of the concrete roadway. During these works, DEM also created a terrace in front of the bathhouse and planted trees into Main Beach, to carry the tree line around, and provide some shade on the beach, all introducing a major visual and aesthetic transformation to the beach area.[33]

[29] DCR interview, July 5, 2007. WPBOD interview, July 9, 2007. WPSR interview, July 13, 2007.

[30] Mass Audubon interview, July 3, 2007. DCR interview, July 5, 2007. WPBOD interview, July 9, 2007. WPSR interview, July 13, 2007.

[31] Mass Audubon interview, July 3, 2007. DCR interview, July 5, 2007. WPSR interview, July 13, 2007.

[32] WPBOD interview, July 9, 2007. WPSR interview, July 13, 2007. Environmental historian telephone interview, July 24, 2007.

[33] Mass Audubon interview, July 3, 2007.

2.5.3 Bear Garden Hill and Brister's Hill Under Threat; the Thoreau Country Conservation Alliance and the Walden Woods Project Respond

Though he could not have known of the real estate battle that would be waged in Walden Woods in the mid- to late-1980s, Thoreau foretold the specter of development in *Walden*: 'The ornamented grounds of villas which will one day be built here may still preserve some trace of this' (Thoreau 2004a: 180). Nearby White Pond had already succumbed to residential development along its shores by the early-twentieth century, and now commercial developers had their sights firmly on Walden.

In 1984, Boston Properties acquired an 18.6-acre site in the northeast part of Walden Woods, near the Hapgood Wright Town Forest (elsewhere named Fairyland Pond by Thoreau, e.g. 1906), for a 147,000-sq. ft. office development, with parking for more than 500 cars. The site of the proposed Concord Office Park was Brister's Hill—the area of Walden Woods where Thoreau had developed his theories of forest succession and seed dispersal. This part of Brister's Hill had already been quarried for sand and gravel in the mid-twentieth century, but by the 1980s, had slowly begun to regenerate.

Elsewhere in Walden Woods, developer Phillip DeNormandie, together with John M. Corcoran & Co., had acquired a 25-acre site on Bear Garden Hill, for 251 townhouses. While Boston Properties had received many of the required permits and approvals from the Town Planning Board, DeNormandie had been less successful—until the addition of affordable housing in the proposed development. A smaller development was negotiated and approved by the Town Planning Board—authorizing the 139-condominium Concord Commons project, with 42 affordable units (including seven for low income families).

These are not the first commercial development threats to Walden Woods. A meatpacking plant was proposed on a parcel that included Thoreau's cabin site in 1902, a gas station was proposed on the former bean field land in December 1938, and a hotel was proposed near the bean field and Brister's Hill in the mid- to late-1970s (Maynard 2004; Dean 2005; Wheeler 2005)—but these earlier developments were quashed, defeated. But with these threats, a precedent was set for Walden conservationists and restorationists.

The meetings and negotiations between Boston Properties, DeNormandie, and the Town Planning Board did not go unnoticed by local conservationists. Botanist and ecologist (and active WFW member) Edmund A. Schofield and Thoreau scholar Thomas Blanding (the incoming Thoreau Society president) were concerned about the projected scale of development, and the possible sacrifice of more of Walden Woods. In their view, Walden Woods—together with Estabrook Woods (described by Thoreau as 'Easterbrook Country')—were the last two of Thoreau's 'great wild tracts' (Thoreau 1906 *V*, also Blanding 1988) remaining in Concord (with the Great Fields and a tract on the old Marlborough road already fragmented and/or developed).

In an effort to assuage concerns, the Town Planning Board enlisted Blanding to confirm that Brister's Hill and Bear Garden Hill held limited import to Thoreau's philosophy and writings. The Board's argument lay primarily in the dearth of references to either site in the index of the 1906 edition of Thoreau's *Journals*. But Blanding's rebuttal pointed to the inconsistencies in the indexing and instead argued that both Bear Garden Hill and Brister's Hill are mentioned throughout the *Journals*, as well as in other writings. Bear Garden Hill is the backdrop for Thoreau's 'Moonlight' papers, such as the 1863 essay 'Night and Moonlight,' and much of the *Walden* chapter 'Former Inhabitants, and Winter Visitors' is centered on Brister's Hill (Blanding 1988). Bear Garden Hill also represents Thoreau's favorite huckleberrying ground, while one of Thoreau's earliest pleas for conservation came after witnessing the felling of pine stands on Brister's Hill.

The Thoreau Country Conservation Alliance (TCCA) was formed in 1988 by Blanding and Schofield with Jack Borden, J. Walter Brain, Vidar Jorgensen, and Robert Allen Schledwitz to challenge the commercial development of Bear Garden Hill and Brister's Hill. By this time, the Fairhaven Preservation Association was already fighting the condo plans for Bear Garden Hill. It seems fortuitous that a conservation campaign should erupt on Bear Garden Hill, granted that a 'bear garden' historically describes a scene of uproar. The commercial development battle over Bear Garden Hill (and Brister's Hill) stands as a harbinger of late-twentieth-century protest and dissent by Walden restorationists (see Blanding 1988 on the impact, importance of Bear Garden Hill, and Fairhaven Hill and the Cliffs, to Thoreau's Walden imagination).

The use of a 'Thoreau Country' marker can be traced back to the late-nineteenth century, appearing in its earliest use shortly after Thoreau's death (Maynard 2004) and was later popularized by Herbert Wendell Gleason, especially his photographs and maps of Walden (Gleason 1917, 1975). Thoreau Country comprises lands across Concord, Lincoln, Walden Pond, Walden Woods, and Estabrook Woods, as well as on Mount Wachusett, Cape Cod, and in Maine's North Woods. The TCCA name derived from a belief that Walden's fate was a much bigger issue than just confined within the Town of Concord[34] (cf. 'Mapping Thoreau Country,' tracking Thoreau's excursions across Massachusetts, at: https://www.mappingthoreaucountry.org).

In paid advertisements in local newspapers—especially *The Concord Journal*—the TCCA, working with WFW and the Thoreau Society, sought to draw attention to the developments planned in Walden Woods. Headlines declared: 'Help Save Walden Woods.' Walden Woods was also politicized as 'Historic Walden Woods'[35] (Blanding 1988, also Henley and Marsh 1991). Throughout the TCCA's campaign, it was the 'idyll of Thoreau,' the 'spirit of Thoreau' that was frequently invoked in defense of the woods, and it was here that Walden-as-monument, Walden-as-shrine was politicized once more. Although the Concord Historical Commission (CHC) was an ally of the TCCA, announcing that it should be Thoreau's voice to resound in Walden Woods, not that of developers,[36] a seachange would come about two decades later as CHC denounced other conservation efforts to prevent the construction of playing fields in the woods. Preserving Walden's 'historic integrity' became *les mots du jour* for Walden restorationists.

[34] FOTC interview, July 27, 2007.

[35] TCCA. Help Save Walden Woods. *The Concord Journal* (July 7, 1988): 5, TCCA. 'Help Save Walden Woods.' *The Concord Journal* (November 3, 1988): 5 [Walden Woods Project Collections—TCCA Archives, at the Thoreau Institute at Walden Woods].

[36] qtd. in TCCA. Help Save Walden Woods. *The Concord Journal* (July 7, 1988): 5 [Walden Woods Project Collections—TCCA Archives, at the Thoreau Institute at Walden Woods].

To further support their case, Blanding and Schofield set out to map the boundary of Walden Woods. Both asserted that 'Walden Woods' describes a 2,680-acre expanse of woods, bogs, ponds, and meadows. Thoreau's famous declaration in *Walden* that 'I went to the woods...' (Thoreau 2004a: 90–91) had, in an earlier draft, appeared as 'I went to the pond...' And two months before his death, Thoreau had requested that *Walden*'s subtitle, *or, Life in the Woods*, be removed from any future printings (Thoreau 1958, see also Blanding 1988; McGregor 1997).[37] What both these vignettes reveal is Thoreau's own reluctance to describe Walden as either *-Pond* or *-Woods*, and an attempt to unite the two. Yet *Walden* (Thoreau 2004a) retains the admission that Thoreau lived on the shore of the *pond* in *Concord, Massachusetts*, further fueling and reinforcing the Concord-Lincoln-Walden confusion.

In 'Historic Walden Woods,' Blanding (1988) unpacks and posits how the term 'Walden Woods' has a well-established and traceable history throughout the nineteenth century—in public records and legal documents, newspapers and magazines, diaries, letters, and other correspondence, and across volumes of history, literature, and science. In particular, he spotlights references to the landscapes of Walden in numerous works by the Alcotts, Emerson, Hawthorne, Thoreau, William Ellery Channing, Franklin B. Sanborn, and other Concord writers. The tracts connecting Brister's Hill and the Cambridge Turnpike, and Walden Pond and Fairhaven Bay, as well as Fairhaven Hill, Bear Garden Hill and the Boiling Spring, and the Lincoln woods receive a particularly close reading.

Schofield (1989, also 1993) meanwhile pioneered the 'Walden ecosystem' construct and presents Walden as a discrete landscape recognizable because of its geology, hydrological regime, soils, and vegetation. In Schofield's (1989) *The Walden Ecosystem*, it is the section entitled 'Indicator Species of the Northern Pine-Oak Forest' that is of particular interest to Walden restorationists. Here, Schofield (1989) reproduces John C. Kricher's (1988) lists of indicator species from *A Field Guide to Eastern Forests, North America*—trees, shrubs, herbaceous species and vines, birds, mammals—but then proceeds to correlate the species with

[37] FOTC interview, July 27, 2007.

occurrences and references in Thoreau's works. Schofield turns primarily to *Walden* (Thoreau 2004a) and the *Journals* (Thoreau 1906), alongside works by later Walden natural historians and ornithologists William Brewster, Ludlow Griscom, and others, to provide confirmation of Walden's status as a northern pine-oak forest unit amidst oak-chestnut-hickory forest, and in doing so, sets down a robust foundation for a Walden reference model. Blanding's (1988) and Schofield's (1989) combined studies in the late-1980s firmly established the geographical extent and boundaries of Walden Woods—and sought to restore the idea of a 'Walden Woods' beyond the administrative and political confines and prescriptions of the WPSR designation (a marked departure from WFW, which had focused its restoration efforts entirely within the boundary of the WPSR).

The TCCA's persistence and campaigning in elevating the idea of a 'Walden Woods' was rewarded with the bittersweet inclusion of 'Walden Pond and Woods' on the National Trust for Historic Preservation's list of 'Eleven Most Endangered Historic Places' in both 1990 and 1991 (see also WWP 2002). A rebuttal to those who denied that Walden Woods existed as a mappable, bounded ecological unit. (A couple of years earlier, in 1989, the TCCA had also requested that the NPS revise the 1962 Walden Pond National Historic Landmark boundary, to more accurately reflect the 'Walden ecosystem' definition.)

To Blanding and Schofield, the Walden ecosystem had been fractured following the Concord-Lincoln town line in 1754 (running from Fairhaven Bay to the Cambridge Turnpike), the completion of the Fitchburg Railroad in 1844, the designation of the Walden Pond State Reservation boundary in 1922 (and the associated assumption that all of Walden Woods was thereafter contained in the now 462-acre reservation), and the construction of Route 2 through the north of the woods in 1934. Further complicating the coherence and unity of the Walden ecosystem is that 1,180 acres of the woods (including the 61-acre pond) are in Concord, and 1,500 acres of the woods are in Lincoln (Blanding 1988; Schofield 1989).

But the 2,680-acre 'Historic Walden Woods' or 'Walden ecosystem' boundary proposed by Blanding (1988) and Schofield (1989) is not autonomous. The boundary is further politicized—and negotiated—by and within the wider administrative context of the towns of Concord and Lincoln. A persistent, albeit malleable, fluid definition of Walden Woods in the early- to mid-nineteenth century, as described by Blanding (1988), is amplified in the Open Space and Recreation Plans (OSRP) for Concord and Lincoln. Both town's OSRPs capture different sides of the same Walden (Open Space and Recreation Plan Committee 2015; Open Space and Recreation Plan Advisory Committee 2017), and so sit in slight conflict or tension with, and are sometimes inconsistent with, the boundary prescribed by Blanding (1988), Schofield (1989), and the TCCA. In Concord's OSRP, Walden is often described as 'Walden Woods/Town Forest,' revising and consolidating the perimeter of the forest unit (Open Space and Recreation Plan Committee 2015). But 'Walden Woods/Town Forest' and 'Estabrook Woods Area' are listed as two of nine large natural areas in Concord, supporting Blanding's (1988) and Schofield's (1989) defense of Thoreau's 'wild tracts' of Concord. Both town administrations have historically approached the conservation of Walden Woods differently, making it harder to reconcile a unified—and protected—Walden Woods (Brooks 1976; McGregor 1997; Foster 1999; Donahue 2004). The zoning history of the two towns adds a further complication to efforts to conserve this suburban forest. Lincoln has a long history of being a conservation town, and a strong reputation of conservation advocacy throughout the township, including in Walden Woods (Brooks 1976). Yet most of the incursions in Walden Woods from the twentieth century onwards—Route 2, the high school campus, the town landfill, gravel operations, and the playing fields—have occurred on the Concord side of the town line.

In November 1989, as Boston Properties was awaiting final approval on its environmental impact report (which was granted the following month), a CNN segment aired on the Walden Woods development issue, featuring Blanding, Schofield, and the work of the TCCA. The segment would later mark the start of a new phase in efforts to preserve Walden—and with it, the institutionalization of an ecological restoration rhetoric in the woods. Musician Don Henley, a founding member of the Eagles, had long been inspired by the philosophy and writings of

Thoreau and Emerson, and after watching the CNN story, began a high-profile fundraising campaign—through donations, concerts, marches—to support the TCCA and halt the two developments.[38] Henley's campaigning is credited with bringing national attention to the Walden issue.

In April 1990, Henley, working with Kathi Anderson, former staffer to Mass. Senator Edward Kennedy, founded the nonprofit Walden Woods Project (WWP). With assistance from the Trust for Public Land and other organizations (including the TCCA), the Walden Woods Project successfully led the opposition to the commercial development projects. Almost immediately, the WWP began fundraising, and set about acquiring the two sites threatened by development (resurrecting the 'bear garden' motif once more). To further complement and accompany the studies of Blanding (1988) and Schofield (1989), the TCCA and the WWP produced a map of Walden Woods, building on and revising Gleason's maps from the early-1900s (e.g. Gleason 1917, 1975), to locate the boundary and major landmarks in Thoreau's Walden, in the heart of Thoreau Country (Fig. 2.6).

DeNormandie sold the Bear Garden Hill property to the WWP in late-1990 at a loss. But with this land purchase, the WWP also committed to providing an alternative site for the provision of the 42 affordable housing units proposed in the Concord Commons development (as an aside, whereas Thoreau's cabin had cost him $28.12½ to build in 1845, average house prices in 1990s Concord were closer to $300,000). Boston Properties was initially reluctant to sell the Brister's Hill property and instead offered to donate $100,000 to the WWP to support its mission. An impasse, stalemate between WWP and Boston Properties over the 18.6-acre site was eventually resolved in mid-1993. Following these two land acquisitions, the mission of the Walden Woods Project expanded to conservation—and to acquiring further threatened parcels. The TCCA and the WWP worked together to secure these early

[38] Henley established the nonprofit Mulholland Tomorrow in 1982, protecting land in the Santa Monica Mountains near Los Angeles, California, and his commitment to environmental activism continues with his establishment in 1992 of the Caddo Lake Institute in Shreveport, Louisiana, which works to protect the ecological, cultural, and economic integrity of Caddo Lake, on the Texas-Louisiana border.

Fig. 2.6 Map of Walden Woods by TCCA and WWP. Courtesy of the Walden Woods Project and Thoreau Institute at Walden Woods

land protections (on the TCCA-WWP relationship, see also Maynard 2004).

In 1991, Henley, with Dave Marsh, co-edited *Heaven Is Under Our Feet*, a collection of more than 60 short essays contributed by environmentalists, politicians, musicians, and actors to support the early work of the WWP. If Thoreau's (2004a) *Walden* is a work of literary activism retroactively, then Henley and Marsh's (1991) *Heaven Is Under Our Feet* is of-the-moment literary activism at its most energized. Within its pages, *Walden* the book and Walden the place are a springboard for calling out the necessity of environmental protections in Walden Woods and beyond. It is at once a book of hope, courage, and respect, and a rallying cry to the late-twentieth-century environmental movement.

In late-1996, the Walden Woods Project moved its headquarters from Boston to Lincoln, Mass., to the heart of Walden Woods. The WWP campus is located a half-mile from Walden Pond on Pine Hill, on the 22-acre former Higginson estate[39] (the Higginson House was added to the National Historic Landmark register in 2005, following nomination by the WWP). The Walden Woods Project began a collaboration with the Thoreau Society to build a repository for research collections (to replace the Thoreau Society's Thoreau Lyceum in Concord). The Thoreau Institute at Walden Woods, part of the Walden Woods Project, was opened by President Bill Clinton in June 1998, and provided the WWP with a platform to strengthen its position in research, education, and advocacy.

By the Walden Woods Project's 30th anniversary in 2020, the organization had protected nearly 170 acres across more than a dozen sites in Walden Woods. The WWP also holds conservation restrictions (CR)—or easements—on a number of other parcels in Walden Woods.[40] Nonprofit land conservation organizations own or manage substantial parcels of Walden Woods, with some of the largest landowners including WWP, Concord Land Conservation Trust, Lincoln Land Conservation Trust, and the Conservation Commission in Lincoln (see also Donahue

[39] The mansion was built by Henry Lee Higginson, founder of the Boston Symphony Orchestra, in 1906. Later, it was the home of descendants of Presidents John Adams and John Quincy Adams.

[40] WWP interview, June 19, 2007, also *Notes from Walden Woods: The Annual Newsletter of the Walden Woods Project & Thoreau Institute*, 2001 through 2020.

2013). Almost 80 per cent of Walden Woods is protected (a near reversal from Thoreau's October 1860 estimation in Thoreau 1906 *XIV*) (Fig. 2.7).

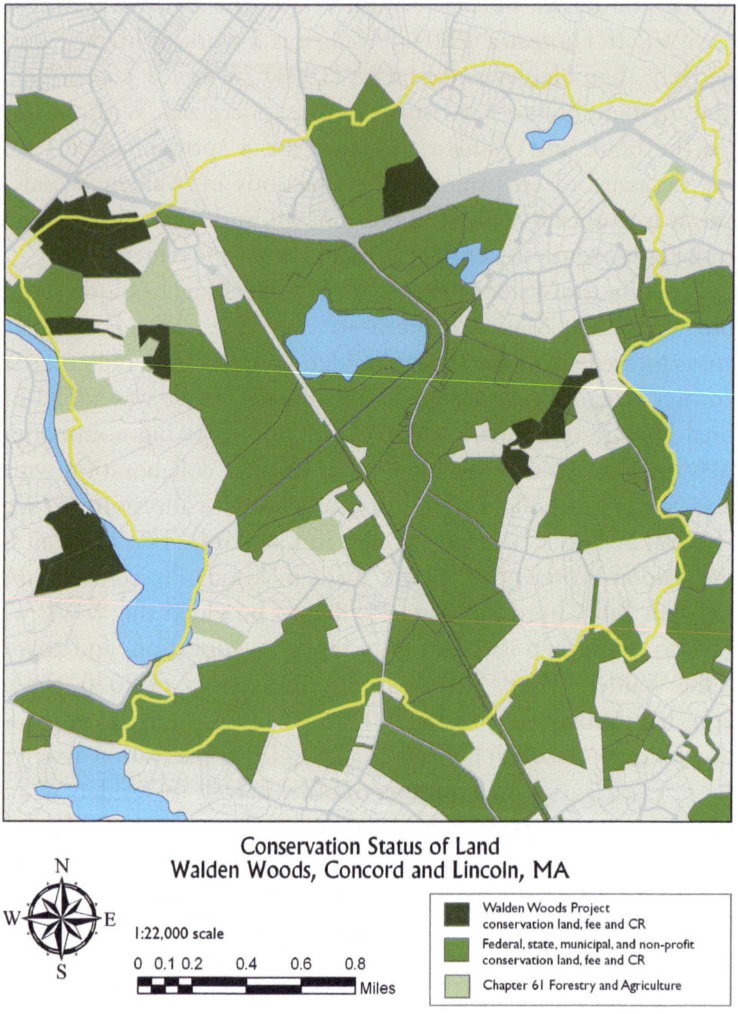

Fig. 2.7 Land protected in Walden Woods. Courtesy of the Walden Woods Project and Thoreau Institute at Walden Woods

2.5.4 Restoration of the Former Town of Concord Landfill

At the same time as the Save Walden Committee was campaigning to restore Red Cross Beach, another tract of Walden Woods was lost in 1958 to the new 35-acre Concord municipal landfill, to hold material excavated during construction of the Concord-Carlisle Regional High School campus. The landfill was located between Goose Pond to the east and Walden Pond to the west, and south of Brister's Hill, in the south-eastern corner of the Route 2-Route 126 intersection, adjacent to the Walden Breezes Trailer Park (which opened in 1936, was purchased by DNR in the 1960s, and was phased out by the early-2000s). The landfill was intended to be temporary, opened under the condition that it would eventually be returned to conservation (Collins et al. 2000). The WWP joined with the TCCA to raise concerns over plans in 1992 to expand the landfill, but it was not until the 1994 annual Concord Town Meeting that the town voted to close the landfill, and cap all used phases.

Following the 2001 Town Meeting, a Landfill Study Committee was appointed by Concord's Board of Selectmen to analyze and evaluate reuse options—including conservation—for the closed and capped landfill. The Massachusetts Audubon Society's (hereafter, Mass Audubon) Ecological Extension Service had completed an ecological inventory of the landfill (and Brister's Hill) property on behalf of the Walden Woods Project, to advise on restoration (Collins et al. 2000). The inventory proposed three Vision Plans for the site: (i) maintain the capped area as open grassland, (ii) establish rotating succession areas (demonstrating Thoreau's theory of succession), and (iii) allow part of the cap to reforest—recommending the grassland option. (Mass Audubon also drew upon its Grassland Birds Program to promote a soil composition and mix of grass species for the landfill that complemented the wider Walden ecosystem.) The Walden Woods Project—with a team of consultants that included Mass Audubon's EES, landscape architects Sasaki Associates, Inc., Rutgers restoration ecologist Steven Handel, and others—became involved in the landfill planning following the Concord Public Works' (CPW) design for closure. The approach proposed by town officials was clinical, distilled to, 'It was landscape architecture practiced by a civil engineering firm hired to be landscape architects—stick some trees along it. It had nothing to do with restoration of a grassland habitat.'[41]

The WWP-Mass Audubon-Sasaki program, by comparison, proposed restoring the site to a native grassland habitat and included plans to establish an interpretive trail system with connections to Goose Pond, Walden Pond, and Walden Pond State Reservation. The grassland would reintegrate, reconnect the property with the rest of Walden Woods, and appease the abrupt ecological and historical disruption, interruption, incongruity caused by the former landfill. It was this proposal, tendered by the Walden Woods Project-led team that was approved by the town.

With a grant from Time Warner, Inc., the capped landfill was graded and seeded with native grasses and wildflowers, while native trees (including oak, pine, maple) primarily provide a buffer screen for methane vents. Although the grassland might be read as another ecological disruption on the former landfill property, the decision to introduce a grassland habitat is twofold: (i) insufficient soil depth on the capped landfill prevents the dense planting of trees, and (ii) with grassland habitat diminishing in New England, any new grassland habitat creation is valued by Mass Audubon.[42] Much like Thoreau's favoring of atmosphere and ambience over fastidiousness, meticulousness (Thoreau 2001a, 'Dark Ages' in 2001b), the grassland restoration and trails places functionality over fidelity—an example of restoration *sensu lato*, rather than *sensu stricto* (after Aronson et al. 1993). It is a pragmatic restoration, rather than a purist restoration. David R. Foster (1999) notes that many of Massachusetts' best (and indeed largest) examples of grassland have been established on artificial landscapes, including former landfill sites. Grassland also connects Walden Woods back to Concord's revolutionary past, particularly the meadow landscapes of Paul Revere (and echoes the restoration of meadowland at Concord's Minute Man

---

[41] Sasaki Associates, Inc. interview, July 23, 2007.

[42] WWP interview, June 19, 2007. Mass Audubon interview, July 11, 2007. Sasaki Associates, Inc. interview, July 23, 2007.

Fig. 2.8 Grassland habitat on the former Town of Concord landfill property. Top: Panoramic view of the capped and planted site. Bottom left, center, and right: Grassland plantings and trails. Photographs by the author, August 2007

National Historical Park).[43] Jeffrey Collins et al.'s (2000) ecological inventory celebrates the landfill property (and the Brister's Hill property) as a cultural landscape (Fig. 2.8).

Together with the town, the WWP advocated for Mass Highways to install a guard rail along Route 2 to restrict off-road vehicle access to the landfill and Brister's Hill. With the guard rail in place, the landfill planting (and Brister's Hill planting) could extend to the edge of the highway.[44] To connect the landfill-grassland with the Brister's Hill area of Walden Woods, and beyond into the Town Forest, the WWP sponsored a study for a pedestrian and wildlife overpass across Route 2—Mullin et al.'s (2007) *Walden Passage Feasibility Study*.[45] The option for a corridor at Goose Pond would play to Concord's philosophical and cultural ecology histories and would help restore the coherence of the 'Walden ecosystem' sought by TCCA members Blanding and Schofield (Mullin et al. 2007, also Collins et al. 2000). But the wildlife corridor remains confined to the page.

[43] Minute Man NHP also connects to Concord's intellectual, literary history, especially at The Wayside, the former home of both Louisa May Alcott and Nathaniel Hawthorne.

[44] Sasaki Associates, Inc. interview, July 23, 2007.

[45] WWP interview, June 19, 2007.

And yet, despite the capping and restoration of the landfill, the property lingers as the single unprotected parcel of land surrounding Walden Pond (Landfill Study Committee 2003, see also the long-term land use of the capped landfill in Open Space and Recreation Plan Committee 2015). This precarity is confirmed in the agreement among town officials to allow the capping and replanting precisely because it would not inhibit or restrict any future land uses. For a number of years, the WWP has lobbied for a possible state-town land swap that would transfer the landfill property to the WPSR (a recommendation repeated by the Landfill Study Committee 2003). The WWP has tried on a number of occasions to purchase a CR for the landfill property. The WWP and Mass Audubon's EES also proposed alternate properties in the township to relocate Concord Public Works' municipal yard operations.[46] At the annual Town Meeting in April 2013, there was town support for the sale of a CR to the WWP (Article 12), but the motion did not receive the two-thirds majority required for approval. The WWP did, however, succeed in stopping development of a bus depot (Article 13) on the landfill property (WWP 2013).[47] The WWP's effort to secure the conveyance of a conservation restriction on the landfill continues.

2.5.5 Installing Thoreau's Path on Brister's Hill

By the end of the 1960s, Brister's Hill had been severely disturbed by gravel operations, with much of the topsoil also excavated and removed from the area. The site was further disturbed by off-road vehicle use and trash dumping. Little vegetation remained, other than two small remnant stands of trees (Collins et al. 2000, also WWP 2002), but processes of natural regeneration were underway. In the late-1980s, the Brister's Hill parcel was the anticipated location of Boston Properties' Concord Office Park, and the ensuing conservation battle over Brister's Hill and Bear Garden Hill led to the founding of the Walden Woods Project. In

[46] WWP interview, June 19, 2007. Mass Audubon interview, July 11, 2007. WWP interview, July 11, 2007. Sasaki Associates, Inc. interview, July 23, 2007. WWP interview, July 23, 2007.
[47] WWP online interview, July 30, 2020.

1993, the 18.6-acre parcel was acquired by the WWP and placed under permanent conservation.

Brister's Hill was a regular, fabled haunt of Thoreau's on his saunters through Walden (see, e.g., description in Thoreau 2004a). This area is the lynchpin of Thoreau's theorizing on forest succession and seed dispersal, but the history of the area also contributed to his thinking on social reform and social conscience—and his essay 'Civil Disobedience' (in Thoreau 2001b). For Brister's Hill commemorates not just Thoreau's Walden legacy, but also another 'former inhabitant' of the woods. The area is named after Brister Freeman, a freed Concord slave who served in the Continental Army during the American Revolution, and was one of the town's first Black landowners. For three decades, Freeman lived with his family on what is now known as Brister's Hill and planted an orchard on the land (see Lemire 2009). Perhaps more so than any other parcel, Brister's Hill tells of the intersection of environmental histories and cultural histories in Walden Woods.[48] (A bench dedicated to Brister Freeman was installed at Brister's Hill in 2013 as part of the Toni Morrison Society's Bench by the Road Project.)

Reflecting on the origin story of Thoreau's Path on Brister's Hill, Walden restorationists note, 'the expectation coming into this was a restoration.'[49] The restoration reference model for Brister's Hill looks to both the environmental history and the cultural history of the site, vividly and evocatively capturing the storying—restorying—restoring framework that underpins this book. A March 18, 1861 journal passage by Thoreau on the importance of history, story, and storytelling informs and underwrites the WWP's restoration program for Brister's Hill: 'Wherever men have lived there is a story to be told, and it depends chiefly on the story-teller or historian whether that is interesting or not.

[48] An early abolitionist, Thoreau was active in Concord's anti-slavery movement. In July 1846, he spent a night in jail after refusing to pay his poll tax, arguing it in part upheld the institution of slavery. Thoreau also opposed the 1850 Fugitive Slave Act, and later that decade, would speak out in support of John Brown's 1859 raid at Harper's Ferry in West Virginia. Thoreau served as a conductor on the Underground Railroad, and his mother's boarding house was used as a station on the Underground Railroad (e.g. Thoreau 1906 *V*), where Thoreau would help organize passage for escaped slaves on their way to Canada. (His mother, Cynthia Thoreau, had founded the Concord Female Anti-Slavery Society in 1837.)

[49] Sasaki Associates, Inc. interview, July 23, 2007.

You are simply a witness on the stand to tell what you know about your neighbors and neighborhood' (Thoreau 1906 *XIV*: 330, also qtd. in WWP 2002: 20). So part of the restoration centered on showcasing the in situ succession of plant communities on a sand plain, on a 'pitch-pine plain,' and processes of natural regeneration. But another part of the restoration sought to celebrate the place of Brister's Hill in Thoreau's writings on social responsibility and environmental stewardship. As with the landfill, the restoration of Brister's Hill is very much restoration *sensu lato* (after Aronson et al. 1993), a cultural landscape restoration.

Working in collaboration with the Department of Conservation and Recreation (DCR, formerly DEM), Mass Audubon's EES, Sasaki, and others (and supported by grants from Time Warner, Inc., and the NPS), the Walden Woods Project began work in 2002 to install Thoreau's Path on Brister's Hill. Thoreau's Path honors Thoreau's legacy as a Transcendentalist philosopher, social reformer, ecologist, conservationist, surveyor, writer, teacher, and inventor. But it also plays to its location, to provide an opportunity to observe the processes of forest succession recounted by Thoreau across the site. Different segments of the one-mile interpretive loop trail align with and articulate important contributions made by Thoreau—progressing through the entry meadow (conservation), to Brister's orchard (social reform and commentary), to the sand plain (teaching and observation), to forest succession (science), culminating at the Reflection Circle (Thoreau's philosophy and spirituality, and his influence on others) (see WWP 2006, 2007).[50] The trail is marked by passages from Thoreau's writings on social responsibility and environmental stewardship, incised in granite and cast in bronze. The markers call attention to parts of the landscape that correspond to, or are otherwise illustrative of, Thoreau's observations.

'I wish to speak a word for Nature' ('Walking' in Thoreau 2001b: 225) is the passage inscribed underfoot at the threshold onto the path. The path connects, *in place*, the lyricism of Thoreau-as-poet with the rigor of Thoreau-as-scientist. The path encourages a scrutiny and rigor in observation and discovery—a slow, careful, close, attentive, 'deliberate' reading of the land, modeled after Thoreau's own practices. Indeed, one marker

[50] WWP interview, July 23, 2007. Sasaki Associates, Inc. interview, July 23, 2007.

features his question: 'Will you be a reader, a student merely, or a seer?' (Thoreau 2004a: 111, cf. WWP 2006, 2007). Two frames placed beside the path further embody this—a horizontal square frame in the grassland draws attention to the terrain underfoot, while elsewhere a vertical frame, a transect, offers a view through the woods, through succession, from lichen to birch to oak to mature forest. At the Reflection Circle, the apex of the loop trail, Thoreau's words are complemented by those of environmental and social reformers, activists, and thinkers influenced by Thoreau, including Ralph Waldo Emerson, Emily Dickinson, John Muir, Chief Luther Standing Bear, Mohandas Gandhi, Aldo Leopold, Rachel Carson, John F. Kennedy, Martin Luther King Jr., E. O. Wilson, and Wendell Berry (Fig. 2.9).

As well as being placed in situ throughout the property, Thoreau's writings also inform the plant material reintroduced on Brister's Hill. The scientific baselines that Thoreau established on Brister's Hill in the 1850s were instrumental in the restoration of the same site over a century-and-a-half later (WWP 2002, also Blanding 1988; Schofield 1989). Already present on the parcel was a successional plant community that could be

Fig. 2.9 Thoreau's Path on Brister's Hill. Top left: Entrance to Brister's Hill from Walden Street/Route 126. Top center: 'I wish to speak a word for Nature' (Entry Meadow). Top right: 'What though the woods be cut down…' (Sand Plain). Bottom left: 'I see in the open field…' (Forest Succession). Bottom center-left: Forest trail. Bottom center-right: 'Heaven is under our feet…' (Entry Meadow). Bottom right: Reflection Circle. Photographs by the author, August 2007

traced through lichens and mosses, to the successional stages of a pioneer woodland dominated by white birch and pitch pine, to a later-stage oak forest (Collins et al. 2000; WWP 2002). A legacy from the prior sand and gravel mining operation was a landscape condition consistent with Thoreau's observations of the railroad deep cut (in e.g. Thoreau 2001a), as well as forest openings and clearings resulting from felling. The geology and ecology of the site are used to explore, retrace, and interrogate Thoreau's experiences, observations, and philosophy, alongside the social and cultural history of this part of Concord.

Restorationists only intervened in the landscape to encourage succession. Grasses, shrubs, and trees were planted to support the restoration of the existing plant community, alongside the selective removal of invasive plants. The plant material used in the Brister's Hill restoration includes arrowwood viburnum, highbush blueberry, low blueberry, huckleberry, pitch pine, white birch, gray birch, and sweet fern. A number of heirloom apple trees, including Roxbury Russet, Blue Pearmain, and Baldwin, were also planted along the entrance path, because of Thoreau's mention of Brister Freeman's apple trees in *Walden* (e.g. Thoreau 2004a: 257–258).[51] The intricacies and intersections of literature, politics, and landscape at Brister's Hill mean that 'you can't hire a lawn service to go out there.'[52]

The planting of pitch pine trees presents another theme, motif in the story of the Brister's Hill restoration. At the start of the restoration work, white pine stands dominated Brister's Hill, having succeeded earlier oak and pitch pine stands. Thoreau wrote extensively on Walden's white pine and pitch pine (Thoreau 2000), and on the succession of pine stands on Brister's Hill (Thoreau 1906 *XIV*, 1993, also Foster 1999), and his observations and records have contributed to the restoration program on Brister's Hill.[53] With the loss of topsoil to gravel operations, Brister's Hill resembles and supports a sand plain ecosystem. The restoration program for Brister's Hill prioritized the reintroduction of pitch pine stands, as

---

[51] WWP interview, June 19, 2007. Sasaki Associates, Inc. interview, July 23, 2007. WWP interview, July 23, 2007.

[52] Sasaki Associates, Inc. interview, July 23, 2007.

[53] Looking beyond Walden, Thoreau's observations on pitch pine, scrub oak vegetation patterns—and succession—have been consulted in, e.g. Foster and Motzkin (1998).

pitch pine is a classic species associated with sand plains. But two factors historically have combined to inhibit its dominance in the landscape: (i) pitch pine grows at a slower rate than white pine, and (ii) fire is less frequent in the woods, and no longer acts as a regulator—pitch pine has a better resistance to fire (and to hotter fire), but is now outcompeted by white pine. Restorationists made the decision to encourage pitch pine and remove many of the smaller white pine from Brister's Hill, in an attempt to mimic the outcome of earlier fires.[54] But another factor drives pitch pine reintroduction: pitch pine sand plain—just like open grassland—is another landscape that is diminishing in Massachusetts.

2.5.6 Concord-Carlisle Regional High School, Friends of Thoreau Country, and the Battle for Deep Cut Woods

While the Walden Woods Project was completing its restoration program on its Brister's Hill parcel, another tract of Brister's Hill emerged as a new battleground between conservationists and developers. It was the Battle for Brister's Hill 2.0. But rather than an office park (or a condo complex), this time the threat to the woods came from high school playing fields. The playing fields issue proved incredibly divisive in Concord (and was the subject of a Special Town Meeting in June 2007 after agreement, consensus on the issue could not be reached at the annual Town Meeting), but the Concord Board of Selectmen authorized the construction of two all-purpose playing fields at the CCHS campus in the summer of 2007. (A few years earlier, the capped landfill parcel had even been suggested as a possible site for the playing fields.)

The objection from conservationists lay in the proposed location of the fields. Although part of the high school campus, the nine-acre site near Brister's Hill Road was also a part of Walden Woods known as Deep Cut Woods. The bottom line in the fight for Deep Cut Woods was whether the woods were part of Walden Woods or not (stirring up Blanding's 1988 and Schofield's 1989 contested definition once more).

[54] Mass Audubon interview, July 11, 2007. Sasaki Associates, Inc. interview, July 23, 2007.

Conservationists, led by Friends of Thoreau Country (FOTC) which had formed in August 2006 in direct response to the issue, argued that the area was part of Walden. The Thoreau Society wrote letters to the editor of *The Concord Journal* defending the woods. The WWP was not part of this campaign, as the parcel was not for sale. FOTC—a group of Concord residents and Thoreau scholars (including Schofield)—argued once more for recognition of Brister's Hill as an important landscape in the development of Thoreau's theories on succession and seed dispersal. A 'Thoreau Country' motif was politicized once more. It was Thoreau who had named Deep Cut Woods on the western part of Brister's Hill, following the construction of the railroad, and it was one of several sites on Brister's Hill—along with Warren's Wood, Thrush Alley, and Laurel Glen—that informed, and contributed to, his botanical theorizing (Blanding 1988; Maynard 2004). In recent decades, Deep Cut Woods had gained a renewed ecological importance as the last remaining wildlife corridor (at the railroad underpass) connecting the northern and southern parts of Walden Woods severed, interrupted by Route 2 (Mullin et al. 2007; Open Space and Recreation Plan Committee 2015).

But for proponents of the playing fields—the Town of Concord, the Concord-Carlisle Regional High School District, the Friends of the Concord-Carlisle Playing Fields—the area was not part of Walden Woods. Arguments from town officials (including CHC) fell into two camps: the area was not part of Walden (i) because of its location north of Route 2, and (ii) because the historical integrity of the site had already been compromised by earlier development and subsequent fragmentation,[55] only to be compounded by discrepancies and inconsistencies in the location of historic landscape features.

This was not a long, protracted battle, but instead played out over a three-month period in mid-2007. FOTC filed for a temporary restraining order against the town and the school district to halt the felling of Deep Cut Woods and the removal of glacial deposits, but following a hearing, the request for injunctive relief was denied.[56] With the ruling, FOTC as a voice for Walden conservation faded from the

[55] FOTC interview, July 27, 2007.

[56] FOTC interview, July 27, 2007.

barricade. But it also joined a pantheon of retired grassroots conservation groups that had challenged threats and incursions in Thoreau's Walden—the Save Walden Committee of the Thoreau Society, Walden Forever Wild, and the Thoreau Country Conservation Alliance. Each group drew on different constituencies, but ultimately shared in the goal of protecting Thoreau's legacy in Walden Woods. The Walden Woods Project is the only organization still active in defending Walden Woods.

But this woods versus playing fields scenario was also unfolding elsewhere in Concord over the same summer. And the battleground was another landmark of Thoreau Country (and the American Revolution)—Estabrook Woods. This time, it was the Estabrook Woods Alliance (EWA, formerly Middlesex Graduates for Estabrook Woods [MGFE])—a grassroots coalition of students, parents, alumni, and local residents—up against the Middlesex School Board of Trustees. Estabrook Road, which passes through the woods, was used by the Minutemen on April 19, 1775, at the start of the American Revolutionary War, and is the last remaining Minuteman road. And Thoreau regularly walked in, and wrote about, the 'Easterbrooks Country,' with more than 160 journal entries and over 50,000 words devoted to the woods (Thoreau 1906).[57] Estabrook Woods has a long and celebrated history of preservation, conservation, and stewardship (see Open Space and Recreation Plan Committee 2015). Most notably, Estabrook Woods is the site of Harvard University's Concord Field Station and ecology study area (see Foster 1999 for further discussion on how Harvard's land preservation program is guided by Thoreau's observations on natural history and archeology).[58] Estabrook Woods was designated a Forest Legacy Area by Congress in 1998, and a Core Habitat by the Massachusetts Office of Environmental Affairs in 2001, with EWA also lobbying for an Area of Critical Environmental Concern designation.[59] Nevertheless, the Middlesex School Board of Trustees authorized development on 11 acres of Estabrook Woods for sports grounds, including tennis courts and artificial soccer fields.

---

[57] EWA email interview, November 30, 2007.

[58] Middlesex School and Harvard University are the largest landowners in Estabrook Woods.

[59] EWA email interview, November 30, 2007.

* * *

Thoreau's justification at the end of *Walden* for why he left the woods in the fall of 1847 also works as a coda to explaining the importance of *Walden* to restorationists working in Walden Woods. For like Thoreau, *Walden* the book still has many more lives to live (Thoreau 2004a: 323)—in the defense of Walden the place. In Thoreau's case, he left Walden Pond because Emerson asked him to take care of his family while he was on a lecture tour in England and Scotland. But perhaps more interesting—and surprising—is that Thoreau left Walden Pond five weeks before Emerson's departure, more than a month before he had to.[60]

2.6 Beyond Walden: 'Ktaadn,' RESTORE The North Woods, and the Katahdin Woods and Waters National Monument

On August 24, 2016, on the eve of the centenary of the National Park Service, President Barack Obama signed Presidential Proclamation 9476, authorizing the establishment of the 87,500-acre Katahdin Woods and Waters National Monument in northern Maine. With the designation, the forests and watersheds east of the Penobscot River—where Thoreau, in *The Maine Woods* essays 'Chesuncook' and 'The Allegash and East Branch,' had most explicitly vocalized the need for 'national preserves,' wilderness preservation, and a forest preserve ethic (Thoreau 2004b)—joined the national parks and monuments system. Thoreau's early call for 'national preserves' (later echoed by John Muir, President Theodore Roosevelt, and others) had gone full circle, returning to its ancestral roots. The designation brings Thoreau's views on preservation into sharp relief. Although the designation of Katahdin Woods and Waters is not an example of ecological restoration in Maine's North Woods, it is nonetheless richly illustrative and evocative of Thoreau's

[60] WWP online interview, July 23, 2020.

restoration sensibility, and how Thoreau's words would again come to be politicized by conservationists who had long campaigned for a Maine Woods national park.

This time, it is Thoreau's (2004b) *The Maine Woods* that has another life to live. A paragraph in President Obama's proclamation is devoted to Thoreau's 1857 excursion (Proclamation No. 9476, 2016) and is immediately followed by similar paragraphs dedicated to other prominent figures who drew attention to (and knowledge and inspiration from) Maine's North Woods—chiefly, former President Theodore Roosevelt, and former Maine governor Percival P. Baxter, with another paragraph pulling together the contributions of John James Audubon, Frederic Edwin Church, George H. Hallowell, and Carl Sprinchorn.

For more than two decades, conservationists, philanthropists, and local communities have fought for federal protection of the forests along the East Branch of the Penobscot River (although as Austin 2015 argues, proposed conservation efforts in the Katahdin region stretch back to the late-nineteenth century, with Mount Katahdin later protected by the designation of Baxter State Park in 1931). Within the various campaigns, references to Thoreau's three excursions to the region have been front and center. Two campaigns in particular have sought to draw attention to Thoreau's Maine Woods legacy.

RESTORE The North Woods, a conservation nonprofit established in 1992, first proposed its national park idea in 1994, with the launch of a bold—and controversial—3.2-million-acre Maine Woods National Park and Preserve (MWNP) campaign. If designated, it would become the largest national park in the lower 48 states—equivalent in area to the state of Connecticut, and larger than Yellowstone and Yosemite combined (see Botkin 2001; Austin 2015; Williams 2016 for an overview of RESTORE's early proposal, Maine politics, and concomitant challenges).[61] The commercial development that threatened Walden in the 1980s is writ large in the working forests of Maine's North Woods. RESTORE politicizes Thoreau's 'national preserves' idea to protect the

61 RESTORE online interview, August 14, 2020.

North Woods from resource extraction and development.[62] In name alone, RESTORE The North Woods capitalizes on (both typographically and metaphorically) an ecological restoration rhetoric, as perhaps a play on Thoreau's exposition of forests, development, and wilderness, or as an acknowledgment that a Maine Woods national park is a second chance (maybe even a last chance?) to honor Thoreau's legacy. Alongside its MWNP bid, RESTORE is also beginning a New National Parks campaign, and has identified 100 areas that might form the next wave of national parks—some as revivals of earlier proposals or expansions of existing parks, but most as new.[63] There is at least one new park idea proposed in every U.S. state. RESTORE's MWNP idea is in many ways the campaign's poster child.

But another powerful and impassioned voice would join that of RESTORE The North Woods at the close of the twentieth century in calling for a Maine Woods national park, and would become a leader, figurehead in the eventual designation of the Katahdin Woods and Waters National Monument—Roxanne Quimby, cofounder of Burt's Bees (Quimby joined the board of RESTORE in October 2000, and worked closely with RESTORE for a number of years, before resigning in 2003). (Phyllis Austin's 2015 biography of Quimby presents an excellent account of Quimby's conservation philanthropy, but was published 14 months before the Katahdin Woods and Waters designation.) Beginning in the summer of 2000, Quimby's advocacy for a national park was confirmed in—and realized through—substantial land purchases. Initially, many of these acquisitions complemented, fell within the proposed perimeter of RESTORE's Maine Woods National Park and Preserve.[64] But a few years later, Quimby split with RESTORE and set up her own landholding company, Elliotsville Plantation Inc. (EPI). RESTORE and Quimby differed on what constituted a 'national park'—Quimby's vision was much more restrictive.

---

[62] RESTORE often collaborates with the WWP on speaker programs and more. Maine Woods Forever, another conservation nonprofit, created the Thoreau-Wabanaki Trail in 2007 to celebrate Thoreau's three excursions, and the Penobscot guides who traveled with him.

[63] WWP online interview, July 30, 2020. RESTORE online interview, August 14, 2020.

[64] RESTORE online interview, August 14, 2020.

For more than a decade, Quimby purchased tens of thousands of acres across the Great North Woods region in Maine, predominantly to the east of Baxter State Park, along the Penobscot East Branch.[65] For Quimby's nonprofit Elliotsville Plantation Inc., the National Park Service's centennial year (and the final year of the Obama administration)—2016—became the deadline for securing a national park designation. When EPI submitted its park proposal for a 75,000-acre national park (and a nearby 75,000-acre national recreation area) with a multimillion dollar endowment in mid-2015, Congress deemed it too controversial and denied the request for a national park (Austin 2015). Up until that point, Quimby's conservation philanthropy and advocacy had largely focused on securing a *national park* designation in Maine's North Woods, but she began campaigning for *national monument* status for the region. In proposing a national monument designation, EPI could bypass Congress and instead appeal to the Obama administration that the land was worthy of federal management (and retaliate against the ambivalence of the Maine congressional delegation toward establishing a national park).

Unlike a national park, a national monument designation does not require an act of Congress for approval. The 1906 Antiquities Act (which predates the National Park Service Organic Act by a decade) invests the President with the authority to designate and protect landscapes with archeological, cultural, scientific, or ecological value. (A brief history of the Antiquities Act appears in Chapter 3, while Chapter 6 unpacks some of the recent challenges to the Antiquities Act by the Trump administration—centered on the Grand Staircase-Escalante and Bears Ears national monuments in southern Utah.) Quimby, RESTORE, and others saw a national monument designation as a placeholder, almost, for a future Maine Woods national park. And so EPI set to work gaining the attention of the Obama administration and met with NPS director Jonathan B. Jarvis when he traveled to the region in the spring of 2016.

Quimby's donation (through her nonprofit Elliotsville Plantation Inc.) of more than 87,500 acres to the federal government in late-August 2016 set in motion the establishment of the Katahdin Woods and Waters

[65] RESTORE online interview, August 14, 2020.

National Monument just two days later, and beating the NPS centenary deadline by a day. But the new Katahdin Woods and Waters monument is not without controversy. It was included in the Trump administration's review of national monument designations in April 2017 (and was the only national monument smaller than 100,000 acres to be included), following lobbying by Maine governor Paul LePage, a vocal opponent of the national monument. The administration also explored opening the national monument up to logging. Even with all these challenges in its first five years, the designation of Katahdin Woods and Waters has added to the park boosterism and advocacy of RESTORE. The national monument is a small part (but the first part) of the proposed national park. Katahdin Woods and Waters might not be a national park in title, but it is managed by the NPS—it is one of 423 units in the parks system and comes under the purview of the 1916 Organic Act, which directs how park units should be managed.[66]

Works Cited

Abbey, Edward. 2004 [1975]. *The Monkey Wrench Gang*. London: Penguin Books Ltd.

Angelo, Ray. 1983a. Thoreau as Botanist. *The Thoreau Quarterly* 15: 15–31.

Angelo, Ray. 1983b. Botanical Index to the Journal of Henry David Thoreau. *The Thoreau Quarterly* 15: 32–203.

Aronson, J., C. Floret, E. LeFloc'h, C. Ovalle, and R. Pontanier. 1993. Restoration and Rehabilitation of Degraded Ecosystems in Arid and Semi-Arid Lands. I. A View from the South. *Restoration Ecology* 1 (1): 8–17.

Arsić, Branka. 2016. *Bird Relics: Grief and Vitalism in Thoreau*. Cambridge, MA: Harvard University Press.

Austin, Phyllis. 2015. *Queen Bee: Roxanne Quimby, Burt's Bees, and Her Quest for a New National Park*. Thomaston, ME: Tilbury House Publishers.

Baym, Nina. 1965. Thoreau's View of Science. *Journal of the History of Ideas* 26 (2): 221–234.

66 RESTORE online interview, August 14, 2020.

Blanding, Thomas. 1988. Historic Walden Woods. *The Concord Saunterer* 20 (1/2): 3–74.

Botkin, Daniel B. 2001. *No Man's Garden: Thoreau and a New Vision for Civilization and Nature*. Washington, DC and Covelo, CA: Island Press/Shearwater Books.

Brooks, Paul. 1976. *The View from Lincoln Hill: Man and the Land in a New England Town*. Boston, MA: Houghton Mifflin Company.

Buell, Lawrence. 1995. *The Environmental Imagination: Thoreau, Nature Writing, and the Formation of American Culture*. Cambridge, MA: Harvard University Press.

Case, Kristen. 2013. Thoreau's Radical Empiricism: The Kalendar, Pragmatism, and Science. In *Thoreauvian Modernities: Transatlantic Conversations on an American Icon*, ed. François Specq, Laura Dassow Walls, and Michel Granger, 187–199. Athens, GA and London: University of Georgia Press.

Chura, Patrick. 2010. *Thoreau the Land Surveyor*. Gainesville, FL: University Press of Florida.

Collins, Jeffrey, with Bill Giezentanner, Stephen Handel, and Christa Hawryluk. 2000. *Ecological Inventory and Conservation Management Plan for Brister's Hill and the Concord Landfill, Concord, Massachusetts*. Lincoln, MA: Mass Audubon Ecological Extension Service.

Cronon, William. 1996. *Uncommon Ground: Rethinking the Human Place in Nature*. New York, NY: W. W. Norton.

Couture, Cindy Hill. 1993. Walden Restoration: Legal and Policy Issues. In *Thoreau's World and Ours: A Natural Legacy*, eds. Edmund A. Schofield and Robert C. Baron, 272–280. Golden, CO: North American Press.

Dean, Bradley P. 2000. Introduction. In *Wild Fruits: Thoreau's Rediscovered Last Manuscript*, eds. Henry David Thoreau and Bradley P. Dean, ix–xvii. New York, NY and London: W. W. Norton.

Dean, Bradley P. 2005. Rediscovery at Walden: The History of Thoreau's Bean-Field. *The Concord Saunterer: Special* Walden *Sesquicentennial Issue: Walden the Place and* Walden *the Book*. New Series, 12/13: 87–137.

Dean, Bradley P. 2007. Natural History, Romanticism, and Thoreau. In *American Wilderness: A New History*, ed. Michael Lewis, 73–89. Oxford: Oxford University Press.

Department of Conservation and Recreation. 2013. *Resource Management Plan: Walden Planning Unit*. Boston, MA: DCR.

Doctorow, E.L. 1991. Remarks at the Walden Woods Project Press Conference—Boston, Massachusetts, April 25, 1990. In *Heaven is Under Our*

Feet: A Book for Walden Woods, eds. Don Henley and Dave Marsh, 37–39. Stamford, CT: Longmeadow Press.

Donahue, Brian. 2004. *The Great Meadow: Farmers and the Land in Colonial Concord*. New Haven, CT and London: Yale University Press.

Donahue, Brian. 2013. Rewilding Walden Woods and Reworking Exurban Woodlands: Higher Uses in Thoreau Country. In *Landscape and the Ideology of Nature in Exurbia: Green Sprawl*, eds. Kirsten Valentine Cadieux and Laura Taylor, 94–120. New York, NY: Routledge.

Elliot, Robert. 1982. Faking Nature. *Inquiry: An Interdisciplinary Journal of Philosophy* 25 (1): 81–93.

Epsilon Associates, Inc. in association with Vanasse and Associates, Inc. 2001. *Environmental Notification Form: Walden Pond State Reservation Gateway Improvements*. Maynard, MA: Epsilon Associates, Inc.

Finley, James S., ed. 2017. *Henry David Thoreau in Context*. Cambridge: Cambridge University Press.

Foster, David R. 1999. *Thoreau's Country: Journey Through a Transformed Landscape*. Cambridge, MA and London: Harvard University Press.

Foster, David R., and Glenn Motzkin. 1998. Ecology and Conservation in the Cultural Landscape of New England: Lessons from Nature's History. *Northeastern Naturalist* 5 (2): 111–126.

France, Robert L., ed. 2008. *Healing Natures, Repairing Relationships: New Perspectives on Restoring Ecological Spaces and Consciousness*. Sheffield, VT: Green Frigate Books.

Garber, Frederick. 1977. *Thoreau's Redemptive Imagination*. New York, NY: New York University Press.

Gardiner, Richard A., and Associates. 1974. *Walden Pond Restoration Study, Final Report for the Middlesex County Commissioners …and the Walden Pond Restoration Committee*. Cambridge, MA: Richard A. Gardiner and Associates.

Gatta, John. 2004. *Making Nature Sacred: Literature, Religion, and Environment in America from the Puritans to the Present*. New York, NY: Oxford University Press.

Giblett, Rod. 1996. *Postmodern Wetlands: Culture, History, Ecology*. Edinburgh: Edinburgh University Press.

Gleason, Herbert W. 1917. *Through the Year with Thoreau: Sketches of Nature from the Writings of Henry D. Thoreau, with Corresponding Photographic Illustrations*. Boston, MA and New York, NY: Houghton Mifflin and Company.

Gleason, Herbert W. 1975. *Thoreau Country: Photographs and Text Selections from the Works of H.D. Thoreau*, ed. Mark Silber. San Francisco, CA: Sierra Club Books.
Golemba, Henry L. 1990. *Thoreau's Wild Rhetoric*. New York, NY and London: New York University Press.
Harding, Walter. 2015 [1982]. *The Days of Henry Thoreau: A Biography*. New York, NY: Dover Publications, Inc.
Henley, Don, and Dave Marsh, eds. 1991. *Heaven Is Under Our Feet: A Book for Walden Woods*. Stamford, CT: Longmeadow Press.
Higgs, Eric S. 1997. What Is Good Ecological Restoration? *Conservation Biology* 11 (2): 338–348.
Higgs, Eric S. 2003. *Nature by Design: People, Natural Process, and Ecological Restoration*. Cambridge, MA: MIT Press.
Higgins, Richard. 2017. *Thoreau and the Language of Trees*. Oakland, CA: University of California Press.
Jordan, William R. III. 1989. Restoration at Walden Pond. *Restoration and Management Notes* 7 (2): 65–69.
Jordan, William R. III. 1993. Renewal and Imagination: Thoreau's Thought and the Restoration of Walden Pond. In *Thoreau's World and Ours: A Natural Legacy*, eds. Edmund A. Schofield and Robert C. Baron, 260–271. Golden, CO: North American Press.
Katz, Eric. 1992. The Big Lie: Human Restoration of Nature. *Research in Philosophy and Technology* 12: 93–107.
Landfill Study Committee. 2003. *The 2001 Landfill Study Committee Final Report, May 2003*. Concord, MA: Landfill Study Committee.
Lemire, Elise. 2009. *Black Walden: Slavery and Its Aftermath in Concord, Massachusetts*. Philadelphia, PA: University of Pennsylvania Press.
Loewer, Peter. 1996. *Thoreau's Garden: Native Plants for the American Landscape*. Mechanicsburg, PA: Stackpole Books.
Marx, Leo. 2000. *The Machine in the Garden: Technology and the Pastoral Ideal in America*, 35th Anniversary Edition. Oxford and New York, NY: Oxford University Press.
Maser, Chris. 1988. *The Redesigned Forest*. San Pedro, CA: R. & E. Miles.
Maynard, W. Barksdale. 2004. *Walden Pond: A History*. Oxford and New York, NY: Oxford University Press.
McGregor, Robert Kuhn. 1997. *A Wider View of the Universe: Henry Thoreau's Study of Nature*. Urbana and Chicago, IL: University of Illinois Press.
Miller, Daegan. 2018. *This Radical Land: A Natural History of Radical Dissent*. Chicago, IL: University of Chicago Press.

Mullin, John R., Jack Ahern, Paige Warren, Ethan Carr, Sergio Breña, Scott Civjan, Zenia Kotval, Scott Jackson, Noah Charney, Beth Fenstermacher, Lee Jennings, and Jinglan Wang. 2007. *Walden Passage Feasibility Study. December 2007*. Amherst, MA: Department of Landscape Architecture and Regional Planning, UMass Amherst.

Nabhan, Gary Paul. 1993. Foreword: Learning the Language of Fields and Forests. In *Faith in a Seed: The Dispersion of Seeds and Other Late Natural History Writings*, eds. Henry David Thoreau and Bradley P. Dean, xi–xviii. Washington, DC and Covelo, CA: Island Press/Shearwater Books.

Nash, Roderick Frazier. 2014. *Wilderness and the American Mind*, 5th ed. New Haven, CT: Yale University Press.

National Park Service. 2002. *Walden Pond and Woods: Special Resource Study, September 2002*. Boston, MA: NPS Boston Support Office, Planning and Legislation.

Open Space and Recreation Plan Advisory Committee. 2017. *Open Space & Recreation Plan. Town of Lincoln, Massachusetts. March 2017*. Lincoln, MA: Open Space Recreation Plan Advisory Committee, Town of Lincoln.

Open Space and Recreation Plan Committee. 2015. *2015 Open Space and Recreation Plan. Concord, Massachusetts*. Concord, MA: Open Space and Recreation Plan Committee, Town of Concord.

Petrulionis, Sandra Harbert, and Laura Dassow Walls, eds. 2007. *More Day to Dawn: Thoreau's Walden for the Twenty-first Century*. Amherst and Boston, MA: University of Massachusetts Press.

Pipkin, John. 2010. *Woodsburner. A Novel*. New York, NY: Anchor Books.

Primack, Richard B. 2014. *Walden Warming: Climate Change Comes to Thoreau's Woods*. Chicago, IL and London: University of Chicago Press.

Proclamation No. 9476, 3 C.F.R. 9476 (2016).

Raden, Audrey. 2017. *When I Came to Die: Process and Prophecy in Thoreau's Vision of Dying*. Amherst and Boston, MA: University of Massachusetts Press.

Richardson, Robert D., Jr. 1986. *Henry Thoreau: A Life of the Mind*. Berkeley, Los Angeles, CA and London: University of California Press.

Richardson, Robert D., Jr. 1992. Thoreau and Science. In *American Literature and Science*, ed. Robert J. Scholnick, 110–127. Lexington, KY: University Press of Kentucky.

Richardson, Robert D. Jr. 1993. Introduction: Thoreau's Broken Task. In Henry David Thoreau, *Faith in a Seed: The Dispersion of Seeds and Other Late Natural History Writings*, ed. Bradley P. Dean, 3–17. Washington, DC and Covelo, CA: Island Press/Shearwater Books.

Robinson, David M. 2004. *Natural Life: Thoreau's Worldly Transcendentalism.* Ithaca, NY and London: Cornell University Press.

Sattelmeyer, Robert. 1990. The Remaking of *Walden.* In *Writing the American Classics*, eds. James Barbour and Tom Quirk, 53–78. Chapel Hill, NC: University of North Carolina Press.

Sattelmeyer, Robert. 2000. Depopulation, Deforestation, and the Actual Walden Pond. In *Thoreau's Sense of Place: Essays in American Environmental Writing*, ed. Richard J. Schneider, 235–243. Iowa City, IO: University of Iowa Press.

Schneider, Richard J., ed. 2000. *Thoreau's Sense of Place: Essays in American Environmental Writing*. Iowa City, IO: University of Iowa Press.

Schneider, Richard J. 2016. *Civilizing Thoreau: Human Ecology and the Emerging Social Sciences in the Major Works*. Rochester, NY: Camden House.

Schofield, Edmund A. 1989. *The Walden Ecosystem: '… a perennial spring in the midst of pine and oak woods …'* [From the Walden Woods Project Collections—Edmund A. Schofield Collection, at the Thoreau Institute at Walden Woods].

Schofield, Edmund A. 1993. The Ecology of Walden Woods. In *Thoreau's World and Ours: A Natural Legacy*, eds. Edmund A. Schofield and Robert C. Baron, 155–171. Golden, CO: North American Press.

Schofield, Edmund A., ed. 2004. *Words for Nature: Spirit, Wildness, and the Sublime in the Writings of Henry D. Thoreau.* [From the Walden Woods Project Collections—Edmund A. Schofield Collection, at the Thoreau Institute at Walden Woods.]

Schofield, Edmund A., and Robert C. Baron, eds. 1993. *Thoreau's World and Ours: A Natural Legacy*. Golden, CO: North American Press.

Schofield, Edmund A., and Mary Bush-Brown. n.d. *Plants of the Thoreau Institute: A Sampling of Thoreau's Favorites*. Lincoln, MA: Walden Woods Project.

Shanley, J. Lyndon. 1957. *The Making of Walden, with the Text of the First Version.* Chicago, IL: University of Chicago Press.

Thoreau, Henry David. 1906. *The Writings of Henry David Thoreau, Journals*, 14 vols., eds. Bradford Torrey and Francis H. Allen. Boston, MA and New York, NY: Houghton Mifflin and Company. [Available online via the Walden Woods Project's Thoreau Institute Library: https://walden.org/what-we-do/library/thoreau/the-writings-of-henry-david-thoreau-the-digital-collection/.]

Thoreau, Henry David. 1958. *The Correspondence of Henry David Thoreau*, eds. Walter Harding and Carl Bode. Washington Square, NY: New

York University Press. [Available online via the Walden Woods Project's Thoreau Institute Library: https://walden.org/what-we-do/library/thoreau/the-writings-of-henry-david-thoreau-the-digital-collection/.]

Thoreau, Henry David. 1993. *Faith in a Seed: The Dispersion of Seeds and Other Late Natural History Writings*, ed. Bradley P. Dean. Washington, DC and Covelo, CA: Island Press/Shearwater Books.

Thoreau, Henry David. 2000. *Wild Fruits: Thoreau's Rediscovered Last Manuscript*, ed. Bradley P. Dean. New York, NY and London: W. W. Norton.

Thoreau, Henry David. 2001a [1849]. *A Week on the Concord and Merrimack Rivers*. Mineola, NY: Dover Publications, Inc.

Thoreau, Henry David. 2001b. *Collected Essays and Poems*, ed. Elizabeth Hall Witherell. New York, NY: The Library of America.

Thoreau, Henry David. 2004a [1854]. *Walden*. 150th Anniversary Edition, ed. J. Lyndon Shanley. Princeton, NJ and Oxford: Princeton University Press.

Thoreau, Henry David. 2004b [1864]. *The Maine Woods*, ed. Joseph J. Moldenhauer. Princeton, NJ and Oxford: Princeton University Press.

Thoreau, Henry David. 2004c [1865]. *Cape Cod*, ed. Joseph J. Moldenhauer. Princeton, NJ and Oxford: Princeton University Press.

Thoreau Farm Trust, ed. 2017. *What Would Henry Do? Essays for the 21st Century*. Concord, MA: Thoreau Farm Trust Inc.

Thorson, Robert M. 2014. *Walden's Shore: Henry David Thoreau and Nineteenth-Century Science*. Cambridge, MA and London: Harvard University Press.

Thorson, Robert M. 2017. *The Boatman: Henry David Thoreau's River Years*. Cambridge, MA and London: Harvard University Press.

Thorson, Robert M. 2018. *The Guide to Walden Pond: An Exploration of the History, Nature, Landscape, and Literature of One of America's Most Iconic Places*. Boston, MA: Houghton Mifflin Harcourt.

Van Noy, Rick. 2003. *Surveying the Interior: Literary Cartographers and the Sense of Place*. Reno, NV: University of Nevada Press.

Walden Woods Project. 2002. *Brister's Hill Interpretive Site*. Lincoln, MA: Walden Woods Project/Watertown, MA: Sasaki Associates, Inc.

Walden Woods Project. 2006. *Thoreau's Path on Brister's Hill in Walden Woods, Concord, Massachusetts. Dedication Ceremony May 15, 2006*. Lincoln, MA: Walden Woods Project.

Walden Woods Project. 2007. *A Guide to Thoreau's Path on Brister's Hill*. Working Draft. Lincoln, MA: Walden Woods Project.

Walden Woods Project. 2013. Explore Walden Woods: Vote Yes on Article 12. *Walden Woods Project.* https://walden.org/Explore/Vote_YES_On_Article_12.

Walls, Laura Dassow. 1995. *Seeing New Worlds: Henry David Thoreau and Nineteenth-Century Natural Science*. Madison, WI: University of Wisconsin Press.

Walls, Laura Dassow. 2000. Believing in Nature: Wilderness and Wildness in Thoreauvian Science. In *Thoreau's Sense of Place: Essays in American Environmental Writing*, ed. Richard J. Schneider, 15–27. Iowa City, IO: University of Iowa Press.

Walls, Laura Dassow. 2009. *The Passage to Cosmos: Alexander von Humboldt and the Shaping of America*. Chicago, IL: University of Chicago Press.

Walls, Laura Dassow. 2016. "The Value of Mutual Intelligence:" Science, Poetry, and Thoreau's Cosmos. In *Thoreau at 200: Essays and Reassessments*, eds. Kristen Case and K.P. Van Anglen, 185–199. Cambridge: Cambridge University Press.

Walls, Laura Dassow. 2017. *Henry David Thoreau: A Life*. Chicago, IL: University of Chicago Press.

Walls, Laura Dassow. 2018. The Corner-stones of Heaven: Science Comes to Concord. In *The Cambridge Companion to the Literature of the American Renaissance*, ed. Christopher N. Phillips, 221–234. Cambridge: Cambridge University Press.

Weinreb, Stuart. 1985. *Walden Pond State Reservation Bank Restoration Project: Report on the Approach and Methodology*. Boston, MA: Massachusetts Department of Environmental Management.

Wheeler, Joseph C. 2005. Saving Walden. *The Concord Saunterer. Special* Walden *Sesquicentennial Issue: Walden the Place and* Walden *the Book*. New Series, 12/13: 195–203.

Whitford, Kathryn. 1950. Thoreau and the Woodlots of Concord. *The New England Quarterly: A Historical Review of New England Life and Letters* 23 (3): 291–306.

Whitney, Gordon G., and William C. Davis. 1986. From Primitive Woods to Cultivated Woodlots: Thoreau and the Forest History of Concord, Massachusetts. *Journal of Forest History* 30 (2): 70–81.

Williams, Terry Tempest. 2016. *The Hour of Land: A Personal Topography of America's National Parks*. New York, NY: Sarah Crichton Books/Farrar, Straus and Giroux.

3

'No Holier Temple': John Muir, the Hetch Hetchy Valley, and Restore Hetch Hetchy

3.1 Into the Range of Light with John of the Mountains

On March 5, 1867, while working at the Osgood, Smith & Co. carriage materials factory in Indianapolis, Indiana, John Muir (1838–1914) was injured in an industrial accident that left him temporarily blind in his right eye. Within days, nerve shock caused his eyesight to also fail in his left eye. Four weeks of bedrest in a darkened room provided Muir with the opportunity to think about the South American voyage of German scientific explorer Alexander von Humboldt, as well as California's Yosemite Valley, and this pause set him on a new professional and intellectual trajectory. With his eyesight restored, he resigned his post at the carriage factory and committed to pursuing what he so nearly lost the faculty to do—the study of natural history, or what he termed the 'inventions of God' (in Badè 1924a).

From Indianapolis he traveled to Jeffersonville, then crossed the Ohio River (and the state line) to Louisville, Kentucky. On September 2, 1867, six months after his accident, Muir began a 1,000-mile 'glorious walk' and botanical expedition through the five states of Kentucky, Tennessee,

L. Smith, *Ecological Restoration and the U.S. Nature and Environmental Writing Tradition*, https://doi.org/10.1007/978-3-030-86148-3_3

North Carolina, Georgia, and Florida—from the Mississippi Valley to the Gulf of Mexico—arriving in Cedar Key on October 23 (see *A Thousand Mile Walk to the Gulf* in Muir 2001b, also Badè 1924a). Muir's plan had been to continue on to South America to the Orinoco and Amazon river basins, retracing Humboldt's 1799–1804 journey across the South American continent. He took a job at Richard W. Hodgson's sawmill in Cedar Key, while he waited for a lumber schooner to Galveston, Texas. But after contracting malaria in Florida, Muir amended his plans 'to be a Humboldt' (Muir 2019), replacing South America with San Francisco, California, and from there, the Yosemite Valley, in an attempt to restore his health. After a three-month convalescence with the Hodgson family, Muir left Florida in January 1868. His ocean voyage to San Francisco took him via Cuba and New York to Panama, and across the Isthmus of Panama by rail to the West Coast. It would be more than four decades before Muir concluded his Humboldtian journey, touring South America (and Africa) for eight months in 1911 and 1912.

This chapter begins with Muir's arrival in San Francisco in the spring of 1868, and lingers on his tenure amidst the forests and mountains of Yosemite through the late-1860s to mid-1870s, before turning to his botanical studies and expeditions, and burgeoning eco-theological philosophy—and the genesis of his restoration sensibility. It is Muir's ecclesiastical descriptions of Yosemite, and particularly his descriptions of the Hetch Hetchy Valley, that would infiltrate political rhetoric on the development of Hetch Hetchy in the opening decades of the twentieth century.

Running alongside Muir's introduction to Yosemite is the origin story of conservation protections in Yosemite. This chapter traces the conservation histories, politics, and controversies of this northern California valley system—beginning with President Abraham Lincoln's signing of the Yosemite Grant in 1864, protecting the Yosemite Valley and Mariposa Grove of Big Trees, four years before Muir first traveled to Yosemite. Muir enters Yosemite's conservation story when he and *The Century Magazine* associate editor Robert Underwood Johnson launched a (literary) campaign for a Yosemite National Park in 1889, to protect the watersheds around the Yosemite Grant land—and including the Hetch Hetchy Valley. President Benjamin Harrison designated Yosemite

National Park just over one year later, in October 1890. Two years afterward, Muir established the Sierra Club to fight for Yosemite and Sierra Nevada protections, and continued campaigning for recession of the Yosemite Grant, and inclusion of the Yosemite Valley in Yosemite National Park. The Yosemite Valley and Mariposa Grove were incorporated into Yosemite National Park in 1906.

But 1906 would emerge as a pivotal year not just for Yosemite protections (and eventual un-protections), but also for the American conservation movement more widely. A devastating earthquake struck San Francisco in the early hours of April 18. President Theodore Roosevelt signed the Antiquities Act on June 8, and the recession of the Yosemite Grant three days later, on June 11. At this conservation juncture, this chapter examines how a remote valley in Yosemite National Park, northwest of the Yosemite Valley, became the theater for the biggest battle in American conservation to that time.

Yosemite's Hetch Hetchy Valley emerged as the lynchpin in political—and public—debates over water politics and water rights-claims in San Francisco in the aftermath of the earthquake. Muir's earlier nature religion framing of Yosemite and the Sierra Nevada would be politicized in defense of the valley. But the damming and flooding of the Hetch Hetchy Valley was authorized by President Woodrow Wilson with the signing of the Raker Act in 1913, and the construction of the O'Shaughnessy Dam on the Tuolumne River began in 1919. The legacy of Hetch Hetchy is told and retold in the signing of the National Park Service Organic Act in August 1916 and the creation of the National Park Service (NPS), and in conservation campaigns against dams on western rivers throughout the 1950s and 1960s, and in many other ways. Still Muir's hope of a restored Hetch Hetchy persists. Secretary of the Interior Donald Hodel first formalized the idea in 1987, and before the end of the decade, the Sierra Club established a Hetch Hetchy Restoration Task Force (which became the nonprofit Restore Hetch Hetchy (RHH) in 2000). This chapter closes by considering the evolution and friction of a Hetch Hetchy restoration rhetoric in California politics, alongside the restoration pathology of RHH, and its technical and political arguments for restoring the Hetch Hetchy Valley.

3.1.1 Yosemite Residencies and Excursions, 1868–1874 and Beyond

One of Muir's earliest mentions of the seven-mile-long Yosemite Valley appears in an April 1867 letter to friend, mentor, and confidante Jeanne C. Carr, dictated shortly after his factory accident as he convalesced, when he reflects, 'I read a description of the Yosemite Valley last year and thought of it most every day since' (Muir 2019: 11). After just one day in San Francisco, Muir and Joseph Chilwell (a young Englishman he met aboard the Panama steamer) took a ferry across the Bay to Oakland, and set out on another 'glorious walk' across the Central Valley, headed for the Sierra Nevada range and Yosemite 180 miles away, arriving six weeks later (Muir 1872c, 1890a, 1896, 2001b, also Badè 1924a). Soon after his arrival, he wrote again to Carr that, 'It is by far the grandest of all of the special temples of Nature I was ever permitted to enter. It must be the *sanctum sanctorum* of the Sierras' (July 26, 1868 letter in Muir 2019: 22).

Muir's first visit into the Sierra Nevada in the spring of 1868 lasted a week-and-a-half, and included his first view of the Yosemite Valley, and the Mariposa Grove of Big Trees, where he met Galen Clark.[1] He wrote very little about this first Yosemite excursion, and soon left to find work in the San Joaquin Valley. But he returned the next year, in June 1869,[2] as a shepherd co-steering more than 2,000 sheep—repeatedly described as 'hoofed locusts' and 'wooly locusts' by Muir (2001a, 2003a, 2018a)—to the headwaters of the Merced and Tuolumne rivers and the alpine Tuolumne Meadows in the High Sierras for Patrick Delaney (his edited journals from this summer posting were serialized in *The Atlantic Monthly* in January–April 1911, and later published as *My First Summer in the Sierra* in 1911, see Muir 2003a, also 2019; Badè 1924a; Wolfe 1966). Muir's shepherd post with Delaney ended in late-September 1869, and he returned to planning his Humboldtian

[1] In 1857, Clark was the first Anglo-American to enter the Mariposa Grove, and was dedicated to its preservation. He was later appointed the first Guardian of the Yosemite Grant.

[2] Also in 1869, Major John Wesley Powell led the first Anglo-American expedition down the Green and Colorado rivers, and through the Grand Canyon (see Chapter 6).

botanical expedition, but in the fall, decided to remain full-time in the Yosemite Valley.

Looking for work, Muir and friend Harry Randall were taken on by James Mason Hutchings. Hutchings, a journalist and editor-publisher of *Hutchings' Illustrated California Magazine* (1856–1861), was central to early Yosemite tourism—but he was also the first permanent Anglo-American resident in the valley, and proprietor of the Hutchings House hotel, one of the earliest hotels in the Yosemite Valley. In the spring, Muir designed, built, and operated a sawmill[3] for Hutchings. As a sawyer and carpenter, Muir also renovated the hotel, and helped Hutchings construct cabins for the early tourist trade, as well as working as a part-time guide for artists, photographers, and small tourist parties into the higher Yosemite.[4] In November 1869, Muir and Randall built for themselves a small, one-room cabin of sugar pine and cedar shakes near the sawmill on Yosemite Creek, yards from Lower Yosemite Falls, where Muir lived for 11 months (Badè 1924a; Wolfe 1966, 2003; Wilkins 1995; Duncan 2013). Much like Henry David Thoreau's Walden cabin, Muir's Yosemite cabin was a cabin of the woods.

Muir left Yosemite with Randall later in 1870, and headed to the San Joaquin plain again to work for Delaney, before returning in January 1871 for 22 months—his longest Yosemite sojourn. In early-1871, he had to give up his cabin, after it was sequestered by Hutchings for use by his sister (Fox 1981; Wilkins 1995; Wolfe 2003). Muir built and moved into a small loft study in the sawmill—his 'hang nest'—again facing Yosemite Falls, where he resided until mid-July, when he finally,

[3] In the heat of the Hetch Hetchy campaign in the early-twentieth century, Muir would be accused of being a hypocrite in seeking to halt industry in Yosemite National Park, but the timber from the sawmill was sourced only from trees uprooted in storms—there was no hand-felling (Badè 1924a; Cohen 1984; Turner 2000; Wolfe 2003; Colwell 2014).

[4] Soon after Hutchings hired Muir and Randall at his sawmill, he left Yosemite Valley for Washington, DC, not returning until May 1870—embroiled in a legal fight with co-claimants James C. Lamon and two others for compensation for his land under preemption laws following the signing of the 1864 Yosemite Grant. Hutchings had acquired the hotel just six weeks before the Yosemite Grant was signed. He appealed first against the Yosemite Board of Commissioners, then the Supreme Court in *Hutchings v. Low*, but his claims were denied in 1872. Hutchings was formally evicted in May 1875 (and again in the fall, after a last stand) (Badè 1924a; Fox 1981; Runte 1990; Wilkins 1995; Turner 2000; Simpson 2005; Worster 2008; Duncan 2013; Binnewies 2015).

and acrimoniously, severed all ties with Hutchings (Muir 2019; also Badè 1924a). Muir was invited by A. G. Black and his wife to make Black's Hotel near Sentinel Rock his base in the valley, where he would serve as caretaker for at least two winters. In the spring of 1872, Muir built another cabin on the Merced River, below the Royal Arches in the upper end of the valley, close to the log cabin, orchard garden, and farm of the valley's first Anglo-American settler James C. Lamon in 1859 (Badè 1924a; Wolfe 1966; Duncan 2013; Binnewies 2015).

From his base in the valley bottom, Muir would regularly head into the Yosemite high country. He spent many nights outdoors, favoring pine and fir boughs for his bedding and blanketing. Favorite Yosemite Valley camps were Sentinel Dome, and the Sunnyside Bench ledge to the east of Lower Yosemite Falls (Wolfe 1966). He would often spend weeks solivagant amidst the canyons, forest groves, mountains, and glaciers of the Sierra Nevada. Other excursions were made in the company of friends and acquaintances, scientists, artists—and two U.S. presidents. Muir's scientific and popular dispatches from the California mountains told of glaciers and glaciology, mountains, trees, and Yosemite flora and fauna, epitomized in his seven-part 'Studies in the Sierra' serialized in *Overland Monthly* in 1874–1875 (collected in Muir 1950), among dozens more. Muir's reputation as a wilderness provocateur began to ascend early in his Yosemite years, so much so that he was fictionalized in Therèse Yelverton's (1872) novel, *Zanita: A Tale of the Yo-semite* as the rugged, wild mountain man Kenmuir, the subject of the opening chapter, published while in the midst of his Yosemite sojourn. Yelverton (1872) presents one of the earliest intimate portraits of Yosemite resident Muir. When 'John of the Mountains' (after Wolfe 1966) eventually departed the Yosemite Valley in the fall of 1875, it was to concentrate on the preservation of the forest groves and watersheds of Yosemite.

3.1.2 Muir as Botanist, Eco-Theologian

To try and define or pigeonhole Muir is to experience something similar to the calamity Muir (2003a) describes in *My First Summer in the Sierra* when he, with sheep-owner Delaney, and other drivers and dogs, tried to coax their caravan of stubborn, cautious sheep across Yosemite Creek. For Muir was at once technologist and inventor, botanist and naturalist, geologist and glaciologist, mountaineer, rancher and fruit farmer, philosopher, writer and campaigner, environmentalist, wilderness polemicist, early preservationist, proto-restorationist. Indeed, Muir struggles to summarize his profession, before settling on a 'poetico-trampo-geologist-bot. & ornith-natural, etc. etc. !—!—! !' (in a September 13, 1889 letter to Robert Underwood Johnson, see Muir 1944: 50). The two greatest contributions to Muir's ecological restoration sensibility lie jointly in his botanical studies and his personal spirituality and mysticism.

The genesis of Muir's passion and enthusiasm for botanizing lies amidst his childhood rambles on the harsh, rugged cliffs and fields surrounding his hometown of Dunbar, 30 miles east of Edinburgh, on Scotland's North Sea coast. But Muir's curiosity in natural history both vibrated against and was irrevocably molded by the evangelical Presbyterianism of his father, Daniel Muir. It was Daniel Muir's religious conviction and zealotry that led the family to abruptly emigrate to the U.S. in 1849—he was drawn to the austere biblical literalism practiced by the Campbellite Body of the Disciples in Christ colonies in the U.S. and Canada. Settling in Wisconsin, Daniel Muir purchased 80 acres of open woodland surrounding Fountain Lake (soon to be known as Muir's Lake, but today named Ennis Lake) near the Fox River, in Buffalo Township, Marquette County. With help from local laborers, he raised first a temporary bur-oak shanty, then a farmhouse, during that first summer (Muir, an older sister, and younger brother accompanied their father in the spring, while their mother and four other siblings followed later in the fall, and Muir's youngest sister was born in Wisconsin). Muir settled at Fountain Lake Farm two years after Thoreau ended his Walden sojourn. The family would live at Fountain Lake Farm for eight years, soon doubling the original acreage to a quarter-section of 160 acres, before raising another half-section of 320 acres as the Hickory Hill farmstead, five miles to the southeast (on Muir's Scotland and Wisconsin years, see especially Badè 1924a; Stanley 1995; Wilkins 1995; Holmes 1999; Turner 2000; Simpson 2003; Wolfe 2003; Worster 2008; Colwell 2014).

Daniel Muir was a formidable patriarch and did not allow any nonconformity to a stifling and strict religious doctrine. As the eldest son, much of the farm labor fell to Muir. From age 11 to 22, Muir worked as an unpaid farm laborer for his father, cultivating and harvesting both the Fountain Lake and Hickory Hill farmsteads. Any rare free time was spent exploring the forests, meadows, fields, and lakes of the farmsteads. At Hickory Hill, Muir regularly began to rise at one o'clock in the morning, and in the pre-dawn hours, split his time between reading and crafting his own mechanical inventions in the cellar of the farmhouse, with wood and metal salvaged from the farm. He began by crafting crude tools, before moving on to timepieces and labor-saving instruments (described in Muir 2018b).

In September 1860, encouraged by neighbor William Duncan, 22-year-old Muir bundled together a selection of his 'whittlings'—two hickory clocks and a thermometer—and left Hickory Hill Farm, traveling to Madison, Wisconsin, to exhibit his inventions at the State Agricultural Society Fair, held near the state university campus (the University of Wisconsin-Madison). Muir's mechanical inventions (including his early-rising bed-tipping machine, assembled using one of his clocks) received much critical acclaim, and he was awarded an honorarium (Muir 2018b; cf. Badè 1924a; Fox 1981; Wilkins 1995; Ehrlich 2000; Turner 2000; Wolfe 2003; Worster 2008). It was at the state fair that Muir first met Jeanne Carr, one of the exhibition judges, beginning a friendship and lively correspondence inspired by a shared passion for botany that would last until Carr's death in December 1903 (on the Muir-Carr friendship, see Muir 2019, also Badè 1924a; Wilkins 1995; Turner 2000; Wolfe 2003; Worster 2008; Colwell 2014). Carr's husband, Ezra Slocum Carr, was chair of natural science at the university, and he too would become a lifelong friend and mentor to Muir.

Muir's technical acclaim also translated into job offers in machine shops, and eventually admission to UW-Madison, a land grant university, in the spring of 1861. Muir completed five semesters at the state university, but did not receive a degree[5] (for an excellent account of

[5] Muir was awarded an honorary degree from UW-Madison 34 years after he left, in June 1897. (He also received honorary degrees from Harvard in June 1896, Yale in June 1911, and UC Berkeley in May 1913.)

Muir's UW-Madison years, see especially Badè 1924a; Worster 2008). An impromptu botany lesson with classmate Milton Griswold beneath a black locust tree on campus near North Hall in June 1863 served as Muir's introduction to plant genealogy, soon to become a lifelong curiosity and passion (Badè 1924a, see also Gisel 2008).

After leaving the university in 1863, Muir, encouraged by Ezra Carr, intended to enter medical school at the University of Michigan in Ann Arbor. In the meantime, the summer would mark his first botanical and geological ramble—spending three weeks in late-June following the Wisconsin River valley through southwestern Wisconsin and into Iowa to the Mississippi River with two UW-Madison classmates (Badè 1924a; Fox 1981; Wilkins 1995; Wolfe 2003; Colwell 2014). But when his number was not called in the Civil War draft, he revised his plans, and in early-March 1864 set out to explore Canada, variously botanizing and working in factories and mills (sometimes alongside his youngest brother Daniel), where he remained until the end of the war.

Back in the U.S. in mid-March 1866, Muir found work first as a millwright and sawyer before quickly working up to supervisor at the Osgood, Smith & Co. carriage factory in Indianapolis. He had struggled to align his twin passions of botany and mechanical invention, and settled on Indianapolis as a railroad center with machine shops and factories, surrounded by expansive deciduous oak and hickory hardwood forests. It was his eye injury at the carriage factory in March 1867, and the panic that he may no longer be able to explore the American wilderness, that ultimately converted Muir-the-industrialist to Muir-the-wilderness-prophet, and he departed Indianapolis after 16 months. His Gulfward walking tour and Yosemite explorations that followed signaled Muir's fracture with the paternal Christianity of his youth, and his quest for a Muirian theology, a religious communion with nature.

Muir was intensely religious, and his adoption of and adherence to a religious vocabulary to describe nature is intentional and deliberate. But it speaks to a personal theology aligned more with mid-nineteenth-century New England Transcendentalist philosophy (Chapter 2), and with pantheism and biocentrism, than with the zealous evangelism of his father (on the influence of Emerson and Thoreau on Muir, see Wolfe 1966; Fox 1981; Cohen 1984; Albanese 1990; Worster 2008; Nash

2014). Muir had just returned for a second year at the state university when Thoreau died in May 1862, but he would meet Ralph Waldo Emerson a decade later, in the Yosemite Valley in May 1871.[6] Muir's essay, 'The Forests of the Yosemite Park,' published in *The Atlantic Monthly* in April 1900 (and included as a chapter in Muir's *Our National Parks* the following year, see Muir 2018a,[7] also Badè 1924a; Wolfe 1966), recounts the meeting between 33-year-old Muir and 68-year-old Emerson during Muir's second spring in Yosemite (Fig. 3.1).

Before the end of the third paragraph in his first book, 1894's *The Mountains of California*, Muir declares that a more apt name for the Sierra Nevada cordillera is the Range of Light (Muir 2001a, the 'Range of Light' motif is also present throughout e.g. Muir 1890a, 2003a, b, 2018a; Badè 1924a). For Muir, the mountains are bathed in a divine light, 'God's holy light' (after Muir 2002), and this natural religion framing sets the tone for much of his California writing (on a Muirian eco-spirituality, or eco-theology, see Fox 1981; Cohen 1984; Albanese 1990; Wilkins 1995; Williams 2002; cf. Bilbro 2015; Stoll 2015).

Perhaps the strongest religious accent is through the analogy Muir draws with Gothic cathedrals, churches, and especially temples—whether in celebrating Yosemite's majestic granite mountain temples, its sylvan, arboreal temple groves, or its glacier temples (Muir 1872a, 1890a, 1896, 1907, 1908a, 1909b, 2001a, 2003a, b, 2018a; Wolfe 1966). Yosemite was a grand, sublime 'Sierra temple' (Muir 2001a, 2003a), a 'Sierra Cathedral' (Muir 1890a, 2003b, 2019), its forest groves silent cathedrals of trees. Muir (2003a: 336) writes during his first Yosemite summer, that, 'This I may say is the first time I have been at church in California.' Muir's first article on forest preservation, inspired by his 1875

[6] Muir visited Concord 22 years later, with *Century* editor Robert Underwood Johnson, in June 1893. The pair had dinner with Emerson's son Edward, and visited Old North Bridge, Author's Ridge in Sleepy Hollow cemetery where Emerson and Thoreau are buried, Walden Pond, and the former homes of several Transcendentalists, including Emerson, Thoreau, and Nathaniel Hawthorne, before Muir departed for a three-month trip to Scotland, England, and continental Europe (Badè 1924b; Muir 2018a, also Fox 1981; Wilkins 1995; Turner 2000; Simpson 2003; Wolfe 2003; Colwell 2014).

[7] Muir's *Our National Parks*, published in 1901 (reprinted as Muir 2018a), is a collection of 10 articles that appeared in *The Atlantic Monthly* between 1897 and 1901, and will be the default reference for these articles henceforth.

Fig. 3.1 John Muir quotation in the Reflection Circle of Thoreau's Path on Brister's Hill, in Walden Woods, Concord, Massachusetts. 'I only went out for a walk,...' Photograph by the author, June 2007

survey of California's sequoia belt, was published in the *Sacramento Daily Record-Union* on February 5, 1876, under the title, 'God's First Temples' (Muir 1876, with the 'God's first temples' motif also featuring in e.g. Muir 2003a).[8] To Muir, the holiest of Yosemite temples was the Hetch Hetchy Valley—this North Tuolumne Church (Muir 2003a) was 'one of Nature's rarest and most precious mountain temples' (Muir 2003b: 255). It is this motif above all that would be politicized in the early decades of the twentieth century as Muir and his Sierra Club fought to keep a dam and reservoir project out of Hetch Hetchy.

Muir extends the Yosemite-as-temple analogy to also describe tree and rock spires, rock altars, aisles of forest groves, congregations of mountains and of waters (Muir 1872a, 1890b, 2003a, 2019, also Badè 1924a,

[8] When Muir wrote this article, American forestry was still in its infancy. The Forestry Division of the Department of Agriculture was established the same year.

b; Wolfe 1966, cf. a palm congregation in Muir 2001b, congregations of glaciers in Muir 2002). There are references to baptism (e.g. Muir's own 'baptism' into the Sierra wilderness at a Twenty Hill Hollow 'font' in Muir 1872c, 2001b, 2019, and in the Wisconsin wilderness in Muir 2018b; Badè 1924a; cf. Cohen 1984; Bilbro 2015), choirs (especially the Illilouette Fall in Muir 1890a, 2003b), and sermons and gospel told in storms, trees, and rocks (Muir 2002, 2003a, 2018a, 2019; Badè 1924b; Wolfe 1966). The sugar pine—one of Muir's favorite trees—is described as the 'priests of pines,' and as 'apostles' preaching to the rest of the forest (as in Muir 2001a, 2003b, 2018a; Badè 1924b, cf. Muir as nature apostle in Wilkins 1995). Muir describes trees bowing as if in worship during wind-storms (Muir 2001a, b, 2003b, 2019), 'preaching God's forestry fresh from heaven' (Muir 2018a: 213), and writes of 'tree scripture,' 'tree hymns and prayers,' and 'psalm-singing' trees and waterfalls (e.g. Muir 2002, 2018a; Badè 1924b; Wolfe 1966).

While Muir's ecological restoration sensibility would bloom around his impassioned defense of Yosemite's Hetch Hetchy Valley in the last decade of his life, its origins lie in his intensive botanical excursions and studies across the Yosemite Valley. But there are other vignettes found elsewhere in his earlier writings that also suggest a latent restoration pathology. One of these is described well into Muir's account of his 1,000-mile tramp from Kentucky to Florida, when, low on funds and awaiting a money package from his brother, David, he camps for five or six nights beneath the moss-covered live oaks in the Bonaventure cemetery three miles outside Savannah, Georgia.[9] Writing on October 9, 1867, Muir notes,

> Even those spots which are disordered by art, Nature is ever at work to reclaim, and to make them look as if the foot of man had never known them. Only a small plot of ground is occupied with graves and the old mansion is in ruins. [...] It is interesting to observe how assiduously Nature seeks to remedy these labored art blunders. She corrodes the iron and marble, and gradually levels the hill which is always heaped up, as

[9] Incidentally, within the grounds of the cemetery, with its damp underbrush, is most likely where Muir contracted malaria that afflicted him later that fall in Florida (Wolfe 2003; Worster 2008; Colwell 2014).

> if a sufficiently heavy quantity of clods could not be laid on the dead. Arching grasses come one by one; seeds come flying on downy wings, silent as fate, to give life's dearest beauty for the ashes of art; and strong evergreen arms laden with ferns and tillandsia drapery are spread over all—Life at work everywhere, obliterating all memory of the confusion of man.
>
> —Muir (2001b: 32, 34)

Indeed, Muir's entire walk to the Gulf was one of spiritual healing, renewal, and restoration following his industrial accident.

There are a small number of direct references to 'restoration' in Muir's writings, indicative of a nascent ecological restoration pathology. Social restoration or redemption in nature are also prominent themes in Muir's writings. Dayton Duncan (2013) argues that Hutchings, Clark, and Muir—and later Muir allies Frederick Law Olmsted and Theodore Roosevelt (often shortened to TR hereafter)—were all not just redeemed or restored by wilderness, but were also redeeming and restoring Yosemite. Approaching the close of the nineteenth century, there was a shift in environmental narrative away from wilderness redeemed by society toward society redeemed by wilderness (on the oxymoron inherent in the historic 'land redeemed from wilderness' motif, see e.g. Duncan 2013).

3.2 Conserving Yosemite

Muir (2003b) writes in *The Yosemite* and elsewhere of the 'discovery' of the Yosemite Valley by Anglo-Americans in the early-1850s. But the valley had long been home to Native Nations including the Ahwahneechee, part of the southern Miwok. Only in the mid-nineteenth century did the first Anglo-Americans enter the valley, as the California gold rush spread across the foothills of the Sierra Nevada. On March 27, 1851, members of the Mariposa Battalion, led by James D. Savage, camped in the valley while deployed to halt raids on trading posts by Native Nations parties.[10] This Mariposa War (1850–1851), which began the year after the Muir family emigrated to the U.S., and more than a

[10] Although a party led by beaver trapper Joseph Reddefern Walker first traversed the valley rim in late-fall 1833 as part of the Bonneville expedition, it remains unknown whether they

decade-and-a-half before Muir first visited Yosemite, was part of the California Indian Wars, that raged between 1850 and 1880. This conflict is part of a much larger—and frequently violent and bloody—history of disruption, destruction, dislocation, and relocation of Yosemite's Indigenous groups by Anglo-American pioneers, settlers, hunters, miners, and others in the nineteenth century (for more discussion on Yosemite's Indigenous histories, see Simpson 2005; Duncan 2013; Solnit 2014; Binnewies 2015; Runte 2020). The racism in some of Muir's writings describing Yosemite's Indigenous groups is uncomfortable and disquieting—and following the death of George Floyd and the Black Lives Matter marches in 2020, the Sierra Club is confronting the early history of the organization, and some of its founding (and early) members, including Muir (Brune 2020, see also Treuer and Grannan 2021). The dislocations and relocations of Native Nations from national parks lands would be repeated across the U.S. into the twentieth century.

The first tourists began arriving in Yosemite a few years later. By the mid-1860s, as the American Civil War drew to a close and the U.S. entered its Gilded Age, summer visitors from San Francisco almost 200 miles away were making the journey across the Bay by ferry to Stockton, then traveling over several days by train, stagecoach, and on horseback to Coulterville, Big Oak Flat, or Mariposa.

Between 1860 and 1874, state geologist Josiah D. Whitney and his team conducted geological and ecological surveys of California, including Yosemite. In the first of the six-volume *Geological Survey of California*, titled *Geology* (Whitney 1865), Whitney introduced his theory on the origin of the Yosemite Valley—'formed by the process of upheaval,' by a cataclysmic convulsion and subsidence (expanded in Whitney 1869). In the 1870s, Muir would (correctly) challenge Whitney's theory to argue that Yosemite was formed by glacial erosion, although his observations of Yosemite's 'glacial manuscripts' would initially be dismissed (beginning with Muir 1871, his first published article, also 1950, 2003a, b; Badè 1924a; Wolfe 1966).

---

viewed the Yosemite Valley (e.g. Wilkins 1995; Turner 2000; Simpson 2005; Duncan 2013; Solnit 2014; Binnewies 2015).

3.2.1 Yosemite Protections I: The Yosemite Grant

On June 30, 1864, 10 months before the end of the American Civil War, and while Muir was botanizing in southern Ontario, President Abraham Lincoln signed what became known as the Yosemite Grant. Thoreau's *The Maine Woods*, with its preservation-of-wilderness proclamations (Thoreau 2004), and George Perkins Marsh's *Man and Nature*, setting out some of the earliest views on ecological restoration (Marsh 2003) were published the same year. The Yosemite bill—unaffiliated with the war effort—was unprecedented, and ceded the seven-mile-long Yosemite Valley and 2,560-acre Mariposa Big Tree Grove, 15 miles south of the valley (combined nearly 40,000 acres), to the control of the state of California, for 'public use, resort, and recreation' (on the history of the Yosemite Grant, see Simpson 2005; Duncan 2013; Solnit 2014; Binnewies 2015; Runte 2020).[11] The Yosemite Grant was in part a response to the many mills, factories, and hotels surrounding Niagara Falls in upstate New York. But more locally, it was also a response to the felling on June 27, 1853, of the Mammoth Tree, a 300-foot-tall, 1,244-year-old giant sequoia in the Calaveras Grove northwest of the Yosemite Valley (Muir 1920; Wolfe 1966).

On September 28, 1864, the Yosemite Board of Commissioners was appointed by California governor Frederick F. Low to oversee the dual task of preservation and public recreation provision within the state park. Clark was appointed as the first Guardian of the Yosemite Grant—becoming Yosemite's inaugural park ranger (on Clark's Yosemite legacy, see Sargent 2001; Duncan 2013; Binnewies 2015; Runte 2020). The chair of the commission, landscape architect Frederick Law Olmsted (manager of the Mariposa Estate, and earlier, the designer and chief architect of New York's Central Park with Calvert Vaux), foresaw a tension in these twin imperatives, and formalized his concerns a little

[11] Yellowstone was the U.S.'s—and the world's—first national park, designated by President Ulysses S. Grant on March 1, 1872, a decade after Thoreau's death, and eight years on from Congress setting aside Yosemite and granting it to the state of California. When designated in Wyoming Territory, with no state status (Wyoming joined the Union in July 1890), the park came under federal responsibility. At more than 2 million acres, it was bigger than Delaware and Rhode Island combined—and more than 50 times bigger than the Yosemite Grant in California.

less than a year later in *Yosemite and the Mariposa Grove: A Preliminary Report, 1865* (as Olmsted 1993). When Olmsted read his survey report and management plan to the commission during a camping trip in the Yosemite Valley on August 9, 1865, four months after the Civil War ended, he called for strict regulation, and was already looking beyond the confines of the Yosemite Valley and Mariposa Grove set down in the grant to celebrate the wider scenic, aesthetic, and therapeutic enjoyment of the region. But Olmsted's report was suppressed, and never reached the state legislature—remaining unpublished until 1952 (see Olmsted 1993; also Wilkins 1995; Turner 2000; Simpson 2005; Duncan 2013; Solnit 2014; Binnewies 2015; Runte 2020). The California state legislature accepted the Yosemite Grant on April 2, 1866.

When Muir arrived in Yosemite in 1868, four years after the Yosemite Grant was signed, he found many year-round residents in the valley, and soon hotels, saloons, lumberyards, pig farms, corrals, plowed meadows, fenced pastures, hayfields, orchards, and felled trees and stumps (Muir 1896; Badè 1924b; cf. Fox 1981; Cohen, 1988; Wilkins 1995; Duncan 2013; Colwell 2014; Binnewies 2015; Runte 2020). Muir's (1872a) essay, *In the Yo-semite: Holidays Among the Rocks*, features some of his earliest critiques of the Yosemite state park commission for the destruction of Yosemite beauty and grandeur (cf. Badè 1924b; Wolfe 2003). Muir laments a dearth of Yosemite law (see especially Muir 1872b), and reworks this dismay into a clarion call for wilderness protection. Muir is highly critical of the idea of 'improvement' within Yosemite's boundary—arguing that Yosemite is unimprovable, and regards attempts at improvement as vulgar and mercenary (Muir 1872a). Meanwhile, stage and railroad companies were introducing tolls, trails, and other developments into the valley. The Transcontinental Railroad was completed in 1869, and brought more tourists from eastern states. The Yosemite Valley Railroad operated between 1907 and 1945, transporting passengers and freight from Merced to El Portal on the Yosemite National Park boundary (see description in Muir 2003b; cf. Binnewies 2015).

On April 14, 1880, Muir married Louisa (Louie) Wanda Strentzel, who he met at the Carrs, who had moved to California in 1869. Jeanne Carr played matchmaker. Muir moved with Louie into the

Dutch colonial house on her family's 2,600-acre fruit ranch of orchards and vineyards in the Alhambra Valley, near Martinez, in Contra Costa County, California (after the death of Louie's father in October 1890, the family would move into the Italianate-style mansion on the ranch). Muir spent much of the period between 1882 and 1888 concentrating on the Strentzel fruit ranch business, growing grapes, oranges, lemons, quinces, pears, peaches, apricots, cherries, plums, and walnuts, and being a father to their two daughters, Annie Wanda and Helen Lillian. Muir increased productivity at the Martinez farm as he had done at factories in Canada and Indianapolis. But still each July through October was reserved for Muir to 'wander and wonder' in the wilderness (to echo phrasing in Muir 2003b), including trips to Yosemite, Alaska, and more.

3.2.2 Yosemite Protections II: Yosemite National Park, the Sierra Club, and the Recession of the Yosemite Grant

In the spring of 1889, Robert Underwood Johnson, associate editor of the New York-based periodical, *The Century Magazine* (and previously at *Scribner's Monthly*, where Muir had contributed early glacier articles), invited Muir to write for the magazine. A meeting in Johnson's hotel room at the Palace Hotel in San Francisco at the start of June, followed days after by a two-week pack trip to the Tuolumne Meadows, revealed a shared commitment to greater protection for the Yosemite Valley. This soon evolved into a campaign to create a Yosemite National Park in the high country around the state park (closely modeled after Yellowstone National Park, established in 1872), to preserve the forests of the Merced and Tuolumne watersheds from felling and overgrazing (on Muir and Johnson's camping trip, see Johnson 1923; cf. Badè 1924b; Wolfe 1966; Fox 1981; Cohen 1984, 1988; Wilkins 1995; Ehrlich 2000; Turner 2000; Wolfe 2003; Worster 2008; Duncan 2013; Colwell 2014; Nash 2014; Binnewies 2015; Runte 2020). While the Yosemite Valley would be difficult to reclaim from state control, 51-year-old Muir and 36-year-old Johnson would lobby for the surrounding land (including the Hetch Hetchy Valley) to be placed under federal control. Their task as park boosters had begun, counter to the boosters of California towns and cities, and reflected the start of the Progressive conservation movement in the U.S.

Johnson launched their campaign in the January 1890 issue of *The Century Magazine*, under the editorial title, 'The Care of Yosemite Valley' (Johnson 1890, 1923; cf. Binnewies 2015). As Muir began work on a pair of articles for *The Century Magazine*,[12] Los Angeles Congressman General William Vandever introduced a Yosemite National Park bill in Congress, House Resolution (H.R.) 8350, but for a much smaller park than Johnson and Muir envisaged—and which excluded Lake Tenaya and the entire Tuolumne River watershed, and much of the Merced River watershed (Muir 1890b; Fox 1981; Wilkins 1995; Simpson 2005; Worster 2008; Binnewies 2015; Runte 2020).

Muir penned the twin articles 'The Treasures of the Yosemite,' and 'Features of the Proposed Yosemite National Park' in mid-June 1890, before embarking on his fourth Alaska excursion. These two essays, published in the August and September 1890 issues of *The Century Magazine* (Muir 1890a, b), were Muir's inaugural campaigning articles. The first essay celebrates the mountains, rivers, streams, creeks, waterfalls, lakes, meadows, and trees in and around the Yosemite Valley, and calls for protection of 'all the fountain region above Yosemite' (Muir 1890a). The second essay opens in the Upper Tuolumne Valley, and describes the valley meadow and surrounding canyons, mountains, and glaciers. The second half of the essay is devoted to descriptions of the Hetch Hetchy Valley. Muir waits until the closing paragraph of 'Features' to deliver his edict on the need for preservation and recreation: 'Unless reserved or protected the whole region will soon or late be devastated by lumbermen and sheepmen, and so of course be made unfit for use as a pleasure ground' (Muir 1890b). The essay closes with a plea for a 'right relationship' with use of the Yosemite region. Muir (1890b) also includes a map detailing the proposed boundary for the new national park. Across his writings, he frames the proposed Yosemite National Park as two parks—or, a dual park (after Cohen 1988)—a wild 'Tuolumne Yosemite' (Muir 1873b, 1907, 1908a, 1909b; Badè 1924a, b), and an

[12] Almost a decade earlier, in the fall and winter of 1881, Muir had been invited to join with members of the California Academy of Sciences to help draft two bills to be presented before Congress—one to expand the Yosemite Valley and Mariposa Grove grant, the other to create a national park in the southern Sierra to protect sequoia groves (what is now protected by the Sequoia and Kings Canyon National Parks). Both stalled in the Senate public lands committee (Cohen 1984; Wilkins 1995; Ehrlich 2000; Wolfe 2003; Simpson 2005).

'improved,' 'Merced Yosemite,' where the latter could be sacrificed to preserve the former.

When Muir returned to San Francisco from Alaska in September, in time for the fruit harvest at the Martinez ranch, the Yosemite National Park campaign had escalated. With the publicity that followed the two *Century* articles, H.R. 12187 replaced the Vandever bill (Wilkins 1995; Turner 2000; Simpson 2005; Nash 2014), and detailed a park boundary of nearly two million acres that was a facsimile of Muir's (1890b) suggestion in 'Features of the Proposed Yosemite National Park.' The Vandever bill had expanded five times following Muir's recommendations.

On September 30, the Yosemite Forest Reservation bill was passed by both houses of Congress, and President Benjamin Harrison signed the bill into law the following day—October 1, 1890. With the signing, President Harrison created the U.S.'s third national park, protecting the watersheds of the Merced and Tuolumne rivers. At more than 900,000 acres, the new park was 30 times larger than the Yosemite Grant, and included the Hetch Hetchy Valley. Muir was now 52 years old, and more than a quarter-century had elapsed since he first visited the Yosemite Valley.[13] But the 39,000 acres of the Yosemite Valley and Mariposa Grove

[13] But Muir could not replicate his conservation momentum in Yosemite back on his family's first homestead in Wisconsin—Fountain Lake Farm. Before Muir left for Canada in 1864, he attempted to purchase 40 acres of sedge meadow beside Fountain Lake from his sister Sarah and brother-in-law David Galloway (who bought the farm from Daniel Muir in 1856—and sister Margaret and brother-in-law John Reid purchased Hickory Hill Farm from Daniel Muir when the family moved to Portage), to be set aside for conservation, illustrative of the plants, wildflowers familiar to pioneers, but it was sold for pastureland in 1865 and 1866, to the Ennis brothers (who renamed the lake Ennis Lake). Muir repeated his offer from Yosemite in 1871, and again in 1896, but was unsuccessful. Muir also tried to preserve Fern Lake on Mound Hill Farm, where the Galloways moved after selling Fountain Lake Farm. Fern Lake was protected for as long as the Galloways owned Mound Hill Farm. Sam Ennis demanded too high a fee for the Fountain Lake acreage (Muir 1896; Badè 1924a, b; Leopold 1968; Cohen 1984; Holmes 1999; Wolfe 2003; Worster 2008; Nash 2014; Root 2017). In an address to the Sierra Club in San Francisco on November 23, 1895, Muir spoke of his regret at being unable to keep the Fountain Lake Farm 'untrampled' (in Root 2017). During his early years in California, Muir also tried to protect Twenty Hill Hollow, near Snelling. It would be another Wisconsin conservation iconoclast who would eventually step in to protect Muir's Fountain Lake—Aldo Leopold, who in the mid-1930s bought a sand farm near Baraboo, just 15 miles from the farmstead (Wilkins 1995; Simpson 2003). Leopold lectured in wildlife management and land conservation at Muir's alma mater, UW-Madison, in the 1930s and 1940s (see Chapter 4). John Warfield Simpson notes an April 14, 1948 letter from Leopold to Ernest Swift, the director of the Wisconsin Conservation Department, proposing the establishment of a state

were still excluded, and remained under the purview of the state of California and the Yosemite Board of Park Commissioners (Binnewies 2015; Runte 2020). Yosemite was a park within a park. Two more California national park bills introduced by Vandever were also authorized days either side of Yosemite National Park that fall—and both save groves of sequoia. Sequoia National Park and General Grant National Park (Cohen 1988; Duncan 2013). The U.S. now had four national parks.

Muir and Johnson contested the exclusion of Kings Canyon on the South Fork of the Kings River from Sequoia National Park, and launched a new literary campaign (returning to a debate in Muir's 1875 'The Kings River Yosemite') to extend the Sequoia National Park boundary northward to include the Kings Canyon sequoia groves. Muir (1891) made the recommendation in 'A Rival of the Yosemite: The Cañon of the South Fork of Kings River, California,' in *The Century Magazine* in November 1891 (see also Muir's journal entries in Wolfe 1966). Once again, the article described the region and the threats facing it. The article also included a map expanding the Sequoia National Park boundary to include Kings Canyon.[14]

While the boundary of Yosemite National Park fluctuated a little following designation, less than two years passed before the park boundary was severely contested. In the late-spring of 1892, Representative Anthony Caminetti introduced bill H.R. 5764 before Congress, which included a proposal to reduce Yosemite National Park by half, in favor of lumber, grazing, and mining interests. Johnson had tried to persuade Muir to organize a watchdog or advocacy group to protect California's mountain range, 'or at least Yosemite' (Johnson 1923), since the pair first collaborated on the national park idea in 1889 (Jones 1965;

park incorporating at least one of the Muir family farms (Simpson 2003). But Leopold passed away one week later, after a heart attack while tackling a wildfire on a neighbor's property. Community groups took on Leopold's campaign, and after much political haranguing, the John Muir Memorial Country Park was established in 1957, and, almost a quarter-century after Leopold's proposal, the Department of Natural Resources established the Muir Park State Natural Area (within the boundaries of the country park) in March 1972 (Simpson 2003).

[14] Kings Canyon National Park (shortened from a proposed John Muir-Kings Canyon National Park) was eventually designated 49 years later on March 4, 1940 by President Franklin Delano Roosevelt (on Muir's and later the Sierra Club's campaign to secure the national park, see Cohen 1988; Wilkins 1995; Wolfe 2003; Worster 2008; Duncan 2013; Binnewies 2015).

Fox 1981; Cohen 1984, 1988; Wilkins 1995; Ehrlich 2000; Wolfe 2003; Simpson 2005; Worster 2008; Colwell 2014). The Caminetti bill proved the pinch point for Muir—and the Sierra Club was born.

On May 28, 1892, in the Sansome Street office of San Francisco attorney Warren Olney, the Sierra Club was first organized. On June 4, 27 members signed articles of incorporation, and soon the Club listed 182 charter members, including Wanda Muir, Clark, and Hutchings (Jones 1965; Cohen 1988; Wilkins 1995; Ehrlich 2000; Turner 2000; Wolfe 2003; Simpson 2005; Worster 2008; Nash 2014; Colwell 2014; Binnewies 2015). The Sierra Club built its advocacy platform on three issues—improving access to the mountains, the publication and dissemination of information on mountain regions, and forest preservation of the Sierra Nevada—and is today one of the leading nonprofit environmental advocacy organizations in the U.S., with chapters in every state. Muir's (2001a) *The Mountains of California* was published in 1894, two years after the Sierra Club was established, and was quickly adopted by the Club as its chief, definitive text (Cohen 1984). *My First Summer in the Sierra*, published in 1911 (see Muir 2003a), is dedicated to the Sierra Club. Muir served as president until his death in 1914, and Olney as the Club's first vice president. The fight against the Caminetti bill was the Sierra Club's first political campaign (for an excellent history on Muir, the Sierra Club, and the fight for Yosemite, see Badè 1924b; Colby 1938; Jones 1965; Cohen 1988). It would not be its last. The establishment and early fight of the Sierra Club ran parallel to, and echoed, the wider political and social reform shaping the start of the Progressive era in America.

Meanwhile, on March 3, 1891, President Harrison established the first federal forest reserves (precursors to national forests) under the Forest Reserve Act (Section 24), attached as a rider to the General Land Law Revision Act. On March 30, he designated the Yellowstone Park Timberland Reserve, the first forest reserve in the U.S. Two weeks before leaving office, President Harrison used his authority under the Forest Reserve Act of 1891 to designate the Sierra Forest Reserve on February 14, 1893—a 4-million-acre tract at the southern boundary of Yosemite National Park, between the Yosemite and Sequoia national parks, including Kings Canyon. The reserve was part of a total of 13 million acres of watershed protected by Harrison under the Forest Reserve Act (Fox 1981; Wilkins 1995; Simpson 2005; Worster 2008).

A few years later, and based on the preliminary findings and recommendations of a forestry commission Muir served on in 1896, President Grover Cleveland established 13 new and expanded forest reserves in eight Western states totaling 21.4 million acres on February 22, 1897, known collectively as 'Washington's Birthday Reserves' (Badè 1924b; Fox 1981; Cohen 1984; Wilkins 1995; Ehrlich 2000; Turner 2000; Wolfe 2003; Simpson 2005; Worster 2008; Nash 2014; Clayton 2019). But opposition swiftly followed from timber, cattle, and mining syndicates. The forest reserves were suspended, then many reinstated, with challenges to the reserves continuing into the McKinley administration.

Muir's article, 'The National Parks and Forest Reserves,' was published in *Harper's Weekly* on June 5, 1897. The day before, President William McKinley signed the 1897 Forest Management Act, which prioritized forest reserves for their timber supply, opening them up to grazing and mining interests. In the article, Muir summarizes the main findings of the forestry commission survey report, and has praise both for 'wise use' forestry and wilderness preservation ideologies (in Muir 2018a). As opposition continued, Muir again responded with 'The American Forests,' published in *The Atlantic Monthly* in August 1897 (in Muir 2018a). In his first article for the *Atlantic*, the nation's most prestigious magazine, Muir looks back through the history of U.S. forests and forestry policy, before bringing his argument to a climax with his scathing, excoriating declaration that God cannot save America's forest groves 'from fools'—that responsibility lies with the federal government (see also Wolfe 1966). Only federally designated forest reserves will halt ecological destruction. His second *Atlantic* article—'Wild Parks and Forest Reservations of the West'—in January 1898 (in Muir 2018a), challenges the emerging utilitarian framing of conservation, and offers a literary tour of U.S. forest reserves. Muir's rousing call to 'save the trees' at the end of his 'The Sequoia and General Grant National Parks' essay (in Muir 2018a, repeated in 2003b) urged the designation of new parks and reserves, and the expansion of existing forest boundaries (see also Muir 1920). The mid- to late-1890s marked an intense period in Muir's literary environmental activism, writing on forest preservation, the creation of more national forest reserves and parks, the recession of Yosemite Valley, and more.

When the Caminetti bill was dismissed in 1894, the Sierra Club returned to another Yosemite issue. Buoyed by the success of the 1890 Yosemite National Park bill, Johnson encouraged Muir to begin a campaign for the recession of the Yosemite Grant—to unite the state and national parks as one national park (see Muir 1896). Two bills had already floundered (Fox 1981). Muir had first traveled to Sacramento, the state capital, in 1897 to petition for recession, but arrived too late—the state legislature had already voted down recession (Cohen 1984; Duncan 2013). Six chapters of Muir's (2018a) *Our National Parks*, published in 1901, describe Yosemite National Park, and these collected essays led Muir's literary charge for the recession of the Yosemite Valley and Mariposa Grove. All of *Our National Parks* (Muir 2018a) might be read as Muir's manifesto on the need for the 'care and preservation' of America's wild places, for establishing national parks for spiritual, aesthetic, recreational pursuits (see also Muir's journal entries on the need for national park protections in Wolfe 1966). In *Our National Parks*, Muir (2018a) also argues for the creation of national parks to protect the Grand Canyon, and Mount Rainier.

In a March 14, 1903 letter, written the same day he signed an executive order designating Florida's Pelican Island on the Indian River as the first federal bird reserve (now national wildlife refuge), President Theodore Roosevelt personally requested Muir as his escort and guide for a four-day excursion in Yosemite later in the spring (Brinkley 2010; Canfield 2015). To accommodate this unexpected presidential excursion, and in the hope of doing 'some forest good' (in Badè 1924b), Muir rescheduled the start of an intercontinental tour across Europe, Russia, and Asia with forester Charles Sprague Sargent and his son to study trees.

Two months later on the evening of May 14, Muir boarded the President's train in Oakland, across the bay from San Francisco, bound for Raymond. The stop was part of TR's eight-week Great Loop tour, visiting 25 states and delivering more than 200 speeches. The week before, on a tour stop at the Grand Canyon in Arizona, TR remarked on May 6, 'Leave it as it is. You cannot improve on it; not a bit. The ages have been at work on it, and man can only mar it' (qtd. in Gessner 2020: 6–7, also Brinkley 2010; Canfield 2015; cf. Muir 2018a). It was a resounding, clear endorsement of wilderness preservation. Conservation had entered the political and legislative vocabulary in the 1880s and 1890s, but it became a movement under TR. From Raymond, the party boarded stages for Wawona, and arrived in the Mariposa Grove on May 15, pausing in the stage wagons for photographs at the Wawona Tunnel Tree and for an official photograph at the base of the Grizzly Giant (Fig. 3.2).

Once the formalities were concluded, TR dismissed his entourage and party—including his Secret Service agents—to the nearby Wawona Hotel (formerly Clark's Station), requesting only Muir, two park rangers, and a U.S. Army packer and climber (Canfield 2015).

The small party headed north on the Lightning Trail the following day, through the last of the winter snow, entering Yosemite's south entrance, to camp near Sentinel Dome in the meadow back of Glacier Point. Around the campfire that evening, discussion turned first to wilderness, and Yosemite's glacier history, then to the use of forests, timber theft and fraud, and forest preservation. Muir also urged TR to expand Yosemite

Fig. 3.2 President Theodore Roosevelt and party at the Wawona Tunnel Tree in Mariposa Grove in May 1903. Courtesy of Yosemite National Park/NPS

National Park to include the Yosemite Valley and Mariposa Grove, and called for the protection of Mount Shasta, and for other expansions to the national park network. TR, an avid outdoorsman and hunter, became a staunch recessionist after hearing Muir's paean to Yosemite.

Political machinations infiltrated the party's third day, including the now-famous photograph of 44-year-old TR and 65-year-old Muir at Overhanging Rock on Glacier Point, framed by Yosemite Falls behind (Fig. 3.3). The photograph is a defining image of their 'backwoods summit meeting' on wilderness conservation (after Canfield 2015), attended by a cavalry unit, photographers, and spectators. For their final night camped in Yosemite—in Bridalveil Meadow, beneath Bridalveil Fall, with El Capitan and Ribbon Fall opposite, beyond the Merced River—Muir and TR settled down to a final night of conversation strolling through the meadow. This Yosemite meeting continues to stand as one of the most important meetings in the annals of the American conservation movement (on TR's 1903 camping trip with Muir, and his presidential conservation legacy, see especially Roosevelt 1924; Fox 1981; Cutright 1985; Brinkley 2010; Duncan 2013; Binnewies 2015; Canfield 2015; Gessner 2020; Runte 2020). Muir and TR are together celebrated as the founding fathers of U.S. conservation and the national park

Fig. 3.3 President Theodore Roosevelt and John Muir (1838–1914) in Yosemite National Park in May 1903. Left: Roosevelt and Muir at Overhanging Rock on Glacier Point. Right: Roosevelt, Muir, and park rangers on horseback in Yosemite Valley, with Half Dome behind. Courtesy of Yosemite National Park/NPS

system. TR rejoined his official traveling party the following morning. The next day on a tour stop in Sacramento, TR sent a telegram to Interior Secretary Ethan Allen Hitchcock, requesting he consult Chief U.S. Forester Gifford Pinchot (America's first professional forester, see also Chapter 4) and explore expanding the Sierra Forest Reserve boundary northward to include Mount Shasta (Fox 1981; Wilkins 1995; Turner 2000; Brinkley 2010; Duncan 2013; Canfield 2015).

When Muir returned to San Francisco after his 12-month tour of Europe, Russia, and Asia in May 1904, he joined William E. Colby, secretary of the Sierra Club, and other Sierra Club members in preparing a bill to be introduced in the next year's session of the state legislature, to cede the Yosemite Valley and Mariposa Grove back to the federal government. But Muir had done most of his politicking in the mountains—with TR as he had done with Johnson, and would later do with President William Howard Taft—and not in San Francisco, Sacramento, or Washington, DC (Cohen 1984). Political momentum to recede the Yosemite Valley to federal control intensified in the fall of 1904. Support for recession came via the political lobbying of the Sierra Club, essays in Muir's (2018a) *Our National Parks*, several California newspaper editorial boards, and it had the endorsement of California governor George C. Pardee, and President Roosevelt. Colby's bill was introduced in the California state legislature in January 1905. In a two-month period, Muir and Colby made nine trips to Sacramento to testify before state legislature committees. The recession bill eventually passed the California Senate on February 23, 1905, by 21 to 13, and Governor Pardee signed the bill into law (Turner 2000; Wolfe 2003; Simpson 2005; Worster 2008; Duncan 2013; Binnewies 2015; Runte 2020). On March 3, TR signed Senate Joint Resolution 115, which called for federal acceptance of Yosemite Valley. But Congress still had to accept the receded Yosemite Grant.

The battle relocated to the East Coast. But approval was prolonged for more than a year by opposition and boundary disputes. Meanwhile, TR established the U.S. Forest Service (USFS, formerly the Bureau, and before that, Division of Forestry) in February 1905, and designated Shasta National Forest on October 3, 1905. Two months earlier, on August 6, 1905, in the midst of the recession campaign, Louie Muir passed away from pneumonia and lung cancer. The recession bill passed the House in May 1906, before being held up again in the Senate, and was finally passed on June 11, 1906, three days after passage of the Antiquities Act, and almost eight weeks after the devastating San Francisco earthquake (Johnson 1923; Jones 1965; Fox 1981; Cohen 1984, 1988; Wilkins 1995; Ehrlich 2000; Wolfe 2003; Worster 2008; Duncan 2013; Binnewies 2015; Runte 2020). Muir and Johnson's campfire experiment in wilderness preservation was finally concluded 17 years later, as the Yosemite Valley became part of Yosemite National Park (now equivalent in size to Rhode Island, at almost 760,000 acres, see Solnit 2014; Binnewies 2015) (Fig. 3.4). But in the aftermath of the earthquake, San Francisco began to scout Yosemite's Hetch Hetchy Valley as a possible dam and reservoir site. The Sierra Club's biggest fight would follow the recession triumph—defending Yosemite's Hetch Hetchy Valley against San Francisco water interests.

Fig. 3.4 Tunnel View overlook through Yosemite Valley in Yosemite National Park, California. From left to right: El Capitan, Cloud's Rest, Half Dome, Sentinel Rock, Sentinel Dome, Cathedral Rocks, and Bridalveil Fall. Photograph by the author, April 2016

3.2.3 The 1906 Antiquities Act, and Muir Woods National Monument

On June 8, 1906, three days before signing the Yosemite Valley recession bill, TR signed another bill that would go down in the annals of the American conservation movement as a landmark piece of legislation, and would establish TR as one of the great conservation presidents. An Act for the Preservation of American Antiquities—or, simply, the Antiquities Act (on the legacy of the Antiquities Act, see especially Rothman 1994; Harmon et al. 2006; Graham 2015). Sponsored by Representative John F. Lacey (who had earlier given his name to the Lacey Act of 1900, discussed in Chapter 5), the Antiquities Act authorizes the president to declare, by presidential proclamation, federal protection for objects and structures—and landscapes—of historic and scientific interest. TR designated the U.S.'s first national monument, Devil's Tower National Monument, in Wyoming, on September 24, 1906 (Wyoming features the U.S.'s first national park, and its first national monument).

On January 9, 1908, TR established the seventh national monument in the U.S.—the 295-acre Muir Woods National Monument (Proclamation No. 973, 1908), a stand of coastal redwoods north of San Francisco, near Mill Valley in Marin County. Businessman and philanthropist (and later congressman) William Kent had purchased 611 acres in 1905 for $45,000, but when a Sausalito water company announced plans to dam Redwood Creek, flowing from Mount Tamalpais through the stand and down to the Pacific coast, Kent donated a center 295-acre tract of the land to the federal government with the stipulation the forest be named in Muir's honor (Jones 1965; Fox 1981; Cohen 1988; Worster 2008; Brinkley 2010; Nash 2014; Gessner 2020) (Fig. 3.5).

By the close of his presidency in March 1909, TR had established 18 national monuments under the Antiquities Act, five national parks (doubling the number of U.S. national parks), 150 national forests after creating the modern USFS in 1905, plus 51 bird reserves, and four big game preserves (now consolidated as national wildlife refuges), protecting 230 million acres of public lands (Fox 1981; Worster 2008; Brinkley 2010; Duncan 2013; Binnewies 2015; Gessner 2020). Muir encouraged TR to establish Petrified Forest National Monument in Arizona

Fig. 3.5 Muir Woods National Monument, Mill Valley, California. The 295-acre national monument was the seventh to be established by President Theodore Roosevelt, on January 9, 1908. Photograph by the author, April 2016

in December 1906. Only President Barack Obama has exceeded TR's conservation legacy, placing almost 550 million acres of public lands under protection between 2009 and 2017. However, the Antiquities Act would come under assault from the Trump administration 111 years after its signing, as President Donald J. Trump attempted to revoke national monument designations (see Chapter 6).

3.3 The Beginning of the End for the Hetch Hetchy Valley: From the 1906 San Francisco Earthquake to the 1913 Raker Act

Muir's many excursions crisscrossing the Hetch Hetchy Valley, located about 20 miles to the northwest of Yosemite Valley, appear throughout his journals, letters, essays, and books. Other writers, artists, and photographers also documented the Hetch Hetchy Valley at the close of the century—including Thomas Ayres, Charles Leander Weed, Carleton E. Watkins, Albert Bierstadt, William Keith, William Henry Jackson, Charles Dormon Robinson, Harriet Monroe, and later, George Masa, Chiura Obata, Iwao Matsushita, and Ansel Adams (Righter 2005; Solnit 2014). Muir describes the smaller valley as a wilder part of Yosemite National Park—identifying it as the 'Tuolumne Yosemite' ever since his first visit in the fall of 1871 (in e.g. Muir 1873b, 1907, 1908a, 1909a, b; Wolfe 1966). But his March 25, 1873 essay in the *Boston Weekly Transcript* and January 1908 essay in the *Sierra Club Bulletin*, both titled 'The Hetch Hetchy Valley' (Muir 1873a, 1908a) are very different beasts. And the pivot was a 48-second, 7.9-magnitude earthquake that struck the northern California coast (the epicenter almost underneath San Francisco) at 05.12 on the morning of April 18, 1906—the same year all of Yosemite was made a national park. Muir was awoken by the tremors at home in Martinez, 23 miles from San Francisco. While his first essay on Hetch Hetchy (Muir 1873a, and 'Explorations in the Great Tuolumne Cañon,' published five months later, see Muir 1873b; Wolfe 1966) melds geography and personal observation, his second under the same title is much more political and activist. There is something uncomfortable and ominous in Muir's (1873a) Hetch Hetchy essay when he remarks how the spring floodwaters are temporarily held by the narrow canyon, briefly transforming the lower half of the valley floor into a lake.

Muir writes that it was a hunter, Joseph Screech, who 'discovered' the three-and-a-half-mile-long Hetch Hetchy Valley (named by Native Nations for its tall grasses) in 1850, the year before the Mariposa Battalion entered the Yosemite Valley (Muir 1873a, b, 1890b, 1907). It was after finding his first active, living glacier between Red and Black Mountains that Muir made his first excursion into the Great Tuolumne Canyon in September 1871, returning two months later to explore the part of the canyon that opens into the Hetch Hetchy Valley. He would return to the Hetch Hetchy Valley on many camping trips and, later, on Sierra Club summer outings (or, 'High Trips') across the next four decades (Fig. 3.6).

Fig. 3.6 The pre-dam Tuolumne River flowing through the Hetch Hetchy Valley, Yosemite National Park. Courtesy of Yosemite National Park/NPS

In 'Yosemite in Spring,' Muir (1872b) animatedly describes being awoken in the Yosemite Valley in the early hours of March 26, 1872, as an earthquake struck—the 7.4-magnitude Inyo or Lone Pine earthquake—and running from his cabin at Black's Hotel near Sentinel Rock where he was the winter caretaker, to experience the 'noble earthquake' (with the event retold in Muir 2003b, 2018a, also Muir 2019; Badè 1924a, b; Wolfe 1966). In the moonlight, he sets out to first explore the newly formed talus slopes at the base of Eagle Rock a half-mile away, then to see if the direction of flow of the Merced River was affected. Back at Black's Hotel, he sets a bucket of water on the desk in his room to study the tremors and aftershocks rippling through the valley in the subsequent weeks and months. Muir dismisses the concerns of his fellow valley residents who fled to the plains, in fear of another awful convulsion and subsidence that might swallow the valley floor (echoing Whitney 1865, 1869), choosing instead to embrace—and learn from—the occasion. But the earthquake that struck almost 200 miles away 34 years later would condemn Muir's 'Tuolumne Yosemite' to a reservoir site. While the 1906 earthquake would not strike Yosemite National Park, a political aftershock would echo in and forever alter Yosemite's Hetch Hetchy Valley a decade later, made manifest in the construction of the O'Shaughnessy Dam and the Hetch Hetchy Reservoir.

3.3.1 Politics, Water, and Water Politics in Hetch Hetchy; an Earthquake Disrupts

City engineers in San Francisco began scoping out the steep, high-walled Hetch Hetchy Valley as a possible dam site as early as 1882. As the Bay Area population increased, the city needed a cheap and reliable municipal water supply, and wanted to escape from the monopolistic control of the Spring Valley Water Company. While many water rights had been claimed, one site persisted—the Hetch Hetchy Valley on the Tuolumne River, together with Eleanor Creek downstream. It was an ideal site for a dam and reservoir, producing water and hydroelectric power. City officials argued it was cheaper and easier to flood the Hetch Hetchy Valley than find alternative sites. But enclosed in Yosemite National Park since 1890, the valley was protected from water rights claims.

Applications for reservoir rights in Hetch Hetchy and Eleanor Creek were first filed by San Francisco mayor James D. Phelan with the Stockton Land Office as early as 1901, while the Yosemite Valley recession bill was underway in the state legislature. (Meanwhile, TR signed the Reclamation Act in 1902, heralding the dam building enterprise on western rivers, picked up in Chapter 6.) Muir only learned of the plans in 1905 from Colby, and Muir and Colby thought they had won a reprieve when Interior Secretary Hitchcock thrice denied the city's applications, for violation of the 1890 Yosemite Forest Reservation bill—on January 20, 1903, December 22, 1903, and February 20, 1905 (Gray 2000; Righter 2005; Simpson 2005; Worster 2008; cf. Muir 1907, 1908a, 1909b, 2003b). The city began to investigate alternative sites. But this technicality was already weakened by H.R. 11973, introduced by Representative Marion de Vries in May 1900, which passed as the Right-of-Way Act on February 15, 1901, authorizing the construction of water infrastructure—canals, tunnels, pipelines, dams, and reservoirs—in national parks (Jones 1965; Fox 1981; Wilkins 1995; Gray 2000; Turner 2000; Wolfe 2003; Simpson 2005; Worster 2008; Brower 2013). In many respects, this piece of legislation signaled the inevitable loss of Hetch Hetchy. The passage of Resolution 6949 by the San Francisco Board of Supervisors on February 6, 1906 required the city to abandon its Tuolumne River-Hetch Hetchy water system plans, and find additional water options elsewhere (Gray 2000; Simpson 2005; Brower 2013).[15] But then April 18, 1906 arrived. And with it, began Muir's seven-year battle for Hetch Hetchy.

The fight over Hetch Hetchy was a much fiercer battle than that fought for Yosemite National Park, and the recession of the Yosemite Valley and Mariposa Grove. Hetch Hetchy brought Muir's preservation career full circle, back to Yosemite. In the space of a half-century, Yosemite protections had gone from strength to strength, from the 1864 Yosemite Grant, to the U.S.'s third national park designation in 1890,

[15] While the San Francisco-Hetch Hetchy battle was unfolding, an analogous battle was commencing elsewhere in California—in Los Angeles, over the watershed of the Owens Valley.

and recession of the Yosemite Valley in 1906, before spectacularly unraveling with a sideswipe to dam the Tuolumne River in the Hetch Hetchy Valley between 1906 and 1913 (on the San Francisco Public Utilities Commission [SFPUC], Hetch Hetchy Water and Power [HHWP], and the politics of twentieth-century Hetch Hetchy water and hydropower infrastructure and governance, see especially Taylor 1926; Jones 1965; Clements 1979; Cohen 1988; Gray 2000; Hundley 2001; Righter 2005; Simpson 2005; Brinkley 2010; Brower 2013; Binnewies 2015; Runte 2020).

With the earthquake, 28,000 buildings collapsed across the city, and gas lines and water mains ruptured. Fires swiftly followed, and many burned for upward of 96 hours. Over 80 per cent of the city was destroyed, and 3,000 people died. In the early-twentieth century, San Francisco was the largest city on the West Coast. The earthquake dramatically revealed how old and inadequate the water supply to the city was, as many pipes fractured and burst. The earthquake came just weeks before TR signed the Antiquities Act and the Yosemite Recession Grant.

The city of San Francisco had long lobbied for Hetch Hetchy water rights, but its campaign intensified and became much more urgent as city repairs got underway. Just six weeks after the earthquake, Pinchot, the Chief Forester of the USFS, spoke out in favor of a water supply from Yosemite National Park—advising San Francisco city engineer Marsden Manson in a May 28, 1906 letter to renew the city's application, and 'make provision for a water supply' from Yosemite (Badè 1924b). Soon after, the Board of Supervisors of the City and County of San Francisco revisited its Tuolumne River dam and reservoir plans, resubmitting the city's petition in 1907 (Fox 1981; Gray 2000; Wolfe 2003; Simpson 2005; Worster 2008). The fate of the Hetch Hetchy Valley was sealed with the quake. At one time colleagues on a commission (with Sargent and others) surveying the condition of western and northwestern forests in 1896 to recommend federal forest reserves, Pinchot became Muir's great antagonist, and the pair collided over Hetch Hetchy on multiple occasions between 1908 and 1913.

The Hetch Hetchy Valley hosted a political showdown split across two camps—Pinchot's conservation and 'wise use' versus Muir's preservation—and the battle would soon erupt onto the national stage. Dam proponents, including mayors, chambers of commerce, business leaders, and newspaper editors across California, argued that Hetch Hetchy was not a one-of-a-kind valley, and so could be forfeited, turning Muir's earlier praise that, 'Nature is not so poor as to possess only one of anything' (Muir 1950, also 1890b) to now condemn the valley. Proponents also pointed to aesthetic concerns—of a valley floor susceptible to flooding in the spring runoff, and little more than a mosquito bog in the summer, remote from the popular Yosemite tourist tours (Badè 1924b; Lowitt 1995; Gray 2000, 2007). Opponents to the dam project, including naturalists, scientists, mountaineers, and others, argued that water could be obtained elsewhere before turning to the national park and inundating the valley's 'grand landscape garden' (Muir 1907, 1908a, b, 1909b, 2003b). Muir was not against 'a Sierra or even a Tuolumne water supply' (Badè 1924b) for the city—his objection was that constructing a dam and reservoir in a national park betrayed the sanctity of the designation. To Muir, the city had other alternatives before condemning Hetch Hetchy. Hetch Hetchy would become the most visible conservation campaign in the history of the American conservation movement up to that time—and for many decades after.

3.3.2 Politicizing Hetch Hetchy: Defending the Tuolumne Yosemite from 'Temple Destroyers'

When Secretary Hitchcock was replaced by James R. Garfield, son of the former President (and ally of TR and Pinchot's 'wise use' practices), in 1906, the fight for Hetch Hetchy accelerated. In the summer of 1907, while Sierra Club members were in Yosemite and Hetch Hetchy on an annual month-long 'High Trip,' Secretary Garfield held hearings on the dam and reservoir in San Francisco. No opponents attended the July 27 hearings, and when the Sierra Club returned a month later, the issue had moved forward. The Sierra Club passed a resolution opposing the damming on August 30, and submitted the resolution to Secretary Garfield (Righter 2005; Simpson 2005). In October that year, Muir and friend and artist William Keith camped for a week in the Hetch

Hetchy Valley—and the trip inspired Muir's (1907) essay 'The Tuolumne Yosemite in Danger,' published in *The Outlook* on November 2, 1907. At the start of 1908, the *Sierra Club Bulletin* published Muir's revised *Outlook* essay as, 'The Hetch Hetchy Valley' (Muir 1908a).

With the success of his earlier forest reserves and Yosemite National Park articles, Muir turned to the theater of Hetch Hetchy. His 1908 *Sierra Club Bulletin* essay is his inaugural Hetch Hetchy campaign dispatch (Muir 1908a). The article was included four years later as the final chapter in *The Yosemite* (Muir 2003b).[16] It is in this article (and chapter) that Muir's wilderness ethic is politicized, when he argues, 'Everybody needs beauty as well as bread, places to play in and pray in, where Nature may heal and cheer and give strength to body and soul alike' (Muir 2003b: 256, cf. 1907, 1908a, 1909b, 2001a, 2018a, Wolfe 1966). For Muir, the wilderness is restorative. But even more pronounced than this, is his articulation and politicization of a temple analogy to defend the Hetch Hetchy wilderness, decrying, 'These temple destroyers, devotees of ravaging commercialism, seem to have a perfect contempt for Nature, and instead of lifting their eyes to the God of the mountains, lift them to the Almighty Dollar. Dam Hetch Hetchy! As well dam for water-tanks the people's cathedrals and churches, for no holier temple has ever been consecrated by the heart of man' (Muir 2003b: 261–262, also 1907, 1908b, 1909a, b, cf. refrain of money-lenders storming, invading the temple at Shadow Lake in Muir 2001a, of garden destroyers in Muir 2003a). Muir also draws parallels with the Garden of Eden, to lament how, 'and earlier still, the first forest reservation, including only one tree, was likewise despoiled' (Muir 2003b: 257, also 1896, 1907, 1908a, 1909b). Muir's literary environmental activism is political, aesthetic, religious, moral (after Worster 2008, see also Spurgeon 2009 on the crossover of public discourse and literary texts at Hetch Hetchy). Muir is a missionary preaching the Hetch Hetchy gospel, and casts dam supporters as national park despoilers.

Variously across the article and revised and expanded chapter, Muir (1908a, 2003b) draws several topographical, geological, ecological,

[16] *The Yosemite* (Muir 2003b) is dedicated to Johnson, the 'originator of the Yosemite National Park.'

hydrological parallels between the Merced Yosemite and the Tuolumne Yosemite to celebrate Hetch Hetchy (also Muir 1873a, b, 1890b, 1907, 1909a, b, 1910; Wolfe 1966), and outlines the threat facing the Hetch Hetchy Valley from San Francisco. He details the city's plans to dam and flood the valley—summarizing the city's 1903 and 1907 applications to build the dam and reservoir, and includes the dismissals from Secretary Hitchcock. He also offers rebuttals to four misleading claims often made by dam supporters. Muir's (2003b) *The Yosemite* also features an Appendix, containing key Yosemite legislation.

A few months later, on April 21, 1908 (his 70th birthday), Muir wrote to TR urging him to intervene on the issue, and save Hetch Hetchy at the sacrifice of Eleanor Creek (Badè 1924b). This marked the start of a renewed letter and telegram campaign. On April 27, TR forwarded Muir's letter on to Secretary Garfield, and proposed that Garfield reconsider, and restrict the permit to Lake Eleanor.

But already committed, Garfield issued a 'revocable permit' (subject to congressional approval) to San Francisco for reservoir rights at both Lake Eleanor and Hetch Hetchy on May 11, 1908. Following pressure from TR, he added a few provisions, including that Lake Eleanor be developed first as a reservoir (Muir 1909b; Taylor 1926; Fox 1981; Cohen 1984; Wilkins 1995; Gray 2000; Wolfe 2003; Righter 2005; Simpson 2005; Worster 2008; Brower 2013; Duncan 2013; Nash 2014; Binnewies 2015; Runte 2020). TR deferred to Garfield and Pinchot's recommendations, rather than Muir, Johnson, and the Sierra Club. Five days later, House Joint Resolution 184 (H.J.R. 184) called for land exchanges 'between the city and the federal government,' and shifted and reframed the debate to congressional hearings (in Righter 2005).

It was on the Hetch Hetchy battleground that the frictions between Muir's preservation philosophy and Pinchot's (and, to a lesser extent, TR's) conservation philosophy were most acute (on the Muir-Pinchot schism in the fledgling American conservation movement, see especially Pinchot 1998; Miller 2001, 2013, 2017; Clayton 2019, also Fox 1981; Cohen 1984; Wilkins 1995; Ehrlich 2000; Turner 2000; Simpson 2003, 2005; Wolfe 2003; Worster 2008; Duncan 2013; Nash 2014). Muir's rift with Pinchot also split the Sierra Club. Such was the bitterness of this fight, that when TR and Pinchot convened the three-day Governors' Conference on the Conservation of Natural Resources at the White House two days after Garfield granted reservoir rights to San Francisco, some of Hetch Hetchy's staunchest defenders, including Muir and Sargent, were not invited (Fox 1981; Cohen 1984, 1988; Wilkins 1995; Turner 2000; Wolfe 2003; Simpson 2005; Worster 2008). Johnson attended as a member of the press. The Sierra Club issued a statement on the omission of preservationists.

The TR-Pinchot allegiance was burgeoning. Pinchot's 'wise use,' multiple use conservation program was becoming federal forestry policy. A city referendum, or special election, on the Hetch Hetchy water project on November 12, 1908 recorded support for the dam by a 6–1 margin (Fox 1981; Gray 2000; Brower 2013). Muir submitted a memorandum to the House Committee on the Public Lands in Washington, DC when hearings on H.J.R. 184 opened on December 16, 1908 (see Muir 1908b; cf. Righter 2005), and when the Committee on the Public Lands supported the project by a much narrower margin—8–7—in February 1909, the issue was paused (Muir 1909b). But preservation-of-wilderness refrain aside, conservationists returned time and again to the argument that the Hetch Hetchy Valley lay within the boundary of Yosemite National Park.

The battle for Hetch Hetchy would prove controversial for the Sierra Club, triggering internal opposition and dissention. Olney, mayor of Oakland from 1903 to 1905, supported the city. Muir, Johnson, Colby, and others supported preservation. In the spring of 1909, an offshoot group was formed by Colby for those against the dam project—the Society for the Preservation of National Parks, with Muir again president. It was this organization that took the Hetch Hetchy campaign national and made Hetch Hetchy a national issue.

But change was underway once again, as the Hetch Hetchy fight stretched into its second administration—on March 4, 1909, William Howard Taft, previously TR's Secretary of War, was inaugurated as the 27th President of the U.S. Pinchot remained chief of the USFS. The newly appointed Secretary of the Interior was Richard A. Ballinger (Wilkins 1995; Wolfe 2003; Worster 2008). The change in administration temporarily halted the Hetch Hetchy fracas. But dam proponents already had a joint resolution before Congress to allow the City and County of San Francisco to exchange lands beyond the Yosemite

National Park boundary for those inside the national park. Hearings on the joint resolution were held in 1909, and Johnson and other Sierra Club members spoke in defense of Yosemite (Wilkins 1995; Righter 2005; Nash 2014).

TR would not be the only U.S. president Muir would seek to influence on preservation, nor the only president he would accompany through Yosemite. Six months after taking office, President Taft invited Muir on a one-day tour of Yosemite, to the dismay of San Francisco officials. This was Muir's chance to do 'some Hetch Hetchy good' (to riff on Muir's 'some forest good' with TR, after Badè 1924b). On October 6, 1909, in the midst of the debate over the future of the Hetch Hetchy Valley, Muir accompanied President Taft and his party through the valley (on Taft and Muir, see Righter 2005; Simpson 2005; Duncan 2013; Binnewies 2015; Runte 2020). This trip was much different to his excursion with TR six-and-a-half years earlier. Not only was it devoid of camping, but there was a much greater media presence. Muir had very little time alone with Taft, but near the end of the excursion he had the opportunity to once more become Yosemite's great orator, making the case for Hetch Hetchy. He laid out his vision for the future of Yosemite National Park, and introduced his idea for a 'grand circular drive' (similar to Yellowstone)—a series of roads and trails connecting the Yosemite Valley, Tuolumne Meadows, Tuolumne Canyon, and Hetch Hetchy Valley (Wilkins 1995; Wolfe 2003; Worster 2008; Duncan 2013). Taft was enthusiastic about Muir's pragmatic proposal—and also disclosed his opposition to the dam project.

After their Yosemite trip and meetings in the following days, Taft organized for Muir to tour the Hetch Hetchy Valley with Interior Secretary Ballinger, George Otis Smith of the U.S. Geological Survey (USGS), and two federal engineers. Smith had been appointed by Taft to investigate Lake Eleanor's capacity to supply San Francisco.

Overlapping with the 1909 hearings, the literary activism of the Sierra Club continued, with the release of a 23-page pamphlet titled, *Let Everyone Help to Save the Famous Hetch-Hetchy Valley* (Muir 1909b; cf. Jones 1965) in November, to accompany its lobbying. The pamphlet includes historical context, testimonials, and editorials, and an edited version of Muir's (1909a) essay for *The Outlook*, 'The Endangered Valley: The Hetch-Hetchy Valley in the Yosemite National Park,' and ends with instructions on how to help protect Hetch Hetchy and Yosemite National Park. Numerous other clubs joined the Sierra Club and its Society for the Preservation of National Parks in opposing the destruction of Hetch Hetchy, including the American Civic Association, the American Scenic and Historic Preservation Society, the General Federation of Women's Clubs, the American Alpine Club, and the Appalachian Mountain Club of Boston (listed in Muir 1909b; Fox 1981). On December 18, 1909, almost two years on from Muir's Hetch Hetchy essay in the *Sierra Club Bulletin* (Muir 1908a), and after a Sierra Club membership referendum vote passed 589 to 161, with 50 resignations, the board of directors retained its anti-dam position (Colby 1938; Jones 1965; Fox 1981; Cohen 1988; Wilkins 1995; Turner 2000; Righter 2005; Worster 2008; Brower 2013).

After reading Smith's report confirming Lake Eleanor's capacity, Ballinger set a hearing in May 1910 requiring city officials to 'show cause,' and explain why the Hetch Hetchy Valley site could not be removed from the Garfield permit (Badè 1924b; Taylor 1926; Wilkins 1995; Gray 2000; Righter 2005; Simpson 2005; Worster 2008). Ballinger also requested President Taft appoint an advisory board of U.S. Army Corps of Engineers (USACE) to survey and report on other available water sources to meet the needs of San Francisco (with the report published in February 1913, as the congressional debate reached its apex—see Advisory Board of Army Engineers 1913). The San Francisco Board of Supervisors requested a series of extensions to mount its rebuttal, and the battle entered another stalemate (Department of the Interior 1910). (Taft also released Pinchot from his position as Chief Forester of the USFS in March 1910, in a dispute over Ballinger and land conservation policy.)

In an article for *American Forestry* in May 1910, Muir framed the controversy as a 'national question,' noting the uncomfortable precedent for the integrity, sanctity of national parks set at Hetch Hetchy (Muir 1910). City officials were attempting to prorogue the issue until after the 1912 presidential election, in the hope that a new administration would be more sympathetic to its cause (Fox 1981; Wilkins 1995). In March

1911, Ballinger resigned from office and was replaced by Walter Lowrie Fisher. Fisher visited Hetch Hetchy in September and would go on to issue almost a half-dozen continuances during his tenure. Before leaving on his eight-month Humboldtian South America tour in August 1911, 73-year-old Muir completed *The Yosemite* (published the following year, see Muir 2003b), then paused in Washington, DC to lobby Taft, new Interior Secretary Fisher, and others (Worster 2008).

The fight assumed a new intensity on both sides. Both parties launched national campaigns. Both hydraulic engineer John R. Freeman's (1912) *The Hetch Hetchy Water Supply for San Francisco* on behalf of San Francisco, and the Advisory Board of Army Engineers' (1913) *Hetch Hetchy Valley: Report of Advisory Board of Army Engineers to the Secretary of the Interior on Investigations Relative to Sources of Water Supply for San Francisco and Bay Communities*, confirmed the Hetch Hetchy Valley was the best source of water for San Francisco—although for differing reasons. Secretary Fisher convened a week-long hearing in late-November 1912. Three days before he left office in March 1913, Fisher repeated the decision to uphold Ballinger's 'show cause' order, determining he did not have the authority to grant the permit to the city (Taylor 1926; Gray 2000, 2007; Righter 2005; Simpson 2005).

San Francisco's delaying tactics paid off when Woodrow Wilson was elected as the 28th U.S. President on November 5, 1912, and took office on March 4, 1913. As the Hetch Hetchy imbroglio entered its third administration—and crossed the desk of a fifth Secretary of the Interior, Franklin K. Lane—it entered its death throes. For Lane had previously served as city attorney of San Francisco under the Phelan administration, and was a prominent supporter of the dam and reservoir project. But in line with his predecessors, Lane deferred the reservoir decision to Congress, as it affected the future of the national park (Taylor 1926; Righter 2005; Simpson 2005).

By the summer of 1913, the Hetch Hetchy campaign was entering its twilight months. Hearings on Representative John E. Raker's Hetch Hetchy bill, H.R. 7207, before the House Committee on the Public Lands opened on June 25, 1913. The dispute between Muir and Pinchot resurfaced when Pinchot was called as an expert witness for the City and County of San Francisco. Lane, Manson's successor

Michael M. O'Shaughnessy, and Phelan were also called as witnesses for the city commissioners. Only one preservationist was present to testify—Edmund A. Whitman of the Appalachian Mountain Club. But Whitman's testimony failed to impress on the committee the wilderness under threat from the dam project (see Gray 2000; Simpson 2005; Worster 2008; Nash 2014; Binnewies 2015). Johnson also submitted written testimony, and Muir sent a telegram requesting a pause in the hearing to allow preservationists to respond.

In its report submitted following the June hearings, the House Committee on the Public Lands unanimously supported the dam project (see Nash 2014). Muir, suspecting defeat was imminent, intensified his defense of Hetch Hetchy. A slew of letters, telegrams, and newspaper articles on behalf of the Sierra Club followed. The Club offshoot, the Society for the Preservation of National Parks, continued to distribute leaflets, circulars, and letters, as did the recently established National Committee for the Preservation of Yosemite National Park. Leading newspapers and magazines on both the West Coast and East Coast ran editorials and articles, often in support of preservation, and thousands of letters and telegrams in defense of Hetch Hetchy were sent to Congress by outdoors clubs and organizations, scientific societies, college and university faculties, and women's clubs (Fox 1981; Wilkins 1995; Wolfe 2003; Righter 2005; Simpson 2005; Nash 2014). But within California, many editorial boards were openly pro-dam—especially the *San Francisco Chronicle*.

It had become apparent to Hetch Hetchy's defenders—especially Muir, Colby, and Johnson—that the Hetch Hetchy Valley site was being lauded by dam proponents not only for its supply of pure water to San Francisco, but for its capacity for hydroelectric power generation. Suspicions were aroused in Congress, too, with the House adding an amendment to the Raker bill requiring the city to distribute power directly to consumers, and not to first divert it through private utilities companies (Lowitt 1995; Wilkins 1995; Righter 2005; Simpson 2005; Worster 2008).

3.3.3 Muir Refuted

The House Committee on the Public Lands voted on August 5, 1913 to recommend H.R. 7207 to the House. The bill arrived on the floor of the House on August 29, 1913, and Raker's Hetch Hetchy bill passed 183 to 43, with 205 absent, on September 3 (Taylor 1926; Lowitt 1995; Wilkins 1995; Gray 2000; Wolfe 2003; Righter 2005; Simpson 2005; Nash 2014). On September 24, the Senate Committee on Public Lands voted to send the legislation to the floor—the Senate began its consideration of the bill on December 1, and on December 6, the bill passed 43 to 25, with 29 absent or abstaining (on the legislative history of the Raker Act and the 1913 Senate debate, see especially Lowitt 1995; Gray 2000, also Taylor 1926; Clements 1979; Wilkins 1995; Hundley 2001; Wolfe 2003; Righter 2005; Simpson 2005; Brower 2013; Nash 2014; Binnewies 2015; Runte 2020). But still Muir held onto the diminishing hope that President Wilson might veto the bill, and cancel what he would later describe as 'the sham dam lake, the grave of Hetch Hetchy' (in Wolfe 1966: 437).

Pushback to Muir and the Sierra Club's defense of Hetch Hetchy would come from both expected—San Francisco's former and present mayors, newspaper editorial boards across the city—and unexpected political bases. One unexpectedly public and outspoken critic of Muir was William Kent, who had donated the land for Muir Woods National Monument in 1908, but now as a U.S. Representative voted to dam Hetch Hetchy (although Kent would later be the lead sponsor in the House of Representatives on the National Park Service bill). Kent accused Muir of being obstructionist, and a wilderness elitist. And Muir and Pinchot's professional relationship had been in terminal decline since Pinchot endorsed the reservoir in 1906.

Insults were leveled at Muir and the Sierra Club, and anyone opposed to the dam and reservoir project. Prominent members of the Sierra Club supported the dam so as not to be targeted. The issue split the Sierra Club. One editorial cartoon in *The San Francisco Examiner* on December 5, under the title, 'Applying the Crusher,' featured doppelgangers of Muir and other high-profile conservationists being steamrollered beneath 'S.F.' and a 'Hetch Hetchy Petition' (in Brower 2013). Editorial cartoonist Paul

Terry's front-page 'Sweeping Back the Flood' in *The San Francisco Call* on December 13 portrayed Muir as a stout old maid, complete with dress, apron, bloomers, Scots bonnet adorned with wildflowers, and broom, trying and failing to hold back the waters of the 'Hetch Hetchy Project' (Terry 1909).

The loss of wilderness in Hetch Hetchy became a national issue in the weeks between September 3 and December 6. *The San Francisco Examiner* ran an editorial on December 2, as deliberation of the Raker Act was underway, arguing that if Los Angeles could drain the Owens Valley, San Francisco should be able to dam Hetch Hetchy (Taylor 1926; Cohen 1984; Brower 2013). But 13 days after the Senate vote, on December 19, President Wilson signed the Raker Act, approving the Hetch Hetchy grant to San Francisco. The Raker Act transferred the Hetch Hetchy Valley from Yosemite National Park to San Francisco for use as a reservoir. Pinchot's conservation philosophy had trumped Muir's preservation philosophy. San Francisco now had the authority to dam Hetch Hetchy. In a December 27, 1913 letter to friends the Kellogg family, and a January 12, 1914 letter to his daughter Helen, Muir described the decision as a 'dark damn-dam-damnation' (qtd. in e.g. Turner 2000; Righter 2005; Simpson 2005; Worster 2008).

Some consolation for Muir came in a widespread public awakening to the need for wilderness preservation—wilderness was now a national issue. As Muir concedes, 'the conscience of the whole country has been aroused from sleep' (in a January 1, 1914 letter to Johnson, qtd. in e.g. Worster 2008: 458, also Nash 2014, see also Wolfe 1966). After the signing, the city reneged on its requirement to distribute power at low cost to municipal residents, selling a substantial proportion to the privately-owned PG&E Company at wholesale prices, with the company reselling the electricity at retail prices, landing massive profits (on San Francisco, PG&E, the public power controversy, and violations of the Raker Act, see Lowitt 1995; Wilkins 1995; Simpson 2005; Nash 2014; Binnewies 2015).

Muir's fear was that with the fall of Hetch Hetchy, other national parks, national monuments, and forest reserves were not inviolable, and would tumble—succumbing to commercial, industrial exploitation. But the defeat of Hetch Hetchy galvanized the American conservation movement to lobby for a federal agency to protect national parks. One year after the Raker Act was authorized, exhausted, frail, and brokenhearted at the loss, Muir died from pneumonia in the California Hospital on South Hope Street in Los Angeles on Christmas Eve 1914, while visiting Helen in Daggett. But Muir's national park legacy would influence another U.S. president, when President Woodrow Wilson signed the National Park Service Organic Act on August 25, 1916, creating the National Park Service to oversee 14 national parks and 21 national monuments. California Congressmen John Raker and William Kent, and Utah Senator Reed Smoot, were instrumental in the establishment of the NPS. Sierra Club member Stephen T. Mather was appointed as the first director of the NPS. The National Parks Association (now known as the National Parks Conservation Association) was established in 1919, inspired by the Society for the Preservation of National Parks, which had ceased in 1913.

3.3.4 Hetch Hetchy Valley Becomes Hetch Hetchy Reservoir

Work on the Hetch Hetchy water system, designed by engineer John Freeman (cf. the Freeman Report, as Freeman 1912), began soon after the signing of the Raker Act, and continued for almost two decades, from 1915 through 1934. As well as the dam and reservoir, an extensive network of aqueducts and tunnels, and the Hetch Hetchy Railroad and other transportation infrastructure, made up the Hetch Hetchy project to deliver water to the Bay Area peninsula (Righter 2005; Simpson 2005). Muir did not live to see the Hetch Hetchy Valley clear-cut. The early construction coincided with the onset of the First World War in July 1914, and the U.S.'s entry into the war in April 1917. Pouring for the arch-gravity O'Shaughnessy Dam—named for Michael O'Shaughnessy, chief of the project and head engineer—began on August 1, 1919, more than four years after Muir's death, and continued for three-and-a-half years. Seventeen workers were killed in accidents during the dam's construction. The dam was dedicated on July 7, 1923

Fig. 3.7 The O'Shaughnessy Dam and Hetch Hetchy Reservoir on the Tuolumne River in the Hetch Hetchy Valley. Courtesy of Yosemite National Park/NPS

(on the engineering history of the O'Shaughnessy Dam and development of Lake Eleanor, see Taylor 1926; O'Shaughnessy 1934; Righter 2005; Simpson 2005; Brower 2013). A decade-and-a-half later, the dam was expanded, doubling the storage capacity of the Hetch Hetchy Reservoir (Righter 2005; Simpson 2005). Today, the dam stands 430 feet high, and 900 feet wide, and the reservoir extends 7.5 miles through the valley, with a capacity of up to 360,000 acre-feet (af) of water (Wilkins 1995; Righter 2005; Simpson 2005; Worster 2008; Binnewies 2015; Runte 2020). It would be almost two decades before water from the Tuolumne River reached San Francisco household faucets 170 miles away (delivery began with the dedication of the Hetch Hetchy system on October 28, 1934, 12 days after the death of O'Shaughnessy) (Fig. 3.7).

3.4 The Battle for Hetch Hetchy, Reignited

Hetch Hetchy is anything but an historic footnote in the reclamation of the American West.

The ghost of Hetch Hetchy has long persisted in the memory of West Coast conservationists. But it experienced something of a resurrection in the mid-twentieth century when David R. Brower, the Sierra Club's first executive director (1952–1969), led a series of high-profile conservation campaigns—first against dams proposed in Dinosaur National Monument in the 1950s, and later against proposed dams in the Grand Canyon in the 1960s (see Chapter 6). Stephen Fox (1981) describes the post-war period as the Muir Redux. Hetch Hetchy was politicized as a conservation cautionary tale on dams and national parks. This was the first time that the national monuments and parks system had been severely challenged since Hetch Hetchy and the passage of the 1916 National Park Service Organic Act (cf. also plans for dams in Yellowstone in the late-1910s and early-1920s, in Binnewies 2015). The triumphs and troubles of the Echo Park, Glen Canyon, and Grand Canyon battles cast a long, lingering shadow over Hetch Hetchy. The dam decommissioning war headed west, returning to California.

Overnight on July 16–17, 1987, an Earth First! (EF!) activist painted a 40-foot crack down the face of the O'Shaughnessy Dam. Accompanying the 'crack,' and broadcast in five-foot-high lettering, was the painted war-cry, 'FREE THE RIVERS!---J. MUIR.' The creative guerrilla proclamation was painted over later on July 17 by dam officials, and the white paint panel remains on the face of the dam today (see *DamNation* 2014, and this creative 'cracking' of dams resumes in Chapter 6). But it was a resounding—and provocative—declaration, *in place*, of Muir's last great preservation battle.

In August 1987, Donald Hodel, Secretary of the Interior under the anti-environment administration of President Ronald Reagan, announced a study on decommissioning the O'Shaughnessy Dam and draining the Hetch Hetchy Reservoir (Righter 2005; Simpson 2005; Spurgeon 2009; Brower 2013; Binnewies 2015; Runte 2020). San Francisco politicians quickly followed in the footsteps of their predecessors to decry the proposed study. Leading the pro-dam politicians was U.S. Senator and former San Francisco mayor Dianne Feinstein. Battle lines were swiftly retraced. Long-dormant editorial page rumblings were resurrected. Meanwhile, the Sierra Club reaffirmed its anti-dam position, adopting its Hetch Hetchy Resolution on September 27, 1987 (Righter 2005). Published one decade earlier, Christopher Swan and Chet Roaman's (1977) speculative fiction *YV88: An Eco-Fiction of Tomorrow*, set in Yosemite National Park, had already begun to explore restoration *on the page*. *YV88* presents a restor(y)ing of the Yosemite Valley (the YV in the title)—threading themes of preservation, wilderness, sustainability, rewilding, and more through the narrative. The final section considers the decommissioning of the dam and reservoir, and the restoration of the Tuolumne River through the Hetch Hetchy Valley. In the late-1980s, Hetch Hetchy restoration entered political dialogue at both the state and federal level.

Before the end of the year, the NPS (1987) released *Alternatives for Restoration of Hetch Hetchy Valley Following Removal of the Dam and Reservoir*. More studies, by both state and federal agencies, followed the next year—including the Bureau of Reclamation's (1988) *Hetch Hetchy: A Survey of Water & Power Replacement Concepts*, the Department of Energy's (1988) *Hetch Hetchy: Striking a Balance. A Review of the Department of the Interior's Survey of Water & Power Replacement Concepts for Hetch Hetchy*, and the California State Assembly Office of Research's (1988) *Restoring Hetch Hetchy* (on the evolution of a Hetch Hetchy restoration rhetoric, see Righter 2005; Simpson 2005; Brower 2013, also Gray 2007 on the 'paradoxes of restoration' in Hetch Hetchy). 1988 also marked the sesquicentennial of Muir's birth, and President Reagan declared Muir's birthday, April 21, as 'John Muir Day' (Proclamation No. 5794, 1988). Earth Day follows on April 22, first held 18 years earlier.

In the late-1980s, the Sierra Club established the Hetch Hetchy Restoration Task Force. By the close of the twentieth century, the Sierra Club task force was recast as the conservation nonprofit, Restore Hetch Hetchy (RHH).[17] Ron Good was appointed as executive director. Restore Hetch Hetchy works to 'return the Hetch Hetchy Valley in Yosemite National Park to its natural splendor—while continuing to meet the water and power needs of all communities that depend on the Tuolumne River' (RHH 2020, on the history of RHH, see also Righter 2005; Simpson 2005). RHH is pursuing technical and political solutions on water and power system alternatives, to relocate water storage outside Yosemite National Park, advocating restoration *sensu lato* (after Aronson et al. 1993). At the same time as David Brower was helping to launch RHH, he was also appointed as a founding board member of Glen Canyon Institute (GCI), another conservation nonprofit dedicated to decommissioning Glen Canyon Dam and draining Lake Powell on the Colorado River (the work of GCI is examined in Chapter 6). When President Bill Clinton established the 328,000-acre Giant Sequoia

[17] RHH interview, April 7, 2016. RHH online interview, July 20, 2020.

National Monument on April 15, 2000 (Proclamation No. 7295, 2000), his remarks at the signing event borrowed from Muir to note, 'these majestic trees will continue, as John Muir said, to preach "God's forestry fresh from heaven"' (Clinton 2000; cf. Muir 1920, 2018a).

In the first decade of the twenty-first century, another suite of studies continued to explore the feasibility of restoring the Hetch Hetchy Valley. Chief among these are analyses including Sarah E. Null's (2003) *Re-Assembling Hetch Hetchy: Water Supply Implications of Removing O'Shaughnessy Dam* (see also Null and Lund 2006), and water specialists Environmental Defense's (as Rosekrans et al. 2004) *Paradise Regained: Solutions for Restoring Yosemite's Hetch Hetchy Valley*. These studies served as the catalyst for a state review of the restoration issue—in 2004, California governor Arnold Schwarzenegger asked the state's Department of Water Resources (DWR) and Department of Parks and Recreation (DPR) to review state and federal studies prepared in response to Secretary Hodel's 1987 proposal to study Hetch Hetchy restoration (Righter 2005; Simpson 2005; Spurgeon 2009).

Meanwhile, other feasibility studies included Environmental Defense's *The Potential Economic Benefits of Restoring Hetch Hetchy* (Rider 2004), and *Tuolumne Watershed Diversions without Hetch Hetchy Reservoir: Comparison of Interties to Cherry and Don Pedro Reservoirs* (Environmental Defense 2005), A. Bennett et al.'s (2004) *Hetch Hetchy Valley: A Plan for Adaptive Restoration*, and Restore Hetch Hetchy's *Finding the Way Back to Hetch Hetchy Valley: A vision of steps to restore Hetch Hetchy Valley in Yosemite National Park and to replace water and energy supplies* (Meral 2005; cf. Meral 2008). The DWR-DPR review of Hetch Hetchy restoration studies was released in 2006, as the California Resources Agency's (2006) *Hetch Hetchy Restoration Study*. In 2005, *The Sacramento Bee* associate editor Tom Philp won the Pulitzer Prize in Editorial Writing for his 2004 series on restoring Hetch Hetchy (the winning articles are listed on The Pulitzer Prizes 2020). In an echo of this uptick in interest in Hetch Hetchy, the release in February 2007 of the Bush administration's budget for the 2008 fiscal year included 'a line item of $7 million earmarked for assessing the details of restoring the Hetch Hetchy Valley' (Spurgeon 2009: 754).

In a December 7, 2011 letter to Interior Secretary Ken Salazar, California Congressman Dan Lungren endorsed Secretary Hodel's 1987 proposal, and requested that Interior investigate whether the City and County of San Francisco was in violation of the 1913 Raker Act. Political debates on the restoration of Hetch Hetchy resumed once again. A November 6, 2012 ballot measure for voters in San Francisco County, supported by RHH—the San Francisco Hetch Hetchy Reservoir Initiative/Water Sustainability and Environmental Restoration Act, Proposition F—was defeated.[18] Spreck Rosekrans, lead author of *Paradise Regained* (Rosekrans et al. 2004), joined RHH as director of policy in 2012, and at the end of the year, was appointed the organization's third executive director. In the subtitle of his 2013 book on Hetch Hetchy, Kenneth Brower, the son of David Brower, framed the restoration of the valley as 'Undoing a Great American Mistake' (Brower 2013). The book continues the tradition of literary activism to take out the dam and reservoir in Hetch Hetchy.

In the spring of 2015, RHH launched a legal campaign to restore the Hetch Hetchy Valley. On April 21—Muir's 177th birthday—RHH filed a petition in Sonora in Tuolumne County against the City and County of San Francisco, arguing that San Francisco's diversion at Hetch Hetchy violated the California Constitution.[19] By the summer, San Francisco filed an unsuccessful motion to move the trial to San Francisco, and at the end of the year, submitted a demurrer and a motion to strike. One year after RHH first filed its petition, on April 28, a Tuolumne County Superior Court ruling ruled in favor of San Francisco on two of the city's four claims. Soon after the decision, RHH announced its intent to appeal. RHH's appeal was heard in the 5th Appellate District in Fresno, and a series of briefs were submitted to the state appellate court by RHH, San Francisco, and other parties throughout 2016 and 2017. However, in mid-2018, RHH was 'denied "standing" in state court, [and] did not have the right to have a trial under California law. The legislation that allowed the O'Shaughnessy Dam to be built was federal legislation which supersedes state law, although that federal legislation

[18] RHH interview, April 7, 2016.

[19] RHH interview, April 7, 2016.

did include a provision that all of California's water laws would still apply. The judge, however, and an appellate panel as well, saw it differently'[20] (for the history of RHH's 2015–2017 legal campaign, see RHH 2017). Following the ruling, RHH resumed work on its technical argument for restoration.

RHH continues to study engineering solutions, and water supply and hydropower replacement, in the Hetch Hetchy system (see especially Rosekrans 2014; Restore Hetch Hetchy 2018a, b, c; ECONorthwest 2019). RHH is also exploring the commitments made by San Francisco not to interrupt or disrupt national park activities—such as hiking and camping—when lobbying for the dam and reservoir project in the early-twentieth century (e.g. Rosekrans and Laws 2021).[21] On July 22, 2018, RHH met with Interior Secretary Ryan Zinke at O'Shaughnessy Dam, and one of the issues discussed that day was a proposal to introduce quiet, electric, nonpolluting tour boats on the reservoir—revisiting the city's recreational promises in the Hetch Hetchy region. The issue appeared in a report attached to the federal budget at the end of the year, but stated that the restriction on boating on Hetch Hetchy Reservoir would continue.[22] In the days after RHH's meeting, newspaper headlines asked, 'Is Zinke trolling San Francisco with plan to dismantle city's reservoir?' (Greenberg 2018), but also noted there was 'No real worry that Hetch Hetchy will be drained after Zinke's visit' (Fracassa 2018). RHH has also met with Department of the Interior officials, and other officials in Washington, DC to explore the issue of restoration (Fig. 3.8).

* * *

The NPS and Yosemite National Park has no political position on the restoration of the Hetch Hetchy Valley. The dam and reservoir were congressionally authorized under the 1913 Raker Act—and three years before the NPS was established. But the NPS is obligated under the 1916 Organic Act (and later, the 1968 Wild and Scenic Rivers Act) to preserve

[20] RHH online interview, July 20, 2020.

[21] RHH online interview, July 20, 2020.

[22] RHH online interview, July 20, 2020.

Fig. 3.8 Restore Hetch Hetchy's vision of a restored Hetch Hetchy Valley. Courtesy of Restore Hetch Hetchy

the park. Yosemite National Park works closely with the city of San Francisco on Hetch Hetchy—not only on the dam and reservoir in the park, but also on the city's interest in issues including visitor use, water quality, stream flow, watershed protection, and power generation.[23]

The three consecutive years of 2014, 2015, and 2016 are significant in the conservation history of Yosemite National Park. 2014 marked the 150th anniversary of the Yosemite Grant, and the centenary of Muir's death. 2015 marked the 125th anniversary of Yosemite becoming a national park. 2016 marked the centennial of the National Park Service.[24] When President Obama spoke at Sentinel Bridge in Yosemite on June 18, 2016, in an early celebration of the NPS centenary, his remarks noted the conservation legacy of both Muir and TR (Obama 2016).

While the political debate over the restoration of the Hetch Hetchy Valley continues, a number of other ecological restoration programs have been undertaken in Yosemite National Park over the last two decades. Many of the restoration projects are in collaboration with the nonprofit Yosemite Conservancy, the park's philanthropic partner. Often, the projects amplify restoration *sensu lato* (Aronson et al. 1993). In November 2003, restoration crews removed the Cascades Diversion Dam on the Merced River (Righter 2005, and on earlier Yosemite restoration projects, see also Solnit 2014). A multi-year project restoring the area around Lower Yosemite Falls was completed in 2005, including the removal of an asphalt parking lot and old cinderblock restroom, construction of new visitor facilities, restoration of 11 acres of meadow habitat, and improvements to the trail system and a new one-mile accessible loop trail.[25] Other ecological restoration projects have focused on lodgepole pine removal and meadow restoration in Tuolumne Meadows, the removal of an old road and drainage ditch across Ahwahnee Meadow, and the removal of fill dirt from campgrounds, recovering native soils (and seeds). A series of restoration projects within the Merced River

23 YNP online interview, August 3, 2020.

24 YNP online interview, August 3, 2020.

25 YNP online interview, August 3, 2020.

Corridor—focusing on restoring riverbanks, wetlands, and meadows—is scaffolded by the Merced River Plan (YNP 2014), which protects and manages the 81-mile-long Merced Wild and Scenic River (designated in 1987) within Yosemite National Park. Restoration also features in infrastructural projects at Tunnel View, Glacier Point, and Olmsted Point. Wildlife restoration projects have focused on the eradication of predatory American bullfrogs, introduced to the reflection pond outside the Ahwahnee Hotel in the 1950s, and the reintroduction of the California red-legged frog, and Sierra Nevada bighorn sheep.[26]

One of the largest restoration projects in the National Park Service—and the largest in Yosemite—is the Mariposa Grove of Giant Sequoias Restoration Project. The project broke ground on June 30, 2014, the sesquicentennial of the Yosemite Grant, and the Mariposa Grove was closed in the spring of 2015 for three years. The project cost $40 million, with the NPS and Yosemite Conservancy each providing $20 million. Restoration crews restored the natural hydrology and flow of water to the trees. The parking lot, open-air diesel tour trams, and gift shop were removed, roads and trails through the grove were realigned and new accessible trails built, and a new 300-space parking lot and shuttle service to the Mariposa Grove was installed at the South Entrance of Yosemite National Park, two miles away. The restoration program revisited the earlier vision of Olmsted and Muir, but also collaborated with local Native groups to tell some of the histories and stories of the Miwok and other Native communities.[27] The Mariposa Grove reopened in June 2018 (Fig. 3.9).

[26] YNP online interview, August 3, 2020.

[27] YNP online interview, August 3, 2020.

Fig. 3.9 Dedication ceremony for Mariposa Grove of Giant Sequoias Restoration Project, June 14, 2018. Photographs by James McGrew, courtesy of Yosemite National Park/NPS

Works Cited

Advisory Board of Army Engineers. 1913. *Hetch Hetchy Valley: Report of Advisory Board of Army Engineers to the Secretary of the Interior on Investigations Relative to Sources of Water Supply for San Francisco and Bay Communities*. Washington, DC: Government Printing Office.

Albanese, Catherine L. 1990. *Nature Religion in America: From the Algonkian Indians to the New Age*. Chicago, IL: University of Chicago Press.

Aronson, J., C. Floret, E. LeFloc'h, C. Ovalle, and R. Pontanier. 1993. Restoration and Rehabilitation of Degraded Ecosystems in Arid and Semi-Arid Lands. I. A View from the South. *Restoration Ecology* 1 (1): 8–17.

Assembly Office of Research. 1988. *Restoring Hetch Hetchy*. 95814. Sacramento, CA: California State Legislature, State Capitol.

Badè, William Frederic. 1924a. *The Life and Letters of John Muir: Volume I*. Boston, MA and New York, NY: Houghton Mifflin Company.

Badè, William Frederic. 1924b. *The Life and Letters of John Muir: Volume II*. Boston, MA and New York, NY: Houghton Mifflin Company.
Bennett, A., N. Fobes, J. Freund, M. Healy, and J. Belknap Williamson with J. Zedler. 2004. *Hetch Hetchy Valley: A Plan for Adaptive Restoration.* Madison, WI: University of Wisconsin-Madison.
Bilbro, Jeffrey. 2015. *Loving God's Wildness: The Christian Roots of Ecological Ethics in American Literature*. Tuscaloosa, AL: University of Alabama Press.
Binnewies, Robert O. 2015. *Your Yosemite: A Threatened Public Treasure.* Ashland, OR: White Cloud Press.
Brinkley, Douglas. 2010. *The Wilderness Warrior: Theodore Roosevelt and the Crusade for America.* New York, NY: Harper Perennial
Brower, Kenneth. 2013. *Hetch Hetchy: Undoing a Great American Mistake.* Berkeley, CA: Heyday.
Brune, Michael. 2020. Pulling Down Our Monuments. *Sierra Club.* https://www.sierraclub.org/michael-brune/2020/07/john-muir-earlyhistory-sierra-club.
Bureau of Reclamation. 1988. *Hetch Hetchy: A Survey of Water & Power Replacement Concepts.* Sacramento, CA: Bureau of Reclamation.
Canfield, Michael R. 2015. *Theodore Roosevelt in the Field*. Chicago, IL and London: University of Chicago Press.
Clayton, John. 2019. *Natural Rivals: John Muir, Gifford Pinchot, and the Creation of America's Wilderness.* New York, NY: Pegasus Books.
Clements, Kendrick A. 1979. Politics and the Park: San Francisco's Fight for Hetch Hetchy, 1908–1913. *Pacific Historical Review* 48 (2): 185–215.
Clinton, President William Jefferson. 2000. Earth Day 2000 [Giant Sequoia National Monument 4/15/00]. *Clinton Digital Library.* https://clinton.presidentiallibraries.us/items/show/32441.
Cohen, Michael P. 1984. *The Pathless Way: John Muir and American Wilderness.* Madison, WI: University of Wisconsin Press.
Cohen, Michael P. 1988. *The History of the Sierra Club.* San Francisco, CA: Sierra Club Books.
Colby, William E. 1938. Yosemite and the Sierra Club. *Sierra Club Bulletin* XXIII (2): 11–19.
Colwell, Mary. 2014. *John Muir: The Scotsman Who Saved America's Wild Places.* Oxford: Lion Hudson.
Cutright, Paul Russell. 1985. *Theodore Roosevelt: The Making of a Conservationist*. Urbana, IL: University of Illinois Press.
DamNation. 2014. Dir. Ben Knight and Travis Rummel. Patagonia Presents a Stoecker Ecological & Felt Soul Media Production.

Department of Energy. 1988. *Hetch Hetchy: Striking a Balance. A Review of the Department of the Interior's Survey of Water & Power Replacement Concepts for Hetch Hetchy.* Washington, DC: Department of Energy.

Department of the Interior. 1910. *Proceedings Before the Secretary of the Interior in Re Use of Hetch Hetchy Reservoir Site in the Yosemite National Park by the City of San Francisco.* Washington, DC: Government Printing Office.

Duncan, Dayton. 2013. *Seed of the Future: Yosemite and the Evolution of the National Park Idea.* Yosemite National Park, CA: Yosemite Conservancy.

ECONorthwest. 2019. *Valuing Hetch Hetchy Valley: Economic Benefits of Restoration in Yosemite National Park.* Portland, OR: ECONorthwest.

Ehrlich, Gretel. 2000. *John Muir: Nature's Visionary.* Washington, DC: National Geographic Society.

Environmental Defense. 2005. *Tuolumne Watershed Diversions Without Hetch Hetchy Reservoir: Comparison of Interties to Cherry and Don Pedro Reservoirs.* New York, NY: Environmental Defense.

Fox, Stephen. 1981. *The American Conservation Movement: John Muir and His Legacy.* Madison, WI: University of Wisconsin Press.

Fracassa, Dominic. 2018. No Real Worry That Hetch Hetchy Will Be Drained After Zinke's Visit. *San Francisco Chronicle.* https://sfchronicle.com/bayarea/article/No-real-worry-that-Hetch-Hetchy-will-be-drained-13105043.php.

Freeman, John R. 1912. *The Hetch Hetchy Water Supply for San Francisco: On the Proposed Use of a Portion of the Hetch Hetchy, Eleanor, and Cherry Valleys Within and Near to the Boundaries of the Stanislaus U.S. National Forest Reserve and the Yosemite National Park as Reservoirs for Impounding Tuolumne River Flood Waters and Appurtenant Works for the Water Supply of San Francisco, California, and Neighboring Cities. A Report to James Rolph, Jr., Mayor of San Francisco, and Percy V. Long, City Attorney.* Santa Barbara, CA: Rincon Publishing.

Gessner, David. 2020. *Leave It As It Is: A Journey Through Theodore Roosevelt's American Wilderness.* New York, NY: Simon & Schuster, Inc.

Gisel, Bonnie J. 2008. *Nature's Beloved Son: Rediscovering John Muir's Botanical Legacy.* Berkeley, CA: Heyday Books.

Graham, Otis L., Jr. 2015. *Presidents and the American Environment.* Lawrence, KS: University Press of Kansas.

Gray, Brian E. 2000. The Battle for Hetch Hetchy Goes to Congress. *Hastings West-Northwest Journal of Environmental Law & Policy* 6 (2/3): 199–237.

Gray, Brian E. 2007. Hetch Hetchy and the Paradoxes of Restoration. *Hastings West-Northwest Journal of Environmental Law & Policy* 13 (2): 211–222.

Greenberg, Alissa. 2018. Is Zinke Trolling San Francisco with Plan to Dismantle City's Reservoir? *The Guardian*. https://theguardian.com/environment/2018/jul/27/ryan-zinke-trolling-san-francisco-dismantle-hetch-hetchy-dam.

Harmon, David, Francis P. McManamon, and Dwight T. Pitcaithley, eds. 2006. *The Antiquities Act: A Century of American Archaeology, Historic Preservation, and Nature Conservation*. Tucson, AZ: University of Arizona Press.

Holmes, Steven J. 1999. *The Young John Muir: An Environmental Biography*. Madison, WI: University of Wisconsin Press.

Hundley, Norris, Jr. 2001. *The Great Thirst: Californians and Water—A History*, rev. ed. Berkeley and Los Angeles, CA: University of California Press.

Johnson, Robert Underwood. 1890. The Care of Yosemite Valley. *The Century Magazine* XXXIX: 494–495.

Johnson, Robert Underwood. 1923. *Remembered Yesterdays*. Boston, MA: Little, Brown.

Jones, Holway R. 1965. *John Muir and the Sierra Club: The Battle for Yosemite*. San Francisco, CA: Sierra Club Books.

Leopold, Aldo. 1968 [1949]. *A Sand County Almanac, and Sketches Here and There*. New York, NY: Oxford University Press.

Lowitt, Richard. 1995. The Hetch Hetchy Controversy, Phase II: The 1913 Senate Debate. *California History* 74 (2): 190–203.

Marsh, George Perkins. 2003 [1864]. *Man and Nature: Or, Physical Geography as Modified by Human Action*. Seattle, WA: Weyerhaeuser Environmental Books/University of Washington Press.

Meral, Gerald H. 2005. *Finding the Way Back to Hetch Hetchy Valley: A vision of steps to restore Hetch Hetchy Valley in Yosemite National Park and to replace water and energy supplies. Feasibility Study 2005*. Sonora, CA: Restore Hetch Hetchy

Meral, Gerald H. 2008. Beyond and Beneath O'Shaughnessy Dam: Options to Restore Hetch Hetchy Valley and Replace Water and Energy Supplies. *Golden Gate University Environmental Law Journal* 2 (1): 22–68.

Miller, Char. 2001. *Gifford Pinchot and the Making of Modern Environmentalism*. Washington, DC, Covelo, CA and London: Island Press/Shearwater Books.

Miller, Char. 2013. *Seeking the Greatest Good: The Conservation Legacy of Gifford Pinchot*. Pittsburgh, PA: University of Pittsburgh Press.

Miller, Char, ed. 2017. *Gifford Pinchot: Selected Writings*. University Park, PA: Pennsylvania State University Press.

Muir, John. 1871. Yosemite Glaciers. *New-York Tribune*, December 5: 5–6.

Muir, John. 1872a. In the Yo-semite: Holidays Among the Rocks. *New-York Tribune*, January 1.

Muir, John. 1872b. Yosemite in Spring. *New-York Tribune*, May 7.

Muir, John. 1872c. Rambles of a Botanist Among the Plants and Climates of California. *Old and New* 5: 767–772.

Muir, John. 1873a. Hetch Hetchy Valley. *Boston Weekly Transcript*, March 25.

Muir, John. 1873b. Explorations in the Great Tuolumne Cañon. *Overland Monthly* 11 (2): 139–147.

Muir, John. 1875. The Kings River Yosemite. *San Francisco Daily Evening Bulletin*, August 13.

Muir, John. 1876. "God's First Temples." How Shall We Preserve Our Forests? The Question Considered by John Muir, the California Geologist—The Views of a Practical Man and a Scientific Observer—A Profoundly Interesting Article. *Sacramento Daily Record-Union*, February 5: 8.

Muir, John. 1890a. The Treasures of the Yosemite. *The Century Magazine* XL (4): 483–500.

Muir, John. 1890b. Features of the Proposed Yosemite National Park. *The Century Magazine* XL (5): 656–667.

Muir, John. 1891. A Rival of the Yosemite: The Cañon of the South Fork of Kings River, California. *The Century Magazine* XLIII (21): 77–97.

Muir, John. 1896. The National Parks and Forest Reservations (Proceedings of the Meeting of the Sierra Club Held November 23, 1895). *Sierra Club Bulletin* 1 (7): 271–284.

Muir, John. 1907. The Tuolumne Yosemite in Danger. *The Outlook* 87: 486–489.

Muir, John. 1908a. The Hetch Hetchy Valley. *Sierra Club Bulletin* VI (4): 212–220.

Muir, John. 1908b. Hetch Hetchy Damming Scheme. In *San Francisco and the Hetch Hetchy Reservoir: Hearing Held Before the Committee on the Public Lands of the House of Representatives, December 16, 1908, on H.J. Res. 184*, ed. Committee on the Public Lands, House of Representatives. Washington, DC: Government Printing Office. 32–33.

Muir, John. 1909a. The Endangered Valley: The Hetch Hetchy Valley in the Yosemite National Park. *The Century Magazine* LXXVII (3): 464–469.

Muir, John. 1909b. *Let Everyone Help to Save the Famous Hetch-Hetchy Valley and Stop the Commercial Destruction Which Threatens Our National Parks.* San Francisco, CA: Society for the Preservation of National Parks/Sierra Club.

Muir, John. 1910. The Hetch-Hetchy Valley: A National Question. *American Forestry* XVI (5): 263–269.

Muir, John. 1920. Save the Redwoods. *Sierra Club Bulletin* XI (1): 1–4.

Muir, John. 1944. The Creation of Yosemite National Park, Letters of John Muir to Robert Underwood Johnson. *Sierra Club Bulletin* XXIX (5): 49–60.

Muir, John. 1950. *Studies in the Sierra*. San Francisco, CA: Sierra Club Books.

Muir, John. 2001a [1894]. *The Mountains of California*. New York, NY: Modern Library.

Muir, John. 2001b. *A Thousand Mile Walk to the Gulf*. In *The Wilderness Journeys*, John Muir. Edinburgh: Canongate Classics.

Muir, John. 2002 [1915]. *Travels in Alaska*. New York, NY: Modern Library.

Muir, John. 2003a [1911]. *My First Summer in the Sierra*. New York, NY: Modern Library.

Muir, John. 2003b [1912]. *The Yosemite*. New York, NY: Modern Library.

Muir, John. 2018a [1901]. *Our National Parks*. Layton, UT: Gibbs Smith.

Muir, John. 2018b [1913]. *The Story of My Boyhood and Youth*. Edinburgh: Birlinn Origin.

Muir, John. 2019 [1915]. *Letters to a Friend: Written to Mrs. Ezra S. Carr, 1866–1879*. Mineola, NY: Dover.

Nash, Roderick Frazier. 2014. *Wilderness and the American Mind*, 5th ed. New Haven, CT and London: Yale University Press.

National Park Service. 1987. *Alternatives for Restoration of Hetch Hetchy Valley Following Removal of the Dam and Reservoir*. Washington, DC: National Park Service.

Null, Sarah, E. 2003. *Re-assembling Hetch Hetchy: Water Supply Implications of Removing O'Shaughnessy Dam*. Unpublished Master of Arts in Geography Thesis, University of California, Davis, Davis, CA.

Null, Sarah, E., and Jay R. Lund. 2006. Reassembling Hetch Hetchy: Water Supply Without O'Shaughnessy Dam. *Journals of the American Water Resources Association* 42: 395–408.

Obama, President Barack. 2016. Remarks by the President at Sentinel Bridge, Yosemite National Park, June 18, 2016. *The White House, President Barack Obama*. https://obamawhitehouse.archives.gov/the-press-office/2016/06/18/remarks-president-sentinel-bridge.

Olmsted, Frederick Law. 1993. *Yosemite and the Mariposa Grove: A Preliminary Report, 1865*. Yosemite National Park, CA: Yosemite Association.

O'Shaughnessy, Michael Maurice. 1934. *Hetch Hetchy: Its Origin and History*. San Francisco, CA: Recorder Printing and Publishing Company.

Pinchot, Gifford. 1998 [1947]. *Breaking New Ground*, commemorative ed. Washington, DC and Covelo, CA.

Proclamation No. 973, 35 Stat. 2174 (1908).

Proclamation No. 5794, 3 C.F.R. 5794 (1988).

Proclamation No. 7295, 3 C.F.R. 7295 (2000).

The Pulitzer Prizes. 2020. The 2005 Pulitzer Prize Winner in Editorial Writing: Tom Philp of *The Sacramento Bee. The Pulitzer Prizes.* https://pulitzer.org/winners/tom-philp.

Resources Agency, State of California. 2006. *Hetch Hetchy Restoration Study*. Sacramento, CA: Department of Water Resources/Department of Parks and Recreation.

Restore Hetch Hetchy. 2017. Legal Campaign Update. *Restore Hetch Hetchy*. https://hetchhetchy.org/legal_campaign_update.

Restore Hetch Hetchy. 2018a. *Water and Power System Improvements Necessary to Accommodate Restoration of Hetch Hetchy Valley in Yosemite National Park*. Berkeley, CA: Restore Hetch Hetchy.

Restore Hetch Hetchy. 2018b. *Hetch Hetchy Valley Restoration and Hydropower in California*. Berkeley, CA: Restore Hetch Hetchy.

Restore Hetch Hetchy. 2018c. *Hetch Hetchy and California Water Supply*. Berkeley, CA: Restore Hetch Hetchy.

Restore Hetch Hetchy. 2020. Our plan. *Restore Hetch Hetchy*. https://hetchhetchy.org/our-plan.

Rider, Jessica K. 2004. *The Potential Economic Benefits of Restoring Hetch Hetchy*. Oakland, CA: Environmental Defense.

Righter, Robert W. 2005. *The Battle Over Hetch Hetchy: America's Most Controversial Dam and the Birth of Modern Environmentalism*. New York, NY: Oxford University Press Inc.

Roosevelt, Theodore. 1924 [1913]. *Theodore Roosevelt: An Autobiography*. New York, NY: Charles Scribner's Sons.

Root, Robert. 2017. *Walking Home Ground: In the Footsteps of Muir, Leopold, and Derleth*. Madison, WI: Wisconsin Historical Society Press.

Rosekrans, Spreck. 2014. *Restoration of Hetch Hetchy Valley and San Francisco's Water Supply*. Oakland, CA: Restore Hetch Hetchy.

Rosekrans, Spreck, and Ryan Laws. 2021. *Keeping Promises: Providing Public Access to Hetch Hetchy Valley, Yosemite National Park*. Berkeley, CA: Restore Hetch Hetchy.

Rosekrans, Spreck, Nancy E. Ryan, Ann H. Hayden, Thomas J. Graff, and John M. Balbus. 2004. *Paradise Regained: Solutions for Restoring Yosemite's Hetch Hetchy Valley*. Oakland, CA: Environmental Defense.

Rothman, Hal. 1994 [1989]. *America's National Monuments: The Politics of Preservation.* Lawrence, KS: University Press of Kansas.
Runte, Alfred. 2020 [1990]. *Yosemite: The Embattled Wilderness.* Lanham, MD: Lyons Press/Rowman & Littlefield.
Sargent, Shirley. 2001. *Galen Clark: Yosemite Guardian,* 4th ed. Yosemite National Park, CA: Flying Spur Press.
Simpson, John Warfield. 2003. *Yearning for the Land: A Search for Homeland in Scotland and America.* New York, NY: Vintage Books.
Simpson, John Warfield. 2005. *Dam! Water, Power, Politics, and Preservation in Hetch Hetchy and Yosemite National Park.* New York, NY: Pantheon Books.
Solnit, Rebecca. 2014. *Savage Dreams: A Journey into the Hidden Wars of the American West,* 20th anniversary ed. Berkeley and Los Angeles, CA: University of California Press.
Spurgeon, Sara L. 2009. Miracles in the Desert: Literature, Water, and Public Discourse in the American West. *ISLE: Interdisciplinary Studies in Literature and Environment* 16 (4): 743–759.
Stanley, Millie. 1995. *The Heart of John Muir's World: Wisconsin, Family, and Wilderness Discovery.* Madison, WI: Prairie Oak Press.
Stoll, Mark R. 2015. *Inherit the Holy Mountain: Religion and the Rise of American Environmentalism.* New York, NY: Oxford University Press.
Swan, Christopher, and Chet Roaman. 1977. *YV88: An Eco-Fiction of Tomorrow.* San Francisco, CA: Sierra Club Books.
Taylor, Ray W. 1926. *Hetch Hetchy: The Story of San Francisco's Struggle to Provide a Water Supply for Future Needs.* San Francisco, CA: Ricardo J. Orozco.
Terry, Paul. 1909. Sweeping Back the Flood. *The San Francisco Call,* December 13: 1.
Thoreau, Henry David. 2004 [1864]. *The Maine Woods,* ed. Joseph J. Moldenhauer. Princeton, NJ and Oxford: Princeton University Press.
Treuer, David, and Katy Grannan. 2021. Return the National Parks to the Tribes. *The Atlantic.* https://www.theatlantic.com/magazine/archive/2021/05/return-the-national-parks-to-the-tribes/618395/.
Turner, Frederick. 2000 [1985]. *John Muir: Rediscovering America.* Cambridge, MA: A Merloyd Lawrence Book/Perseus Publishing.
Whitney, Josiah D. 1865. *Geological Survey of California. Geology. Volume I. Report of Progress and Synopsis of the Field-Work, from 1860 to 1864.* Philadelphia, PA: Caxton Press of Sherman & Co.
Whitney, Josiah D. 1869. *The Yosemite Guide-Book: A Description of the Yosemite Valley and the Adjacent Regions of the Sierra Nevada, and of the Big Trees*

of California. Geological Survey of California. Sacramento, CA: California State Legislature, State Capitol.
Wilkins, Thurman. 1995. *John Muir: Apostle of Nature*. Norman, OK: University of Oklahoma Press.
Williams, Dennis C. 2002. *God's Wilds: John Muir's Vision of Nature*. College Station, TX: Texas A&M University Press.
Wolfe, Linnie Marsh, ed. 1966 [1938]. *John of the Mountains: The Unpublished Journals of John Muir*. Madison, WI: University of Wisconsin Press.
Wolfe, Linnie Marsh. 2003 [1945]. *Son of the Wilderness: The Life of John Muir*. Madison, WI: University of Wisconsin Press.
Worster, Donald. 2008. *A Passion for Nature: The Life of John Muir*. New York, NY: Oxford University Press Inc.
Yelverton, Therèse. 1872. *Zanita: A Tale of the Yo-semite*. New York, NY: Hurd and Houghton.
Yosemite National Park. 2014. *Merced Wild and Scenic River: Final Comprehensive Management Plan and Environmental Impact Statement*, 3 vols. Yosemite, CA: Yosemite National Park.

4

'On This Sand Farm in Wisconsin': Aldo Leopold, the Leopold Shack, and the Aldo Leopold Foundation

4.1 Leopold, the U.S. Forest Service, and Gila Wilderness Area

When Rand Aldo Leopold (1887–1948) was born in Burlington, Iowa, on January 11, 1887, John Muir and Robert Underwood Johnson had not yet begun their campaign for the designation of Yosemite National Park. The 1906 San Francisco earthquake, and Muir's public disagreement with U.S. Forest Service (USFS) chief Gifford Pinchot over the preservation versus 'wise use' of Yosemite's Hetch Hetchy Valley, was still two decades and more away. Aldo Leopold would be strongly influenced by the conservation rhetoric and philosophies of both Muir and Pinchot, although over the course of his career he would gradually abandon Pinchot's utilitarian forestry in favor of Muir's preservation—despite beginning his career in the Forest Service under Pinchot.

Leopold's late-childhood and teenage years, during the administration of conservation President Theodore Roosevelt (hereafter, TR), were spent exploring (and later hunting) along the Mississippi River near the family home on Prospect Hill. After Leopold transferred to the Lawrenceville School, a preparatory boarding school in Lawrenceville, New Jersey, in

L. Smith, *Ecological Restoration and the U.S. Nature and Environmental Writing Tradition*, https://doi.org/10.1007/978-3-030-86148-3_4

January 1904, he resumed his tramps through local woods, farm fields, and meadows—and again at Yale. In February 1905, seven months before Leopold enrolled at Yale University, TR signed the Transfer Act, placing responsibility for the forest reserves under the Bureau of Forestry in the U.S. Department of Agriculture (USDA). In the July, the Bureau was renamed the U.S. Forest Service, and Pinchot was appointed as Chief Forester. After graduating from the Sheffield Scientific School at Yale in 1908, Leopold remained at Yale and entered the Yale School of Forestry in the fall, which had opened eight years earlier with philanthropic support from the Pinchot family, and was the first graduate school of forestry in the U.S.

This chapter begins with Leopold's tenure with the Forest Service in the American Southwest. Discussion follows Leopold's work on timber crews and his regional office positions, to track the evolution of his fledgling conservation philosophy, especially on game management and game restoration. It was while working in New Mexico's Carson National Forest in the late-1910s and early-1920s that Leopold first articulated his wilderness advocacy, through his work to secure the first wilderness area designation in the national forest system—the headwaters of the Gila River. In the mid-1920s, Leopold and his family returned to his Midwestern roots, after transferring to the USFS's Forest Products Laboratory in Wisconsin as assistant director, before quitting the USFS for the academy, joining the University of Wisconsin-Madison as the inaugural chair of game management in 1933. The chapter broadens its scope at this juncture, to explore the formalization and rise of a wilderness rhetoric in the American conservation movement during the first half of the twentieth century and consider Leopold's place in (together with the legislative history of) wilderness preservation—from Leopold's contribution to the Gila Wilderness designation, to his cofounding the Wilderness Society in 1935, to the signing of the Wilderness Act in 1964. 2014 marked the 50th anniversary of the Wilderness Act and the 90th anniversary of the Gila Wilderness.

It was in Wisconsin that Leopold's interest in ecological restoration found two important outlets—through his work as director of research at the newly created UW Arboretum, and through his family's venture on their sand farm near Baraboo, 50 miles northwest of Madison, on the south shore of the Wisconsin River. The chapter lingers at the family's worn-out and abandoned farmstead to unpack the entanglements of cultural landscape restoration at 'the Shack'—through Leopold's annual pine planting program, and the restoration of prairie on the former cornfield. His sand farm is the theater for many of the essays in his 1949 *A Sand County Almanac, and Sketches Here and There* (see Leopold 1968), published the year after he died fighting a grass fire on a neighbor's farm. While the restoration of the sand farm is quietly woven through the collected essays, it is in *A Sand County Almanac*'s final essay—'The Land Ethic'—that restorationists can also find a powerful and lyrical commentary on a moral-emotional, ethical land conservation philosophy.

The final part of this chapter returns to the farm country of the Leopold Shack to consider the restoration legacy of the author, the book, and the place, *in place*. In the decades since Leopold's death, several restoration programs have variously 'restor(y)ed' the Shack lands. In the late-1960s, the Leopold Memorial Reserve (LMR) was established by a group of private landowners to protect the land around the Shack property from development. Out of the collaboration emerged the Sand County Foundation, which oversaw the management of the LMR for many decades, the Bradley Study Center and its prairie restorations established by Leopold's eldest daughter Nina and her husband, and the Aldo Leopold Shack Foundation (today, simply the Aldo Leopold Foundation), formed in 1982 by Leopold's five children, and which protects the Shack farm and property at the center of the reserve. This chapter concludes by tracing Leopold's ecological restoration legacy in the work of the Aldo Leopold Foundation at the Shack farm and property.

4.1.1 Leopold, Game Conservation, and the National Forests of Arizona and New Mexico

After graduating from Yale Forest School in the February of 1909, Leopold was still required to complete a ten-week field training program at the university's forestry camp in Doucette, Texas, and pass the civil service exams, before he could join the fledgling USFS. Leopold was sent to his first choice of the Forest Service's six districts—District 3 (the Southwest region), covering the 21 forests of the South and Southwest—arriving in Albuquerque, New Mexico, on July 2, 1909. He would remain with the Forest Service for 15 years (for more context and history on Leopold's Southwest years, see his 'sketches' essays in Leopold 1968, also Meine 1988; Flader 1994; Brown and Carmony 1995; Lannoo 2010; Nash 2014; Lorbiecki 2016; Warren 2016). Leopold's early career with the USFS overlapped with the residual momentum of a Progressive conservation movement galvanized by Pinchot and Roosevelt, and exposed him to—and positioned him amidst—some of the early-twentieth-century's major 'wise use' land use policies. His tenure with the Forest Service began a career trajectory that would variously cast Leopold as forester, field biologist, wildlife manager, writer, bow and gun hunter, sportsman, outdoor recreationist, environmental historian, wilderness preservation advocate, college professor, scientist-philosopher, and as a pioneer in conservation biology, restoration ecology, sustainable agriculture, ecological economics, environmental ethics, environmental policy and law, and environmental education, all nascent fields that would expand in the 1970s and 1980s.

Leopold's first post was as a forest assistant in the year-old Apache National Forest in the Arizona Territory (Arizona became the 48th state in February 1912), one of the wildest parts of the district. Arriving in Springerville in mid-July, Forest Assistant Leopold was put in charge of a reconnaissance crew, calculating and drafting section maps, surveying, and timber cruising in the Blue Range and White Mountains of eastern Arizona. The imprint of his Apache forest adventures on his early conservation sensibility would ricochet across his writings years later. It was during this posting that Leopold shot and killed a female wolf, arriving just as 'a fierce green fire' was extinguished in the wolf's dying eyes (recounted in the 1944 essay, 'Thinking Like a Mountain,' in Leopold 1968, cf. 'a fierce green fire' motif adopted by *Green Fire* 2011; Lorbiecki 2016). The essay is one of the most emotive and evocative in Leopold's canon, and has become shorthand for a provocation on wildness and wild land.

After two years, Leopold was assigned to the district office in Albuquerque as a temporary staff officer, but he did not remain there long. Meanwhile, while on this temporary assignment, he met María Alvira Estella Bergere. Leopold's work quickly impressed District Forester Arthur C. Ringland, and in April 1911, Leopold was appointed deputy supervisor of New Mexico's Carson National Forest, north of Santa Fe. The Apache and Carson national forests were established on the same day—July 1, 1908 (Meine 1988). But where the Apache National Forest had experienced relatively light incursions and intrusions, Deputy Supervisor Leopold was confronted by sheep and cattle overgrazing, soil erosion, flooding, vegetation changes, and an absence of game on the Carson National Forest. To address the poor condition of the land, and outline his conservation ideas and programs, Leopold turned to a platform that would recur across his professional career—short essays and reports. Beginning in June 1911, Leopold edited a newsletter, *The Carson Pine Cone*, to communicate his plans to his team of forest rangers (Meine 1988; Flader 1994). Throughout his career, Leopold would publish more than 500 newsletters, articles, reports, handbooks, reviews, and speeches. A similar volume would remain in draft or unpublished during his lifetime (Leopold 1991).

In March 1912, Leopold was promoted to acting supervisor of the Carson (two months before, in early-January, New Mexico became the 47th U.S. state). By the fall, he was appointed the Carson's full supervisor (Meine 1988; Lorbiecki 2016; Warren 2016). Leopold and Estella Bergere married in Santa Fe on October 9, 1912. But Leopold's burgeoning career with the USFS was interrupted when he developed a life-threatening kidney infection—Bright's disease, or acute nephritis—after getting caught in a spring storm while on a range inspection of the Carson's overgrazed Jicarilla unit in mid-April 1913 (Meine 1988; Brown and Carmony 1995; Lorbiecki 2016; Warren 2016). Unable to work for almost 18 months, Leopold had to abandon his stewardship of the Carson. His convalescence in Santa Fe and at his family's home on Prospect Hill in Burlington was spent celebrating the birth of his first son in October 1913—and reading. In particular, his reading of William Temple Hornaday's (1913) *Our Vanishing Wild Life* prompted an interest in developing game reserves, and in game management and conservation (Meine 1988; Flader 1994; Brown and Carmony 1995; Lorbiecki 2016; Warren 2016). Leopold continued to write for *The Carson Pine Cone* and introduced his game program in the *Pine Cone*'s January 1914 issue (Meine 1988; Flader 1994; Lorbiecki 2016).

With the support of Ringland, Leopold resumed his Forest Service career in September 1914, accepting a temporary assignment in the district's Office of Grazing in Albuquerque. The post gave Leopold the opportunity to more closely consider the policies and practices of the USFS. In January 1915, Leopold submitted a memo outlining a policy approach to restoring game in District 3, but the plan was rejected by the Washington, DC office (Meine 1988; Lorbiecki 2016). Ringland continued to support Leopold's creativity and ambitions, and encouraged Leopold's pursuit of game management ideas. In June 1915, after nine months as assistant to the district supervisor, Ringland put Leopold in charge of the new recreational policy for the district. Recreation was emerging as a new forest 'product.' This assignment also tasked Leopold with initiating and coordinating a new fish and game program, and overseeing publicity and public relations work (Meine 1988; Dunlap 1993; Flader 1994; Sutter 2005; Lorbiecki 2016). The Forest Service's timber and grazing management paradigm was echoed in Leopold's approach to game management. The program was soon adopted as a Forest Service model nationwide, and Leopold wrote the Forest Service's first *Game and Fish Handbook* (Leopold 1915). The protection of game had superseded forestry as Leopold's principal focus in his work in District 3. Game *management* and game (later expanding to 'wild life') *restoration* were beginning to coalesce in Leopold's forestry lexicon.

His assignment also saw Leopold meet with sportsmen across the Southwest to organize new local and state cooperative game protective associations (GPAs), to lobby for new state game conservation laws, greater law enforcement, the creation of nonpolitical, administrative state agencies, new game refuges, and greater predator control (Meine 1988; Dunlap 1993; Flader 1994; Brown and Carmony 1995). Leopold served as secretary of the Albuquerque GPA (and later the New Mexico GPA, or NMGPA). His interest in game restoration was further consolidated after meeting Hornaday in Albuquerque in October 1915 on Hornaday's lecture tour (Meine 1988; Flader 1994). To bring together and update the new game protection societies and district foresters, Leopold issued a quarterly bulletin, *The Pine Cone*, the name echoing his earlier Carson

National Forest newsletter, beginning in December 1915, and continuing into 1920 (Meine 1988; Flader 1994). To Leopold, the newsletter and magazine offered a wider audience than USFS memos and reports. Leopold turned down a transfer to Washington, DC to work on forest policy (he would later also decline postings in District 1 covering the northern Rockies, and with the U.S. Biological Survey, predecessor of the U.S. Fish and Wildlife Service).

Leopold's first magazine article was published in the November–December 1916 issue of *Arizona*—'Game Conservation: A Warning, Also an Opportunity' (Leopold 1916). His early bulletin, newsletter, and magazine writings would quickly establish Leopold as an organizer, activist, and policymaker on game management. In January 1917, Leopold received a letter of congratulation from former President Theodore Roosevelt for his game enforcement work with the Albuquerque GPA, and six months later, he (with the Albuquerque GPA) was awarded a gold medal by Hornaday's Permanent Wild Life Protection Fund (Meine 1988; Flader 1994; Lannoo 2010; Nash 2014; Lorbiecki 2016; Warren 2016). Soon after, Leopold was promoted. In the fall of 1917, Leopold also began to compile detailed field journal accounts and hunting records, a practice he would maintain for the rest of his life.

The U.S.'s entry into the First World War in April 1917, and a shift to wartime operations, imposed budget restrictions on the USFS. The Forest Service introduced a vigorous timber and grazing policy on national forest rangelands that largely stalled or reversed the range management progress of the past decade. With these changes—and Ringland's transfer to Washington, DC in April 1916—Leopold reluctantly resigned on January 3 and took a 20-month hiatus from the Forest Service in January 1918, to become secretary of the Albuquerque Chamber of Commerce (Meine 1988; Lorbiecki 2016; Warren 2016). Leopold was excused from the draft because of his age, health, and young family (two more sons and a daughter would be born in New Mexico between 1915 and 1919, and a fifth child, a daughter, would be born in Wisconsin in 1927). Leopold continued to work on and promote his plan for game management and conservation. His first article on wilderness, 'The Popular Wilderness Fallacy: An Idea That Is Fast Exploding' appeared the same month, in *Outers Book—Recreation*

(Leopold 1918). In Leopold's absence, meanwhile, the Forest Service gained a new outdoor recreation competitor in the form of the three-year-old National Park Service (NPS), resulting in a recasting of scenery and outdoor recreation as major 'products' of USFS lands. Leopold returned to District 3 on August 1, 1919, to the second-highest position in the district—Assistant District Forester in Charge of Operations for Ringland's successor, Paul G. Redington. It was a post he would hold for five years. Leopold was responsible for managing 20 million acres, taking him on inspection tours and surveys of national forests across the American Southwest (Meine 1988; Flader 1994; Brown and Carmony 1995; Lorbiecki 2016; Warren 2016).

Although constrained by postwar funding cutbacks, and another change in district forester in November 1919—with Frank C. W. Pooler replacing Redington, and less sympathetic to Leopold's conservation ideas than his two predecessors—Leopold's conservation sensibility flourished during this period. Leopold stepped back from his NMGPA and *Pine Cone* commitments in 1920 to focus on his role as chief of operations (but stayed on at the Albuquerque GPA) and grew increasingly uneasy at the erosion that followed overgrazing and poor logging practices in the forests of the Southwest. 'Forest restoration' joined 'game restoration' in Leopold's professional lexicon. Leopold was emerging as an urgent voice and authority on the Southwest's changing landscapes. This was reflected in many of his essays from the interwar period, with overgrazing,[1] erosion control, and fire ecology central themes (compiled in e.g. Leopold 1991; Brown and Carmony 1995; Meine and Knight 1999; see also Meine 1988; Dunlap 1993; Flader 1994; Lorbiecki 2016). While his interest in game conservation continued, wilderness preservation was entering and metastasizing in Leopold's thinking. Curt Meine (2004) describes this friction of utilitarian conservation and wilderness preservation in Leopold's conservation thinking as 'Leopold's fine line.'

[1] Not until passage of the Taylor Grazing Act in June 1934 was grazing on public lands federally regulated. The 1934 Act also created the Division of Grazing within the U.S. Department of the Interior (renamed the U.S. Grazing Service in 1939, which merged with the General Land Office in 1946 to form the Bureau of Land Management [BLM]).

Leopold was also growing increasingly concerned about road building activity on national forest lands, and the fracturing and loss of wilderness due to motorized recreation (Harvey 2005; Sutter 2005). In several articles written in the early- to mid-1920s, Leopold spoke out against the 'good roads movement' expansion of the road system across Forest Service lands (in Leopold 1991, also 1968; cf. Meine 1988; Harvey 2005; Sutter 2005).

4.1.2 Establishing the First Wilderness Area in the U.S.

With his return to District 3 in the summer of 1919, Leopold began, with fellow foresters Frederic Winn and Ward Shepard, to explore setting aside areas of wilderness in the national forests. In December that year, Leopold traveled to Denver, Colorado, to meet with Forest Service rangers in District 2. A December 6 conversation with the Forest Service's first landscape architect and planner, Arthur H. Carhart, brought together two sympathizers on the need for wilderness preservation, and the value of wilderness and the wilderness experience (Meine 1988; Brown and Carmony 1995; Sutter 2005; Nash 2014; Lorbiecki 2016). The meeting heralded the start of a struggle to recast and redefine the Forest Service's description of its forests beyond timber production and game management, toward the protection of wilderness as a recreation resource. Yet Leopold's work on a wilderness preservation program for the national forests vibrated against—and marked his fracture with—Pinchot's 'wise use' of the forests. Leopold's 'fine line' between utility and preservation (after Meine 2004) was again narrowing. He returned to Albuquerque energized on protecting wilderness. Leopold would become as passionate a wilderness prophet as John Muir (see Chapter 3). Public debate on 'the wilderness idea' was just beginning.

While Carhart and another District 2 forester, Carl J. Stahl, would initiate and introduce a wilderness preservation rhetoric in the Forest Service (securing Trappers Lake in Colorado's White River National Forest as a roadless, undeveloped area in early-1920), it was Leopold who popularized—and expanded—'the issue of wilderness conservation.'

'The Wilderness and its Place in Forest Recreational Policy,' published in the *Journal of Forestry* in November 1921 (Leopold 1921), is Leopold's first formalization of the wilderness question in American professional forestry and articulates the need for a wilderness preservation policy. As Leopold (1921: 720) argues, 'It will be much easier to keep wilderness areas than to create them. In fact, the latter alternative may be dismissed as impossible. Right here is the whole reason for forehandedness in the proposed wilderness area policy.' Leopold (1921) closes his article with the provocation that an undeveloped section of the Gila National Forest in southwestern New Mexico be designated a wilderness reserve.

Early the following year, Pooler instructed Leopold to inspect the Gila National Forest. Working with the supervisor of the Gila National Forest, Fred Winn, in May and June 1922, Leopold drew up a 'General Inspection Report on the Gila National Forest.' He submitted his 'Report on the Proposed Wilderness Area' to Pooler in early-October, mapping the boundary of the first official wilderness area in the U.S. (on Leopold, the Gila National Forest, and wilderness, see Meine 1988; Flader 1994; Sutter 2005; Lannoo 2010; Nash 2014; Lorbiecki 2016). A few decades later, Edward Abbey (Chapter 6) would work as a fire lookout and firefighter in the Gila National Forest. In late-October 1922, one month before the Colorado River Compact was signed (Chapter 6), Leopold and his younger brother Carl set out on a canoe trip, camping and hunting along the Colorado River Delta (recounted in e.g. 'The Green Lagoons' in Leopold 1968).

But Leopold was becoming increasingly frustrated with the conservation ethos of District 3. His 1923 *Watershed Handbook* was adopted as the Forest Service's formal manual for erosion control (Leopold 1923; cf. Dunlap 1993; Brown and Carmony 1995; Meine 1988; Lorbiecki 2016; Warren 2016 on Leopold's USFS legacy of game management, erosion control, and wilderness protection). Leopold and his colleague in the district's Land Office, Morton Cheney, finalized a 'Recreational Working Plan' for a 755,000-acre Gila Wilderness Area in March 1924 (Meine 1988). Designation of the Gila Wilderness was imminent. But his authority to inform and mold Forest Service policies and practices in District 3 was waning. Leopold's conservation ideas were considered too radical by the new district forester, Pooler, and many of his colleagues.

4.1.3 From New Mexico to Wisconsin

Leopold—like John Muir seven-and-a-half decades before him—was a Wisconsin transplant. On May 29, 1924, Leopold and his young family left the Southwest for a USFS post in Madison, Wisconsin. Five days later, Pooler approved Leopold's recreational working plan and designated the 755,000-acre Gila Wilderness Area (Meine 1988; Lorbiecki 2016) (Fig. 4.1). The headwaters of the Gila River became America's first official wilderness area (the 202,016-acre Aldo Leopold Wilderness was designated in the Black Range region of the Gila National Forest in 1980). A decade after the passage of the 1913 Raker Act, and Muir's Hetch Hetchy defeat (Chapter 3), Leopold succeeded in introducing wilderness preservation policy into the national forest system (Leopold

Fig. 4.1 Gila Wilderness Area marker showing dedication to Aldo Leopold, ca. 1970. Courtesy of the Aldo Leopold Foundation and University of Wisconsin-Madison Archives

celebrates the protection of the Gila headwaters as a wilderness hunting ground in Leopold 1925; 1940b; see also Sutter 2005; Nash 2014; Lorbiecki 2016 on the history of wilderness preservation policy in the national forests). Hetch Hetchy marked a pivot in the American conservation movement. But a federal wilderness preservation system was still several decades away.

Although Leopold moved away from the Southwest in the mid-1920s, his personal, intellectual, and professional interest in the region continued. In 1927 and 1929, he returned to New Mexico to hunt deer in Gila National Forest, camped and hunted near Colonia Pacheco and along the Río Galiván in the northern Sierra Madre of Chihuahua, Mexico in 1936 and 1937, and visited Arizona's Kaibab National Forest in 1941 (see accounts in Leopold 1968, 1991, 1993, 2013). The 'unspoiled wilderness' of Chihuahua sat in stark contrast to the 'sick land' Leopold had so far encountered in the Southwest and Midwest, and it was from these Chihuahuan camps that Leopold began to develop his idea of 'land health' (Leopold 1968, 1993; cf. Meine 1988; Flader 1994; Lorbiecki 2016; Warren 2016). With the move back to the Midwest, Leopold continued to develop his ideas on the meaning and value of wilderness preservation, and the need for a federal system of wilderness areas. A series of wilderness essays and articles followed in the latter half of the decade and reflect a wilderness advocacy born in the Gila forest (see compilations in e.g. Leopold 1991; Brown and Carmony 1995; Meine and Knight 1999). Leopold's wilderness philosophy breaks from the Transcendental, spiritual wilderness philosophy of Thoreau and Muir, and aligns with scientific theory.

Leopold had left District 3 to take up the post of associate director of the Forest Products Laboratory, the USFS's center for timber research, beginning in early-July 1924. But the post took Leopold away from the outdoors and his passion for conservation. After four years supervising the laboratory, Leopold, now 41 years old, resigned on June 26, 1928—and left the USFS altogether. But during these four years, Leopold had continued to write on game management and its intersections with forestry, agriculture, and grazing, as well as on forestry and wilderness protection. Through various conservation clubs, Leopold also continued to contribute to the national wilderness debate, and in 1927, he worked to establish (and from 1942, served on) a nonpolitical state conservation commission (Meine 1988; Lorbiecki 2016).

After 15 years with the Forest Service, Leopold worked from July 1928 to 1931 for the Sporting Arms and Ammunition Manufacturers' Institute (SAAMI), conducting a national survey of game conditions, and coordinating game management research projects for increasing game populations. Game management became Leopold's new preoccupation. At the end of the year, the national survey was revised to an eight-state midwestern block—Ohio, Indiana, Michigan, Illinois, Wisconsin, Minnesota, Iowa, and Missouri. With the publication of his survey report, *Report on a Game Survey of the North Central States* (Leopold 1931), in the spring of 1931, game management conclusively joined technical forestry, wildlife protection, and wilderness preservation in Leopold's professional and intellectual repertoire. A game (and wild life) restoration lexicon again infiltrated and molded Leopold's recommendations (on Leopold's surveys and reports for SAAMI, see Meine 1988; Lorbiecki 2016). In September 1928, the University of Wisconsin-Madison offered Leopold office space in the Chemistry Building, and in February and March 1929, he delivered a series of lectures on game management, beginning a connection with the university that would last the rest of his life (Meine 1988; Brown and Carmony 1995; Court 2012; Lorbiecki 2016). When Leopold's funding from SAAMI ended in March 1932, he sought work as a consulting forester. But the period from late-1932 through early-1933 marked some of the worst months of the Great Depression, and Leopold used any time he was unemployed to work on a game management textbook. In October 1932, he secured intermittent work with the state conservation commission to establish game management demonstration projects (Meine 1988).

In April 1933, Leopold's first book, *Game Management*, was published (as Leopold 1986). *Game Management* was scaffolded by Leopold's 1929 lecture series, and also incorporated material from his abandoned 'Deer Management in the Southwest' (earlier, 'Southwestern Game Fields') manuscript (Meine 1988; Lorbiecki 2016; Warren 2016). Leopold regarded the textbook as a companion (and much expanded) volume to his *Report on a Game Survey of the North Central States* (Leopold 1931). Begun during his years in the Southwest, the expansive volume presents the first comprehensive history of, and commentary and studies in, the new science of game management. But most significantly, it demonstrates how, through attention to scientific principles and land use practices and management, wildlife populations and environments could be restored and sustainably managed—by deploying the same tools that earlier led to their ruin—tools such as the 'axe, plow, cow, fire, and gun' (Leopold 1986: xxxi). *Game Management* has become a cornerstone in conservation literature. Revealed in the work is a further shift beyond utilitarian conservation.

The publication of *Game Management* coincided with the inauguration of Franklin Delano Roosevelt (hereafter, FDR) on March 4, 1933, as America's 32nd President. Before the end of FDR's first month in office, Congress authorized the creation of the Civilian Conservation Corps (CCC), placing land conservation as a central part of FDR's New Deal program. That summer, Leopold returned to the Southwest to assist in running the CCC's erosion control program across Region 3 (the USFS Districts were renamed in 1930) (on Leopold's critique of a conservation taxonomy in the CCC program, see Meine 1988; Sutter 2005; Lorbiecki 2016; Warren 2016). While back in New Mexico, Leopold delivered the fourth annual John Wesley Powell Lecture at a meeting of the Southwestern Division of the American Association for the Advancement of Science in Las Cruces on May 1. In his address (published in the *Journal of Forestry* in October as Leopold 1933), he first formalized his 'The Conservation Ethic' that would later underpin 'The Land Ethic' of *A Sand County Almanac* (see Leopold 1968, also Meine 1988; Lorbiecki 2016).

In late-June, while Leopold was consulting on CCC erosion control projects in the Southwest, UW-Madison confirmed his appointment as professor of game management in the College of Agriculture's Department of Agricultural Economics. Leopold joined the university in the fall. The chair was supported by a $8,000-a-year endowment for five years from the Wisconsin Alumni Research Foundation (Meine 1988; Flader 1994). Leopold held the first-ever academic appointment (and ran the nation's first graduate program) in game management (Meine 1988; Lannoo 2010; Lorbiecki 2016). His roster included short courses for local farmers, undergraduate and graduate classes and supervision, research, and outreach—including field demonstrations, monthly radio talks on a university-sponsored station aimed at farmers and rural landowners, and essays in magazines, newsletters, and pamphlets (McCabe 1987; Meine 1988; Lorbiecki 2016; Warren 2016). Leopold's Game Management 118 class began in March 1934. At the same time, Leopold also served on the university's interdisciplinary Arboretum Committee, which was working toward 'reconstructing' or 'rebuilding' a sample of pre-farmland Wisconsin on the southern shore of Lake Wingra.

In the fall of 1938, Leopold switched his title to Professor of Wildlife Management, to reflect his widening thinking in the field. Wildlife Ecology 118 replaced Game Management 118 the same year. One year later, the state legislature approved funding for the continuation of Leopold's post, and Leopold became chair of the new Department of Wildlife Management. Leopold was at once the chair and only faculty member. He held the academic position for the rest of his life. He introduced a new class, Advanced Game Management 179, in the fall of 1946 (Meine 1988). Leopold continued to write scientific papers, technical reports, policy statements, position papers, editorials, popular articles and essays, speeches, serve on national organizations, and keep detailed natural history records. His attention was also shifting more to farm country—to the homesteads, cultivated fields, pastures, woodlots, and hedgerows of the Midwest.

4.2 The Value of Wilderness, the Wilderness Society, and the 1964 Wilderness Act

'There are some who can live without wild things, and some who cannot. These essays are the delights and dilemmas of one who cannot' (Leopold 1968: vii, also 1986). So begins the Foreword to Leopold's *A Sand County Almanac, and Sketches Here and There*, dated March 4, 1948, written just seven weeks before his death. These two sentences are a signal, portend to Leopold's literary wilderness advocacy. A siren call for the preservation of wilderness. But it is the 'delights and dilemmas' that I always pause at—unpicking the tensions and frictions of the ethics and aesthetics of the wilderness idea.

As Leopold's advocacy for a national system of wilderness preserves developed and intensified across the 1920s, 1930s, and 1940s, his writings on wilderness echoed this shift across three principal, overlapping concerns. Leopold variously enters the preservation of wilderness debate to cast wilderness in terms of recreation, roadlessness, and protecting wilderness hunting grounds to critique 'Outdoor Recreation, Latest Model,' as well as to amplify the cultural and historical value of wilderness preservation, and the scientific value of wilderness as a 'land laboratory' to study land health (Leopold 1968, 1991; Meine and Knight 1999, on Leopold and wilderness, see Tanner 1987; Meine 1988, 2004; Nash 2014). Yet Leopold (1968: 101) worries in 'Marshland Elegy' that, 'all conservation of wildness is self-defeating, for to cherish we must see and fondle, and when enough have seen and fondled, there is no wilderness left to cherish.'

When Robert Marshall organized the Wilderness Society at the Cosmos Club in Washington, DC on January 20–21, 1935, Leopold was one of the eight cofounders (John Wesley Powell founded the Cosmos Club in 1878). The Wilderness Society formed to provide political support for (roadless) wilderness preservation, a response to the impacts of motorized recreation, and to campaign for a national wilderness preservation system. Leopold was invited to serve as president, but he opted instead to serve as a senior advisor. Robert Sterling Yard was appointed as the Society's first president (on the early history of the Wilderness Society, and Marshall's legacy, see Meine 1988; Harvey 2005; Sutter 2005; Nash 2014; Lorbiecki 2016).[2] In the lead article in the inaugural issue of *The Living Wilderness* (now simply *Wilderness*), the magazine of the Wilderness Society, Leopold (1935b: 6) addresses the question, 'Why the Wilderness Society?' He opens with the rebuttal that, 'no idea is significant except in the presence of its opposite,' and brings

[2] The Wilderness Society features in a number of conservation campaigns later in this book, including in the Everglades (Chapter 5), and at Dinosaur National Monument, the Grand Canyon, and Bears Ears National Monument (Chapter 6).

his answer to a close by articulating the need for 'intelligent humility' (cf. Leopold 1968) in conservation.

Momentum and support for a national wilderness preservation system had been building since the early-1920s. Wilderness preservation received a boost from FDR's Secretary of the Interior Harold L. Ickes throughout the 1930s, and a wilderness bill was introduced in Congress in 1940, before fading with the U.S.'s entry into the Second World War. In the mid-1940s, following the death of Yard and a leadership reshuffle, Olaus Murie was appointed the Wilderness Society's new director, with Howard Zahniser as executive secretary, Benton MacKaye as president, and Leopold as vice president (Harvey 2005). Zahniser (later appointed director) resurrected the campaign for a wilderness preservation system in the late-1940s, and his campaign received a further boost with the defeat of the Echo Park dam project in Dinosaur National Monument in 1955 (examined in Chapter 6). Zahniser capitalized on the Echo Park wilderness battle in articles and speeches, and in January 1956 began drafting and lobbying for a wilderness bill to be presented before Congress. Later in 1957, Zahniser, in his capacity as president of the Thoreau Society, spoke of wilderness and Thoreau at the Thoreau Society Annual Gathering.[3] That summer was the Red Cross Beach summer (see Chapter 2).

Zahniser's wilderness campaign would last eight-and-a-half years and span three administrations. His wilderness bill would see 66 redrafts or resubmissions. Between June 1957 and May 1964, 19 congressional hearings were held on the proposed bill authored by Zahniser, generating more than 16,000 pages of testimony that frequently cited Thoreau, Muir, Leopold, Marshall, and other conservation figures. Zahniser attended every hearing (Harvey 2005; Nash 2014). Zahniser's wilderness preservation bill passed the Senate with a vote of 73-12 on April 10, 1963, and the House by 373-1 on July 30, 1964 (on Zahniser, the Wilderness Society, and the legislative history of the Wilderness Act, see especially Harvey 2005, 2014, also Nash 2014; Lorbiecki 2016). Zahniser died on May 5, 1964, as the bill headed to the House. After

[3] Zahniser, Howard. 1957. Thoreau and the Preservation of Wildness. *Thoreau Society Bulletin* 60: 1–2 [Thoreau Society Collections—Archives of the Thoreau Society, at the Thoreau Institute at Walden Woods].

some final adjustments in August, President Lyndon B. Johnson signed the Wilderness Act into law on September 3, 1964, establishing the National Wilderness Preservation System, and protecting 9.2 million acres. Not long after its signing, a new wilderness battle erupted in the Grand Canyon (on the Marble Canyon and Bridge Canyon dam proposals, see Chapter 6).

By 2014 and the 50th anniversary of the Wilderness Act, there were 758 areas in the National Wilderness Preservation System (Lorbiecki 2016), protecting upward of 100 million acres. The same year, the Gila Wilderness celebrated its 90th anniversary. In 2015, the Wilderness Society celebrated its 80th anniversary (on the achievements of the Wilderness Society and the Wilderness Act, see Molvar and Nesset 2014; The Wilderness Society 2014).

4.3 Curtis Prairie at UW Arboretum: A World First for Ecological Restoration

More than a year before his appointment as chair of game management in the College of Agriculture, Leopold accepted a position at the end of August 1932 on the Advisory Committee of the new university arboretum. Amidst the Great Depression and Dust Bowl years of the 1930s, the University of Wisconsin's 'Forest Preserve Arboretum and Wild Life Refuge' was established in Madison on April 26, 1932, to protect land bordering the western and southwestern shore of Lake Wingra. In 1932, the Arboretum lands consisted of six parcels totaling 245 acres, and by the turn of the century would expand to more than 1,200 acres, and upward of 500 acres across almost a dozen satellite properties across the state (Court 2012; Lorbiecki 2016). The history of the UW Arboretum is entangled with the history of restoration ecology and ecological restoration in the U.S.

In the fall of 1933, Leopold was appointed research director of the Arboretum, and G. William Longenecker was appointed the Arboretum's executive director. Leopold's research centered on the 'Wild Life Refuge' part in the Arboretum's early title and included a migratory game bird project, a trapping and bird banding program, establishing feeding stations (food-patch plantings), and exploring habitat types to support game cropping. In October 1933, the same month 'The Conservation Ethic' (Leopold 1933) was published, he submitted his 'Wild Life Management Plan' for the Arboretum (Meine 1988; Court 2012; Lorbiecki 2016). Speaking at the dedication ceremony of the 'University of Wisconsin Arboretum, Wild Life Refuge, and Experimental Forest Preserve' on June 17, 1934, Leopold notes, 'This, in a nutshell, is the function of the Arboretum: a reconstructed sample of old Wisconsin, to serve as a bench mark, a starting point, in the long and laborious job of building a permanent and mutually beneficial relationship between civilized men and a civilized landscape' (Leopold 1934a: 5; cf. Leopold 1934b, 1991, and on the tensions between different versions of Leopold's Arboretum address, see Meine 1988; Callicott 1999b; Court 2012; Greenwood 2017). The 'reconstructed sample of old Wisconsin' refrain speaks back to a more restoration *sensu stricto* goal (Aronson et al. 1993). The 'wild life refuge' focus would later be dropped from the Arboretum's title and remit.

The history of the Arboretum and its restored tallgrass prairies, oak savannas, forests, and wetlands has been extensively addressed elsewhere (Sachse 1965; Jordan 1984; Court 2012; Kozik 2017, also Meine 1988; Hall 2005; Cochrane et al. 2007; Warren 2016) and will not be repeated here. Through an interdisciplinary program with colleagues in botany and horticulture, and CCC crews based at the Arboretum, Leopold worked to restore tallgrass prairie and other plant communities across the arboretum land parcels. It was Leopold's work at the Arboretum, and his interest in evolutionary biology and ecology, succession, and ecological restoration, that led to his decision to purchase a worn-out sand farm in the mid-1930s. Ecologists at the UW Arboretum continue Leopold's phenological legacy, through the collection of phenological data 'on over 200 events each year.'[4]

The Arboretum fosters Leopold's 'land ethic' (Leopold 1968), but two sites are particularly closely aligned and intertwined with Leopold's restoration ethic. The 73-acre Curtis Prairie on the southwestern edge

[4] UWMA email interview, August 10, 2020.

of the Arboretum was the Arboretum's first prairie restoration. The 21-acre pinetum of red and white pines, with red maple and white birch, at the southern boundary of the Curtis Prairie was the Arboretum's first reforestation project and is known as the Aldo Leopold Memorial Forest, or the Leopold Pines (Jordan 1984; Court 2012). I pause at the Curtis Prairie, named in 1962 for UW botany professor and Arboretum plant research director John T. Curtis, not only for its Leopold connection, but also for its legacy in the annals of restoration ecology as the first prairie restoration in the U.S. Curtis's work on the prairie restoration was preceded by fellow UW botany professor Norman Carter Fassett's early experiments with transplanting prairie grass sods (Court 2012).[5] In 1936, Theodore Sperry was hired as the first Arboretum ecologist, and Leopold instructed him to 'go build a prairie.'[6] Sperry supervised the Arboretum CCC crews, and between April 1936 and 1941, they collected native sod, plants, seeds, and hay from prairie remnants along the Wisconsin River west of Madison, to plant in the fallow horse pasture.[7] In spring 1939, Sperry introduced experimental controlled burns on limited prairie plots (Court 2012, also Jordan 1984; Meine 1988; Leopold 2016; Kozik 2017). When Curtis was appointed director of plant research in 1940, he directed much of the early research on the prairie, monitored the progress of the restoration (cf. Curtis 1959; Cottam and Wilson 1966), and directed some of the first experiments using prescribed fire (e.g. Curtis and Partch 1948) (Fig. 4.2).

The UW Arboretum has another entanglement with the history of restoration ecology and ecological restoration. The restoration journal *Restoration and Management Notes* began in June 1981 as an Arboretum publication. When the University of Wisconsin Press took over publication in early-1984, the Arboretum retained editorial control (Jordan 1984; Court 2012). In 1988, the Arboretum, its public service coordinator William R. Jordan III, and the *R&MN* journal all had a part in organizing the Society for Ecological Restoration (SER), and SER was allocated office space in the Arboretum Security Office that May

[5] UWMA email interview, August 10, 2020.

[6] UWMA email interview, August 10, 2020.

[7] UWMA email interview, August 10, 2020.

Fig. 4.2 Aldo Leopold (1887–1948) with group participating in controlled prairie burn, UW Arboretum, mid-1940s. Left to right: Pepper Jackson, Aldo Leopold, Jim Hale, and Mary Ellen Helgren. Courtesy of the Aldo Leopold Foundation and University of Wisconsin-Madison Archives

(Court 2012). In January 1998, Joy B. Zedler joined the faculty of UW-Madison as the first Leopold Chair of Restoration Ecology, jointly in the Department of Botany and the UW Arboretum. In the late-1990s, *R&MN* relaunched as *Ecological Restoration.*

As an extension scientist, Leopold also oversaw other field trials and research on restoration beyond the UW-Madison campus, including several long-term game management projects on privately-owned farms. In 1931, Leopold and local farmer Reuben J. Paulson organized the 1,700-acre Riley Game Cooperative in western Dane County, Wisconsin. A game cooperative of farmers and sportsmen, Riley was a demonstration site for game cropping and restoration, and a conservation experiment on private land (on the history of the Riley shooting preserve and hunting ground, see Leopold and Paulson 1934; Leopold 1940a; cf. Meine 1988; Lorbiecki 2016; Warren 2016). The Leopold family would often make hunting trips to the preserve (Leopold 2016). Leopold also worked with private landowners producing game east of Madison near Lake Mills, in Jefferson County. In 1933, the farmers formed the 2,400-acre Faville Grove Wildlife Experiment Area (McCabe 1978; Meine 1988; Court 2012).

When FDR established the U.S. Soil Erosion Service in 1933 (SES, later expanded and renamed the Soil Conservation Service in 1935, and the Natural Resources Conservation Service in 1994), the Coon Creek watershed near La Crosse in west-central Wisconsin was selected as the inaugural soil conservation and watershed restoration demonstration area in the U.S. The Coon Valley project area included a 22-mile segment of the river and extended across 92,000 acres and three counties. The pioneering, community-wide program was a collaboration between the SES, farmers, scientists and technicians, a CCC camp, and others. Leopold joined the project at its inception as an extension advisor, supervising on wildlife management (Leopold 1935a; cf. Brown and Carmony 1995; Court 2012; Lorbiecki 2016; Warren 2016 on the history of the Coon Valley demonstration area). Soil and water conservation were a frequent focus of Leopold's conservation writings from the 1920s onward (included in Leopold 1991; Brown and Carmony 1995; Meine and Knight 1999). The subtitle of Leopold's May 1935 article in *American Forests* describes the five-year Coon Valley project as 'An Adventure in Cooperative Conservation' (Leopold 1935a). Leopold's two eldest sons, Starker and Luna, also worked on the project (Meine 1988; Lorbiecki 2016).

In addition to directing on-the-ground restoration programs, Leopold was also invited to contribute to restoration policymaking. In January 1934, he was appointed to the three-member President's Committee on Wild Life Restoration, serving with Thomas Beck and Jay 'Ding' Darling (Meine 1988; Flader 1994; Court 2012; Lorbiecki 2016; Warren 2016). The committee submitted their restoration report to FDR—a National Plan for Wildlife Restoration—the following month (Meine 1988). Throughout his career, Leopold would advise, serve on, and chair dozens of federal, state, local, and university committees, boards, commissions, organizations, and agencies (cataloged throughout Meine 1988; Flader 1994).

Several federal laws would echo Leopold's wildlife conservation and resource management concepts, principles, procedures, and philosophies—including the 1934 Fish and Wildlife Coordination Act, the landmark 1969 National Environmental Policy Act, the 1973 Endangered Species Act, the 1974 Forest and Rangelands Resources Planning Act, the 1976 National Forest Management Act, the 1976 Federal Land Policy and Management Act, the 1985 Food Security Act, and many more.

1935 emerged as a pivotal year in the development and intersection of Leopold's wilderness *and* restoration philosophies—in January, Leopold cofounded the Wilderness Society with Robert Marshall and others; in May, he closed on the purchase of an abandoned sand farm in Sauk County, Wisconsin, a weekend retreat for his family and the setting for many of the essays in *A Sand County Almanac* (Leopold 1968); and in the fall, he spent three months in Germany on his first (and only) transatlantic trip, to study German forestry and wildlife management practices on a Carl Schurz fellowship.

4.4 A Restoration Experiment at 'the Shack'

Four paragraphs on from Leopold's paean to wilderness and 'wild things' in the Foreword to *A Sand County Almanac* (Leopold 1968) is a subtle directive for restorationists. On the importance of restoring a personal, familial, intimate correspondence with place. It comes not from Leopold's work at UW-Madison, or before that, with the USFS, but from the weekends his family spent at their farm—'On this sand farm in Wisconsin, first worn out and then abandoned by our bigger-and-better society, we try to rebuild, with shovel and axe, what we are losing everywhere. It is here that we seek—and still find—our meat from God' (Leopold 1968: viii). This attention to rebuilding—and *restor(y)ing*—at the Leopold farm property (often hereafter, just 'the Shack') permeates and is central to the almanac section of *A Sand County Almanac*, the book's first and longest section.

Leopold had visited the sand counties of central Wisconsin on hunting and fishing trips during his first years in Madison, and had surveyed the region as part of his *Report on a Game Survey of the North Central States* (Leopold 1931), but it was through his extension work at the university that he formed the strongest correspondences with the area. One day after his 48th birthday, on January 12, 1935, Leopold and friend Ed Ochsner were driving in the Baraboo Hills, north of Madison, searching for farmland for Leopold to lease near the Wisconsin River. What they found was an abandoned sand farm on the floodplain of the Wisconsin River, north of Baraboo, in Sauk County, south-central Wisconsin. After initially leasing the 80 acres of abandoned farmland, Leopold closed on the purchase on May 17, 1935, for $8.00 an acre (Meine 1988; Lannoo 2010; Laubach 2014; Leopold 2016; Lorbiecki 2016). Another 40 acres were added later. The farm was 50 miles northwest of the Leopold family home on 2222 Van Hise Avenue in Madison. This is the Central Sands—or 'sand county'—landscape Leopold would become most closely associated with. The isolated farm property lay just beyond the eastern edge of the Driftless Area, a region of the Upper Midwest (spanning southeastern Minnesota, southwestern Wisconsin, northeastern Iowa, and northwestern Illinois) bypassed by glaciers during the last ice age, and devoid of glacial drift, preserving its rugged topography (see e.g. 'Smoky Gold,' 'Red Lanterns,' 'Marshland Elegy,' 'The Sand Counties' in Leopold 1968, also Flader 1973, 1987; Ross and Ross 1998; Laubach 2014; cf. Meine and Keeley 2017 on the histories and stories of the Driftless Area).

While less than 20 miles as the crow flies separate the Leopold farm from the Muirs' Fountain Lake Farm in Marquette County to the northeast, the intervening 85 years harbored a dramatic decline in the condition of the land. Where Muir (2018) had earlier exulted the unsettled and uncultivated 'glorious Wisconsin wilderness!' surrounding Fountain Lake Farm, Leopold (1968) offers a sharp contrast in his 'The Sand Counties' essay in *A Sand County Almanac*, which describes the sand counties as economically and ecologically poor. The Muirs' pioneer farmstead was untrammeled wilderness. The Leopold farm was depleted and abandoned. Yet it was *because of* the poor condition that Leopold purchased the sand farm (Leopold 1968). It was a humble landscape, where the family could live simply and deliberately (to borrow phrasing from Thoreau 2004). Leopold wanted to observe how the principles laid down in *Game Management* (Leopold 1986), and at the Arboretum, might play out 'on the back forty.'

It was in the early spring of 1935 that Leopold and his wife Estella B. Leopold first brought their three sons and two daughters—22-year-old Starker, 19-year-old Luna, 18-year-old Nina, 16-year-old Carl, and eight-year-old Estella Jr.—to the farm, beginning a tradition that would come to define the next dozen or so years of weekends and vacations. Carl Leopold soon became the Shack photographer, documenting the family's labors on the farmstead. The Dust Bowl droughts and the Second World War stand as silent sentinels at either end of this period.

Leopold had initially intended to buy the farm as a hunting camp, as a place to hunt deer and ducks, to fish, and for his family to practice archery (he crafted bows and arrows in the basement workshop of their Madison home). Five years earlier, Leopold had purchased a small cabin in the Ozarks in Missouri for quail hunts. But very quickly, the farm became the place where Leopold could practice ecological restoration on his own land. A half-century of poor farming practices on the farmstead had diminished the vegetation and soils. The open land disclosed little more than fallow, corned-out fields, with sand blows on the hills. Most white pine stands had been logged by the end of the nineteenth century. The only trace of the farmhouse, burned down by the previous owner, was part of its foundations. Fences along the boundary line had collapsed (Fig. 4.3).

Leopold's (1939a) essay, 'The Farmer as a Conservationist,' opens with the summary statement, 'Conservation means harmony between men and land,' and on the sand farm, Leopold-the-rural-landowner could translate his idea from page to place. Leopold adopted the credo he set down in 'Game Cropping in Southern Wisconsin:' 'Just now, conservation is short of doers. We need plants and birds and trees restored to ten thousand farms, not merely to a few paltry reservations' (Leopold 1927, also in Meine and Knight 1999: 121). It was on this 120-acre farm tract that Leopold developed and articulated his vision of 'land health,' and a 'land ethic' (Leopold 1968, also Leopold 2016; cf. Root 2017). Like the UW Arboretum, dedicated six months before Leopold and Ochsner found the farm, the farm provided instruction in 'reconstructing' the land (cf. Leopold 1934a, b).

Fig. 4.3 Leopold at the Shack, near Baraboo, Wisconsin. Photograph by Carl Leopold, 1936, courtesy of the Aldo Leopold Foundation and University of Wisconsin-Madison Archives

Across Leopold's canon, 'reconstructing,' 'rebuilding,' 'regenerating,' and 'restocking' appear alongside, and are interchangeable with, 'restoration' (on ecological (and game, wildlife) restoration in Leopold's conservation philosophy, see Brown and Carmony 1995; Leopold 1999; Meine and Knight 1999; Zedler 1999; Lorbiecki 2016; Meine 2017, also Woodworth 2013 on 'the possibilities and quandaries of restoration,' with connections to Leopold). As Leopold's youngest daughter Estella B. Leopold Jr. (2016) notes in *Stories from the Leopold Shack: Sand County Revisited*, weekend restoration work at the Shack also proved a 'reciprocal exercise.'

4.4.1 Restoring the Leopold Shack

One of the first tasks for the family was remodeling the abandoned chicken coop, half-filled with frozen manure, into a habitable barn. The manure was excavated with pickaxe and shovel, and used to map out garden plots. At the start of spring 1935, Leopold and his two eldest sons built a fireplace (substantially remodeled in 1936), and in April, Leopold and his sons began to expand the barn—already called 'the Shack' by the family—by attaching a bunkhouse wing (Leopold 2016). The Shack, much like Thoreau's Walden cabin, and Muir's Yosemite Valley cabin, was very much of the land. And of its time—the Shack repairs were undertaken during the Depression. Unlike Thoreau's and Muir's cabins, the original Leopold Shack still stands. The Leopold Shack and Farm National Historic Landmark was designated in 2009.

Many of the beams and boards for rebuilding the Shack were driftwood salvaged from the Wisconsin River. Windows and a door (and more lumber) came from the Madison city dump (screens were added after an aggressive mosquito tempest). Sand from river bars was used to mix cement. Red clay from a road cut was used to build the clay floor (later topped by white pine floorboards). In the early-1940s, the tarpaper roof was replaced with cedar shingles (Leopold 2016). Salvaged lumber was also used to build indoor and outdoor furniture at the Shack. Carpentry projects at the Shack continued a strong woodworking tradition in the Leopold family—Leopold's father was president of the Rand and Leopold Desk Company (later the Leopold Desk Company) in Burlington. Plank benches were built along the back and side walls inside the Shack. A series of bunk beds spanned the new wing. A collection of stools and tables were crafted. A hand pump and fire pit were built in the front yard.

Other weekend projects included building trails through the woods. An early bench of driftwood cottonwood boards was built against an old elm in front of the Shack, and other benches were similarly built into the landscape, including another cottonwood bench built against two elms on 'Slide Hill' above the Shack overlooking the slough. Leopold meanwhile began to construct a series of simple, practical, portable benches—now recognized as Leopold Benches (see description in Leopold 2016)—that were placed in the yard, along trails, and elsewhere across the Shack property. In March 1936, Starker Leopold built an outhouse, referred to by the family as the Parthenon (as Leopold 2016). Trees were later felled for firewood, with the woodpile stacked between two elms next to the Shack.

That first spring, a neighboring farmer, Mr. Gilbert, plowed a garden plot on the east side of the Shack for the family, which was planted with sorghum and hemp as winter food for birds, then potatoes. The following year, another farm neighbor, Mr. Webster, plowed another garden plot west of the Shack, on the gentle eastern slope of Sand Hill, in the apple orchard inherited from the previous farm owners (Leopold 2016). The orchard vegetable garden produced potatoes, onions, corn, tomatoes, and a mulberry tree.

* * *

The Shack model and philosophy has been replicated by each of Leopold's five children across the western U.S. And by later Leopold generations (see Leopold 2016; Lorbiecki 2016; Freeman 2018). Each time, an ecological restoration rhetoric scaffolds the project.

4.4.2 Restoring Oak Savanna, Marshland, and Forest on the Shack Property

Next, attention turned to the land. As Leopold (1993: 165) notes in 'The Round River—A Parable,' 'One of the penalties of an ecological education is that one lives alone in a world of wounds.' The family started projects to triage and heal some of those ecological wounds.

During the first summer at the farm, the family planted out the Shack yard. With the Shack overhaul and garden plantings in progress, and influenced by his work at the university arboretum, Leopold began to consider the restoration of the larger farm property. The historic landscape of central Wisconsin—pre-farm disturbance—was characterized by oak and grass savanna (cf. 'oak openings' in Muir 2018), and Leopold developed a restoration program that fortified and celebrated oak openings and prairie. As Leopold (1993: 147) argues, 'To keep every cog and wheel is the first precaution of intelligent tinkering.' Leopold's ideas on restoring 'land health,' developed at the Shack, correspond with a restoration *sensu lato* goal (Aronson et al. 1993), and he reforested woodland stands and restored prairie on the corned-out fields. Leopold used 'ecological landmarks' (after 'The Ecological Conscience' in Leopold 1991) to inform the reference model at the Shack. As Leopold (2016) notes, the reference model was also guided by Asa Gray's (1848) *A Manual of the Botany of the Northern United States*, and Fassett's (1931) *Spring Flora of Wisconsin*, Leopold's colleague at the Arboretum.

Leopold's ecological restoration program was in two parts. The first part introduced a spring planting program to reestablish the white pine stands that once grew along the Wisconsin River. By the 1930s, the woods contained soft maple, birch, linden, and elm (Leopold 2016). Beginning in April 1936, the Leopold family hosted an annual pine planting, with seedlings ordered from the Wisconsin Conservation District. The pine planting coincided with UW-Madison's spring break each year, and family, friends, and students were invited to join the planting (Leopold also ran Wildlife Ecology 118 field classes at the Shack, as Leopold 2016). In the first year, 1,000 white pine seedlings, and 1,000 Norway or red pine seedlings were planted, along with shrubs including mountain ash, juneberry, nannyberry, cranberry, raspberry, and plum (Meine 1988; Leopold 2016). The first planting season centered on planting pines along the northern margin of the former cornfield, which included Birch Row, southeast of the Shack. The view was kept open southeastward from the Shack across the cornfield toward the marsh (see 'The Deer Swath' in Leopold 1993, also Leopold 2016). Almost all the pine saplings from the first planting season were killed by drought (Meine 1988; Leopold 2016; Freeman 2018). But planting resumed the following year.

Subsequent spring plantings increased sometimes to 2,000 or 3,000 white pine, 1,000 red pine, and 1,000 jack pine, more mountain ash and raspberry, and cedar, juniper, aspen, dogwood, crabapple, witch-hazel, and dewberry (Meine 1988; Leopold 2016; Aldo Leopold Foundation

Fig. 4.4 Pine plantings at the Shack. Left: Aldo and Estella Bergere Leopold planting pines. Right: Leopold examining red pines near the Shack. Photograph by Robert McCabe, 1946, courtesy of the Aldo Leopold Foundation and University of Wisconsin-Madison Archives

2020c).[8] The plantings moved around the farm property—with plantations on top of the clay hill southwest of the Shack near Lake Chapman, on the sand hill to the west, along the row of elms south of the Shack, at the property's eastern boundary, and near the river, in the riverine woods. Tens of thousands of pine trees were planted at the Leopold Shack and farm in the 1930s and 1940s (Fig. 4.4).

The pine plantations[9] were interspersed with native hardwood plantings—hickory nuts, walnuts, and acorns were planted amidst the stands, with sumac and other bushes planted on the periphery. Wildflowers were also added to the woods. The Leopolds also planted hard, or sugar, maples—one in the front yard of the Shack, others above the orchard, near the old farmhouse foundation. The family attempted to plant tamarack in a swamp of alder, river birch, and soft maple at base of Sand Hill, west of the old row of elms, but with limited success (as Leopold 2016).

[8] ALF online interview, October 2, 2020.

[9] It was when Leopold cut fire lanes to protect the pine plantations that he discovered prairie grasses flourishing on the newly burned-over fire lanes (decades later, controlled prairie burns would be introduced into the Leopold Memorial Reserve management plan, coinciding with burns on the Curtis Prairie, see Curtis 1959, also Leopold 2016).

In the shovel and the axe, Leopold sees the capacity to create (or restore) and destroy plant communities—one tool will plant a tree, one will cut one down. And yet it is the axe, according to Leopold, that also best defines the conservationist (and more so than the pen). As Leopold (1968: 68) argues in the almanac's second November essay, 'Axe-in-Hand,' 'It is a matter of what a man thinks about while chopping, or while deciding what to chop. A conservationist is one who is humbly aware that with each stroke he is writing his signature on the face of his land. Signatures of course differ, whether written with axe or pen, and this is as it should be.' The shovel–axe diptych plays out in the convergence of white pine and red birch on the property. Where the two trees compete, Leopold admits that he always fells the birch to protect the pine, and in an effort to deconstruct this 'plant bias,' concedes that, 'I love all trees, but I am in love with pines' (Leopold 1968: 70). But part of this is an 'underdog bias,' too, as white pine is scarce in the township (cf. pitch pine and white pine on Brister's Hill in Walden in Chapter 2). Yet revealed in this 'plant bias' is a 'sensitive index to our affections, our tastes, our loyalties, our generosities, and our manner of wasting weekends' (Leopold 1968: 72). Loyalties and generosities, which, in Leopold's case, extend to the championing of aspen, tamarack, and cottonwood, as well as wahoo, red dogwood, prickly ash, hazel, and bittersweet, where his neighbors might see only brush (Leopold 2016).

The second part of Leopold's restoration program, a summer planting program, focused on restoring a prairie on the former cornfield. Leopold wanted to build up a living collection, a 'living exhibit' (Leopold 1934a), of plants native to south-central Wisconsin on the Shack's restored prairie and savanna. Although not writing about the Shack farm in his 'Conservation' essay, the importance of his prairie restoration task is clear: 'Some day we may need this prairie flora not only to look at but to rebuild the wasting soil of prairie farms. Many species may then be missing. We have our hearts in the right place, but we do not yet recognize the small cogs and wheels' (Leopold 1993: 148).

Already present among the old corn stalks on the experimental prairie site were a number of weedy plants and grasses. Leopold began to scout out and explore local 'idle spots' (as 'Prairie Birthday' in Leopold

1968)—beside railroad tracks, along old road cuts, and in cemetery edgelands—in search of native wildflower patches. No commercial source for native prairie species existed in the 1930s, so Leopold learned the local flora in situ. He dug up prairie wildflowers, as well as prairie grasses, legumes, asters, and perennial prairie herb species for transplant, or collected seeds, for the Shack prairie restoration (Leopold 2016; Aldo Leopold Foundation 2020c).[10] Wildflower propagation and prairie seeding were central to the prairie restoration.[11] For his bachelor's dissertation in botany, Carl Leopold produced an inventory and herbarium of plant species on the Shack land—and it is now stored in the Leopold Center, in a bespoke cypress cabinet he crafted (see Leopold 2016). The Shack prairie restoration is now more than eight decades old, and according to the Aldo Leopold Foundation (2020c) is 'the 2nd oldest prairie restoration in the world.' As a sand county restorationist notes of the Shack prairie, 'we have made additions of species into the Shack prairie—we always pick species that they didn't pick, and so we knew what they did, and we knew what we did. But that was more beautification and enhancement than anything'[12] (Figs. 4.5 and 4.6).

Between 1940 and 1948, Leopold expanded the Shack property. By 1948, the farm property totaled 350 acres (Leopold 2016). On April 14, 1948, the same day Leopold learned that Oxford University Press would publish *Great Possessions* (now known as *A Sand County Almanac*), and just a week before his death, Leopold wrote to Ernest Swift, director of the Wisconsin Conservation Department, to propose the acquisition of Fountain Lake Farm, the boyhood home of John Muir, as a state park—and called for 'restor[ing] the flora' (in Meine 1988; Simpson 2003; Root 2017).

The work Leopold and his family undertook on the farm was one of the earliest tests, experiments in ecological restoration. A remark Leopold

[10] An experiment atop Sand Hill, just west of the orchard—where prairie seed was planted in a one-meter-square quadrant—confirmed that sowing by seed could be successful, and in pure sand (Leopold 2016). Estella B. Leopold Jr. (2016) writes of her later regret at another experiment—Leopold planted dune plant and sweet fern, gifted by graduate student Fran Hamerstrom, on the sand blow, but both species spread, and now sand country restorationists are working to reverse the plant cover and restore the open sand blow.

[11] ALF online interview, October 2, 2020.

[12] ALF online interview, October 2, 2020.

Fig. 4.5 Shack prairie in summer, ca. 1947. Courtesy of the Aldo Leopold Foundation and University of Wisconsin-Madison Archives

Fig. 4.6 A Shack landscape panorama. Photograph by Carl Leopold, 1939, courtesy of the Aldo Leopold Foundation and University of Wisconsin-Madison Archives

(1993: 145) makes in his 'Conservation' essay about game conservation reform could just as easily be applied to his sand farm restorations: 'The job we thought would take five years will barely be started in fifty.' But the Shack venture bestowed another ecological legacy—all five Leopold siblings had ecology or conservation careers, working in universities and federal agencies, and Starker, Luna, and Estella Jr. were elected to the

National Academy of Sciences (on the legacy of the Leopold siblings, see Leopold 2016, also Meine 1988; Lorbiecki 2016).

4.4.3 A Sand County Phenology

Leopold wrote the first entry in the Shack field journal on April 27, 1935. Throughout the 1930s and 1940s, Leopold and his family compiled extensive phenological records, field observations, and other scientific data of the Shack's sand county environs. On summer mornings at the Shack, Leopold would rise before dawn and sit on a bench in the front yard to record the times and light levels as different birds joined the dawn chorus (see Leopold and Eynon 1961). Transferring field observations from a small notebook to the Shack journal was the only writing Leopold did at the Shack (excerpts from the Shack journals, 1935–1948, are included in Leopold 2013). All other writing was done back in Madison, either at home, or in his office on campus.

Revealed in the Shack journals—and later in Leopold's (1968) *A Sand County Almanac*—is not just an intimacy with the oak openings and prairie across the Shack property, but also the privilege of exposure. Through the 22 seasonal essays collected in the opening almanac section of *A Sand County Almanac*, Leopold (1968) shares a biography of place—of his sand county farm, and the 'land community' (the almanac structure is repeated in Ross and Ross 1998; Leopold 2016). These short essays offer descriptive field sketches of the farm property, stories, activisms, testimonies, and ecological cautionary tales, to encourage care of the land. His phenological records underpin or are otherwise quietly present in many of the vignettes. 'Good Oak' condenses and aligns eight decades of Wisconsin history with the transect of tree rings as the family collectively fell an oak tree killed by lightning. 'Come High Water' tells of being marooned by flooding. 'Sky Dance' recounts the drama of woodcock peentings. '65290' describes a returning chickadee banded by the family. Leopold's (1968: 30, also 1999) assertion that, 'Thus, he who owns a veteran bur oak owns more than a tree. He owns a historical library, and a reserved seat in the theater of evolution' could equally be expanded to the farmstead. Restorationists can browse historic library stacks of the Shack in Leopold's meticulous phenological records.

4.5 Restoration and *A Sand County Almanac*'s 'The Land Ethic' Essay

A Sand County Almanac, and Sketches Here and There was composed and compiled during the last decade of Leopold's life, and published posthumously. But it was not always titled so. When Leopold submitted the finished manuscript to Oxford University Press in 1947, it bore the title, *Great Possessions* (echoing the title of the first July essay in the almanac). It was Leopold's second eldest son, Luna (who features in Chapters 3, 5, and 6), who oversaw the change in title and guided the book to publication after Leopold's sudden death in April 1948. But the ephemeral, momentary *Great Possessions* title is fascinating for restorationists, for it speaks back to ideas of stewardship and ecological care, and amplifies in a different way the central themes of the book—ecology, aesthetics, and ethics—that the *A Sand County Almanac* title achieves through motifs of place-making and place attachment. But *Great Possessions* was in fact the third working title Leopold assigned to the book. Leopold had earlier considered *Marshland Elegy—And Other Essays*, then *Thinking Like a Mountain—And Other Essays* (on the history and chronology of *A Sand County Almanac*, see Ribbens 1987; Meine 1988, 2002, 2004).

But the 'Sand County' motif is also interesting for restorationists, for there is no Sand County in Wisconsin. Another early title for the almanac section of the volume was 'A Sauk County Almanac' (see Meine 1988, 2004; Lannoo 2010), which is found in Wisconsin. But this 'Sand County' motif—as with a 'Thoreau Country' (Chapter 2) or 'Abbey's Country' (Chapter 6) motif—serves to intimately align and anchor author with place, place with author.

A Sand County Almanac is organized across three sections—it opens with the almanac centered on seasonal and historical descriptions and stories of the Shack and its environs, then moves to a series of essays, sketches, and episodes from across the U.S. that collectively speak back to conservation ideas, before closing with a quartet of essays, labeled

'The Upshot,' which introduce and consolidate Leopold's assessments of and statements on conservation science and philosophy, wilderness preservation, and land health and a land ethic. Many of the essays were published elsewhere first, often as conservation essays for farmers in the *Wisconsin Agriculturalist and Farmer*, and the *Wisconsin Conservation Bulletin* (Ribbens 1987; Meine 1988). There is a progression within and across these sections from reflective and experiential pieces to more philosophical pieces (for critical commentaries on *A Sand County Almanac*, see Callicott 1987), and sit in contrast to Leopold's earlier technical essays of the 1920s and 1930s. It is the last of the essays presented in 'The Upshot'—'The Land Ethic'—that offers the greatest provocation for restorationists.

'The Land Ethic' essay is in many ways the culmination of *A Sand County Almanac* (Leopold 1968). Not just superficially in terms of its location in the book, but conceptually, theoretically, and philosophically. All other vignettes and essays in the slender volume quietly build to this declaration (as, indeed, does the rest of Leopold's canon). Yet its final placement was an editorial decision by Luna Leopold (Callicott 1987; Meine 1988). 'The Land Ethic' was written during the last year of Leopold's life and is a compilation—and expansion—of three previous essays: 'The Conservation Ethic' (Leopold 1933), 'A Biotic View of the Land' (Leopold 1939b), and 'The Ecological Conscience' (Leopold 1947) (for a chronology and critical exposition of 'The Land Ethic,' see essays in Callicott 1987; Meine 1987, 2004; Tanner 1987, and on the legacy of Leopold's land ethic in environmental ethics, see Callicott 1989, 1999a, 2013).

On at least two occasions in the essay, Leopold observes that there is not yet a land ethic, an ethic connecting society with the land (Leopold 1968). *A Sand County Almanac* is bookended by two connected statements that powerfully unpack and illustrate Leopold's moral precept. The Foreword argues that, 'We abuse land because we regard it as a commodity belonging to us. When we see land as a community to which we belong, we may begin to use it with love and respect. [...] That land is a community is the basic concept of ecology, but that land is to be loved and respected is an extension of ethics' (Leopold 1968: viii–ix), for 'The Land Ethic' then to instruct, 'A thing is right when it tends to preserve the integrity, stability, and beauty of the biotic community. It is wrong when it tends otherwise' (Leopold 1968: 224–225, cf. 1947: 52). The interim builds a carefully crafted case for a land ethic (Fig. 4.7).

'The Land Ethic' discloses Leopold's (1968) conviction on the ethics of conservation and casts conservation as a moral, ethical, and aesthetic issue. Absent from a conservation rhetoric, Leopold argues, is an 'ecological conscience,' that is, individual responsibility in society's relationship with land (cf. Meine 1988). For Leopold, society is part of the 'land community'—with moral responsibilities toward the nonhuman, more-than-human community. Leopold advocates the expansion, or recasting, of 'community,' and of citizenship, and civic responsibility (on Leopold and land ethics, see Meine and Knight 1999; cf. 'civic environmentalism' in Haglund and Still 2005). Tied to this, land health becomes a moral duty, and the endgame of conservation (see collected land health essays in Leopold 1999). Restorationists find in Leopold's land ethic instruction

Fig. 4.7 Aldo Leopold quotation in the Reflection Circle of Thoreau's Path on Brister's Hill, in Walden Woods, Concord, Massachusetts. 'That land is a community...' Photograph by the author, June 2007

in the moral and ethical ecologies of conservation, and, by extension, restoration. As a sand county restorationist notes, 'the biotic community is, both philosophically and practically, the gateway to ecological restoration'[13].

4.6 Building a Land Ethic at the Leopold Shack Property

On April 21, 1948 (on what would also have been Muir's 110th birthday), Leopold died aged 61 after suffering a heart attack while fighting a grass fire on a neighbor's farm. Leopold, Estella, and their youngest daughter Estella Jr. had traveled to the Shack five days earlier for their annual spring planting trip, and to allow Leopold to recuperate from surgery for trigeminal neuralgia (Meine 1988; Lorbiecki 2016; Warren 2016). At the time of Leopold's death, more than 8,000 of the 16,000 pines that he and his family had planted since 1935 had taken hold, and the prairie restoration was established and flourishing (Lannoo 2010).[14]

One month after Leopold's death, on May 21, 1948, Estella Bergere Leopold, and her daughters Nina and Estella Jr., returned to the Shack for the weekend—and resumed the phenological study, adding birdsong, bloom dates, and other observations to the Shack journal (Leopold 2016). Leopold's restoration legacy persists on the sand county farmlands.

4.6.1 The Leopold Memorial Reserve, a Story of Private Land Conservation Around the Shack

Tom and Catherine Coleman, Madison friends of Aldo and Estella Bergere Leopold, purchased an adjoining tract of land south of the

[13] SCF online interview, July 24, 2020.

[14] ALF online interview, October 2, 2020.

Shack in 1937, near Lake Chapman, with views across the great marsh and floodplain. In the mid-1960s, their son Reed Coleman (and the Leopolds' godson), now executor of his family's land, was looking to expand his family's landholding south of the River Road (now Levee Road). But Coleman also wanted to support Leopold's legacy and the land they shared.[15] In March 1965, Coleman and friend and colleague Howard Mead sought to protect the land surrounding the 200-acre Leopold Shack and farm property—through a voluntary commitment between a coalition of private landowners to restrict development, similar to Riley, Faville Grove, and Coon Valley (Laubach 2014; Leopold 2016). The old Gilbert farm had earlier been subdivided for summer homes, and the construction of Interstate 90/94 through south-central Wisconsin in 1962 further opened up the area to developers. Leopold was an advocate of private land conservation and regarded it as the leading conservation challenge of his generation, arguing that, 'Conservation can accomplish its objectives only when it springs from an impelling conviction on the part of private land owners' (qtd. in Meine and Knight 1999: 166; cf. Knight and Riedel 2002 on Leopold and conservation on public and private lands).

With encouragement from Estella Bergere Leopold to protect the Shack and the land around it,[16] Coleman next approached neighbor Frank Terbilcox. The early scaffolding of a coalition was soon in place—the Coleman, Leopold, and Terbilcox families, together with any other interested landowners, would enter into a cooperative conservation agreement on land stewardship and responsible ownership. Land management decisions would be anchored to Leopold's ideas on land health. The group assigned the environmental nonprofit Louis R. Head Foundation, recently established by Coleman and his mother Catherine, to oversee the guiding principles for the reserve drawn up by Coleman, Mead, and Terbilcox (Laubach 2014).

The Leopold Memorial Reserve (LMR) was created in December 1967. Original signatories included the Coleman and Terbilcox families, and neighboring landowners the Anchor and Van Hoosen families. The

[15] SCF online interview, July 24, 2020.

[16] SCF online interview, July 24, 2020.

Leopold family formed a trust—the Aldo Leopold Sand County Trust—and joined the agreement seven months later, in June 1968 (Laubach 2014). The agreement introduced restrictions on landowner activities—signatories agreed to restrict building construction and wetland draining on the property, and the Head Foundation was granted first refusal of any land sale from members (Laubach 2014; Leopold 2016). The 900-acre reserve was jointly owned by five families. During the reserve's first year, the Head Foundation purchased the Turner and Kammerer lands to form a buffer between the LMR and development, with more acquisitions in later years (Laubach 2014). The LMR would eventually expand to 1,600 acres of private land around the Shack (on the history of the LMR, see Laubach 2014, also C. Bradley 1987; Ross and Ross 1998; Leopold 2016; Root 2017) (Fig. 4.8).

In January 1968, the Head Foundation hired Terbilcox as the land manager of the LMR (a post he held until his retirement in 1993). The Head Foundation also hired Robert Ellarson, UW professor of wildlife ecology and former student of Leopold, to write the reserve's first management plan (Laubach 2014). Ellarson would remain involved with the reserve as a consultant. The management plan largely echoed Leopold's methods, to focus on prairie restoration, wetland and flood-plain restoration, controlled burns, and wildlife management, as well as invasive species control, seed harvesting, management of timber stands, and food-patch plantings. Terbilcox also introduced a number of artificial waterfowl ponds (Laubach 2014).

Early in the history of the LMR, two nearby tracts of land owned by Wisconsin Power and Light (WPL) were being explored as possible locations for a new coal-fired power plant. The Head Foundation assembled a scientific advisory team led by UW wildlife ecologist Joseph J. Hickey to consult with the regional power company on alternative sites. When WPL relocated the project, the Head Foundation worked with the company on a management plan for the undeveloped wetlands bordering the new location (Laubach 2014)[17].

As a sand county restorationist notes, of the 200 acres now owned by the Leopold family, the part of the property most heavily influenced by Leopold and his family, the 'cultural legacy,' extends across

[17] SCF online interview, July 24, 2020.

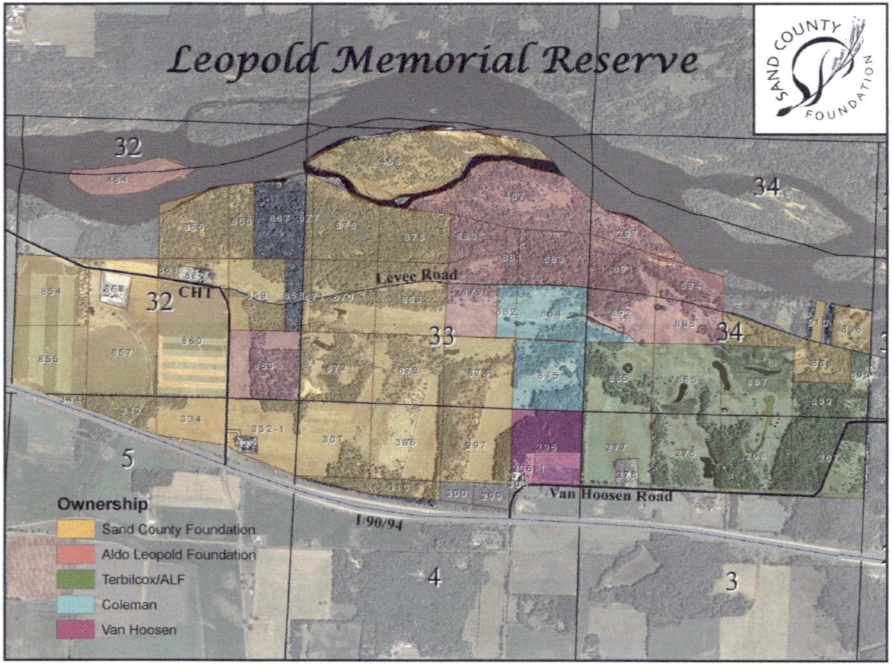

Fig. 4.8 Land ownership of the Leopold Memorial Reserve, Baraboo, Wisconsin. Courtesy of Brent M. Haglund, Ph.D./Sand County Foundation

20 or 30 acres—the Shack, the pine plantations, the prairie restoration, the orchard.[18] Enveloping the Leopold land is 600 acres owned by the Aldo Leopold Foundation, and about 1,000 acres owned by the Sand County Foundation (until 2015)—making up the LMR. That land parcel also forms one of the tracts in the 16,000-acre Leopold-Pine Island Important Bird Area (IBA), alongside other private, state, and federally-owned lands, designated in 2005 (Laubach 2014; Lorbiecki 2016).[19] The Leopold-Pine Island IBA is another cooperative conservation program—organized around bird conservation. Leopold's restoration legacy is variously 'restor(y)ed' through overlapping land conservation designations.

4.6.2 Leopold's Restoration Legacy, and the Aldo Leopold Foundation

In the mid-1970s, conservationist Nina Leopold Bradley and geologist Charles Crane Bradley[20] retired near the Leopold Shack. The Bradleys rented five acres of land in the LMR from the Head Foundation, a mile west of the Shack, and constructed a log house in 1976. Timber was sourced from the Leopold pine plantations, particularly jack pine, planted by the family in the 1930s and 1940s.[21] The Bradley home soon became known as the Bradley Study Center (Leopold 2016). Nina and Charles Bradley worked out of a walkout basement, and their home also featured a greenhouse, office space, and a large garden with vegetable, fruit, and flower beds.

The Bradley Study Center became a hub for Leopold family gatherings (although the Shack is still used by the Leopold family for private events) and was the early headquarters of the Aldo Leopold Shack Foundation in the early-1980s.[22] The Bradleys launched the Leopold Fellows

---

[18] ALF online interview, October 2, 2020.

[19] ALF online interview, October 2, 2020.

[20] Bradley, with his father and brother, had been some of the earliest conservationists to draw attention to the proposed dam projects in Dinosaur National Monument on the Colorado-Utah border, explored in Chapter 6.

[21] ALF online interview, October 2, 2020.

[22] The Aldo Leopold Foundation purchased the Bradley Study Center building from the Sand County Foundation in 2004. Following Nina's death in May 2011, it is now the home of

Program in 1978 through the Head Foundation, working with graduate students on research projects and hosting a weekly summer seminar series (Laubach 2014). The headquarters of the International Crane Foundation (ICF), established in 1973, is also located close to the Shack property. The Bradleys (and ICF) established a plant nursery near the Bradley Study Center for nearly five dozen rare prairie species, with financial support from the Environmental Protection Agency (EPA) and Charles Bradley's family Friendship Foundation (Leopold 2016). Seeds were shared with restoration ecologists in Sauk County and elsewhere.

Back amidst the ecological theater of the sand farm, Nina Leopold Bradley (and Charles Bradley) resumed her family's meticulous phenological studies after a lapse of 29 years, keeping detailed records of about 300 events.[23] When Bradley retired, the task passed to a staffer at the Aldo Leopold Foundation and continues today (the phenological records of both Leopold and Nina Leopold Bradley have featured in several scientific studies, including Leopold and Jones 1947; Leopold and Eynon 1961; Liegel 1982, 1988; Ross and Ross 1998; Bradley et al. 1999; Aldo Leopold Foundation 2020b; cf. Ellwood et al. 2013; Primack 2014 and Thoreau's Walden phenology). The Bradleys began to plant on Leopold family land, and on some of the land rented from the Head Foundation. Charles Bradley also purchased land southwest of the Bradley Study Center, comprising alfalfa fields, black oak, red oak, shagbark hickory, walnut, and white oak (Leopold 2016).

Nina and Charles Bradley introduced two prairie restorations onto the land surrounding the Bradley Study Center. The first, in the fall of 1976, was an experimental prairie restoration that used the spoil pile from a new pond excavation (Leopold 2016).[24] The excavated material was spread onto a worn-out agricultural field, and the Bradleys 'collected seeds from prairie plants, [and] brought in local genotypes of prairie

---

the Aldo Leopold Foundation's second executive director (and former seasonal fellow) Buddy Huffaker and his family.

[23] ALF online interview, October 2, 2020.

[24] ALF online interview, October 2, 2020.

plants'[25] to plant in the soil. More than two dozen species of native prairie seeds were initially collected, and more species were added later. The restored prairie features *Silphium laciniatum*, the compass plant, the focus of the second July almanac essay, 'Prairie Birthday' (Leopold 1968).[26] Annual oat seed was also added to the prairie seeding (N. Bradley 1987; Leopold 2016). The prairie is known as the EBL Prairie to honor the Leopold matriarch and is an 'ecological landmark' on the LMR (Leopold 2016). The second prairie restoration, known as the Too Bare, or Two Bears, Prairie (on the wordplay, see Leopold 2016), looked at introducing prairie on land free of weeds. Building on the experiences on the EBL Prairie, the tract was prepared with a glyphosate spraying in the summer (before environmental toxicity concerns emerged), and then prairie seed was sown in the fall (N. Bradley 1987). The result was what Leopold (2016) describes as 'instant prairie,' and after five years, the site supported prairie cover with no weed, as root competition from prairie plants overwhelmed the weeds.

During the same period, a team including Nina's niece Susan Leopold (Carl Leopold's daughter), a Leopold Fellow, restored a four-acre tract of relict savanna midway between the Shack and the Bradley Study Center. The parcel comprised native grasses and wildflowers, and swamp white oak trees, but it was being invaded by prickly ash. The team introduced a restoration program to cut down the prickly ash, and began a series of prescribed burns to stifle its regrowth, and encourage native prairie species to return. A year later, bottle gentian, absent for decades, bloomed on the tract (told in Freeman 2018).[27]

By the early-1980s, following the death of Estella Bergere Leopold in Santa Fe in 1975, the Leopold family was reviewing the future of the Shack property. Among the five Leopold siblings, there were conversations about donating the Shack land to UW-Madison as a research

[25] SCF online interview, July 24, 2020.

[26] ALF online interview, October 2, 2020.

[27] ALF online interview, October 2, 2020.

site. But younger family members opposed the land gift to the university and argued to retain the land and establish a family foundation to protect the land and legacy (Laubach 2014; Leopold 2016). In December 1982, the environmental nonprofit Aldo Leopold Shack Foundation was founded by the five siblings as a family foundation (the history of the Aldo Leopold Foundation is discussed in Leopold 2016, also Laubach 2014; Lorbiecki 2016). Also in 1982, the Head Foundation changed its name to the Sand County Foundation (cf. 'Thoreau Country' motif and TCCA in Chapter 2, and 'Abbey's Country' motif in Chapter 6). In its early years, the Aldo Leopold Shack Foundation was focused on the preservation of the historic site—the 200 acres owned by the Leopold family.[28] In the 1990s, the foundation began to move away from a family foundation structure, hiring its first executive director in 1995—Charlie Luthin, a former Leopold Fellow. The same year, the foundation shortened its name to the Aldo Leopold Foundation.[29] The Aldo Leopold Foundation (2020a) works 'to foster the land ethic through the legacy of Aldo Leopold, awakening an ecological conscience in people throughout the world.'

Two nonprofit foundations—the Sand County Foundation and the Aldo Leopold Foundation—were now coordinating the conservation approach on the LMR. But the two organizations have quite different conservation philosophies. The Sand County Foundation views the LMR as a model for private land conservation and collaborates on private land conservation and advancing a land ethic and 'healthier land community'[30] with landowners—ranchers, farmers, and foresters—beyond the reserve, in other states and countries. As a brief sidebar, a dam removal rhetoric also infiltrates the restoration work of the Sand County Foundation (cf. Chapters 3 and 6). The Sand County Foundation was involved in the removal of five small dams on the Baraboo River between 1997

[28] ALF online interview, October 2, 2020.

[29] Three seats on the board are reserved for Leopold family members, and Estella B. Leopold Jr., the sole surviving sibling, has a permanent seat.

[30] SCF online interview, July 24, 2020.

and 2001 (Haglund and Still 2005; Jepsen 2009).[31] As of 2015, the Sand County Foundation no longer has a land presence on the LMR—the organization sold all holdings on the reserve to a private landowner with a conservation commitment.[32] The Aldo Leopold Foundation focuses on Leopold's science and philosophy legacy—on land health, the land ethic and the ethics of conservation, ecological restoration, and on the Shack (Laubach 2014). The organization is also collaborating with private landowners on conservation projects, including the Leopold-Pine Island IBA.

The campus of the Aldo Leopold Foundation is located on a moraine near Levee Road and close to where Leopold died firefighting in April 1948. For a number of years, the Aldo Leopold Foundation occupied the basement of the Bradley Study Center. The Aldo Leopold Legacy Center, or Leopold Center, the headquarters of the Aldo Leopold Foundation, opened in 2007. The project broke ground in 2006, and timber from the 2005–2006 winter timber harvest was used in the construction. Five hundred trees were felled, and about 100,000 board-feet of Leopold pines planted by the family decades earlier were used in the building[33]—for flooring, paneling, ceiling paneling, and structural supports (Lorbiecki 2016). Some hardwoods were harvested offsite. The LEED-certified Leopold Center was the 'greenest' building in America in 2007 (Root 2017), and features solar panels, a rain garden, and a geothermal heating and cooling system (Laubach 2014). It includes offices, classrooms, meeting and exhibition spaces, and a library.

By the late-1980s and early-1990s, the land management and conservation approach on the LMR shifted from projects on small tracts and targeted locations to landscape-scale conservation practices. Leopold's theory of land health is writ large on the landscape where he refined and revised it. Today, restoration projects by the Aldo Leopold Foundation and its partners focus on returning natural processes to the landscape, such as predation, fire, and flooding,[34] to enhance and promote native

[31] SCF online interview, July 24, 2020.

[32] SCF online interview, July 24, 2020.

[33] ALF online interview, October 2, 2020.

[34] SCF online interview, July 24, 2020.

plant and animal species, and remove invasive species. A program of timber harvesting introduced over 450 acres on the LMR supports restoration of the woods. Pine plantations (and other forest stands) are thinned every few years. As a sand county restorationist observes, 'what we realized is we could not achieve a savanna-like condition with dominant oak trees in there, and a real dominant herb layer that was more savanna-like, without a timber harvest.'[35] Meanwhile, larger prescribed burn units are being set on the prairie and savanna (Terbilcox had first experimented with controlled burns at the Shack while Curtis was trialing the same at the Arboretum). The goal is restoration *sensu lato* (Aronson et al. 1993): 'It's not a typology we're after, it doesn't have to absolutely look like that everywhere. But it's clear when you're losing to the woody brush and the young trees, and you'll just end up with a forest all over again, and all of that pre-work basically didn't achieve the savanna or prairie you were after.'[36] It is about 'get[ting] back to a maintenance phase.'[37] Restorationists are also investigating how to restore and reshape the floodplain—to 'build an acquaintance with the Wisconsin River'[38]—and use floods to support wildlife habitat.

Works Cited

Aldo Leopold Foundation. 2020a. Who We Are, What We Do. *Aldo Leopold Foundation*. https://aldoleopold.org/about/our-work/.

Aldo Leopold Foundation. 2020b. Phenology Calendar. *Aldo Leopold Foundation*. https://aldoleopold.org/teach-learn/phenology/.

Aldo Leopold Foundation. 2020c. Practicing Ecological Restoration. *Aldo Leopold Foundation*. https://aldoleopold.org/post/practicing-ecological-restoration/.

[35] ALF online interview, October 2, 2020.

[36] ALF online interview, October 2, 2020.

[37] ALF online interview, October 2, 2020.

[38] SCF online interview, July 24, 2020. ALF online interview, October 2, 2020.

Aronson, J., C. Floret, E. LeFloc'h, C. Ovalle, and R. Pontanier. 1993. Restoration and Rehabilitation of Degraded Ecosystems in Arid and Semi-Arid Lands. I. A View from the South. *Restoration Ecology* 1 (1): 8–17.

Bradley, Charles Crane. 1987. The Leopold Memorial Reserve. In *Aldo Leopold: The Man and His Legacy*, ed. Thomas Tanner, 161–164. Ankeny, IA: Soil Conservation Society of America.

Bradley, Nina Leopold. 1987. A Sand County Restoration. *Restoration and Management Notes* 5 (2): 77–79.

Bradley, Nina L., A. Carl Leopold, John Ross, and Wellington Huffaker. 1999. Phenological Changes Reflect Climate Change in Wisconsin. *Proceedings of the National Academy of Sciences of the United States of America* 96 (17): 9701–9704.

Brown, David E., and Neil B. Carmony, eds. 1995 [1990]. *Aldo Leopold's Southwest*. Albuquerque, NM: University of New Mexico Press.

Callicott, J. Baird, ed. 1987. *Companion to A Sand County Almanac: Interpretive & Critical Essays*. Madison, WI: University of Wisconsin Press.

Callicott, J. Baird. 1989. *In Defense of the Land Ethic: Essays in Environmental Philosophy*. Albany, NY: State University of New York Press.

Callicott, J. Baird. 1999a. *Beyond the Land Ethic: More Essays in Environmental Philosophy*. Albany, NY: State University of New York Press.

Callicott, J. Baird. 1999b. 'The Arboretum and the University': The Speech and the Essay. *Transactions of the Wisconsin Academy of Sciences, Arts, and Letters* 87: 5–21.

Callicott, J. Baird. 2013. *Thinking Like a Planet: The Land Ethic and the Earth Ethic*. New York, NY: Oxford University Press.

Cochrane, Theodore S., Kandis Elliot, and Claudia S. Lipke. 2007. *Prairie Plants of the University of Wisconsin-Madison Arboretum*. Madison, WI: University of Wisconsin Press.

Cottam, Grant, and H. Cameron Wilson. 1966. Community Dynamics on an Artificial Prairie. *Ecology* 47: 88–96.

Court, Franklin E. 2012. *Pioneers of Ecological Restoration: The People and Legacy of the University of Wisconsin Arboretum*. Madison, WI: University of Wisconsin Press.

Curtis, John T. 1959. *The Vegetation of Wisconsin: An Ordination of Plant Communities*. Madison, WI: University of Wisconsin Press.

Curtis, John T., and Max L. Partch. 1948. Effect of Fire on the Competition Between Blue Grass and Certain Prairie Plants. *The American Midland Naturalist* 39 (2): 437–443.

Dunlap, Julie. 1993. *Aldo Leopold: Living with the Land*. New York, NY: Twenty-First Century Books/Henry Holt.

Ellwood, Elizabeth R., Stanley A. Temple, Richard B. Primack, Nina L. Bradley, and Charles C. Davis. 2013. Record-Breaking Early Flowering in the Eastern United States. *PLoS ONE* 8 (1): e53788.

Fassett, Norman Carter. 1931. *Spring Flora of Wisconsin: A Manual of Plants Growing Without Cultivation and Flowering Before June 15*. Madison, WI: Department of Botany, University of Wisconsin.

Flader, Susan L. 1973. The Person and the Place. In *The Sand Country of Aldo Leopold*, eds. Charles Steinhacker with Susan Flader, 8–49. San Francisco, CA: Sierra Club Books.

Flader, Susan L. 1987. Aldo Leopold's Sand Country. In *Companion to A Sand County Almanac: Interpretive & Critical Essays*, ed. J. Baird Callicott, 40–62. Madison, WI: University of Wisconsin Press.

Flader, Susan L. 1994 [1974]. *Thinking Like a Mountain: Aldo Leopold and the Evolution of an Ecological Attitude toward Deer, Wolves, and Forests*. Madison, WI: University of Wisconsin Press.

Freeman, Scott. 2018. *Saving Tarboo Creek: One Family's Quest to Heal the Land*. Portland, OR: Timber Press Inc.

Gray, Asa. 1848. *A Manual of the Botany of the Northern United States: From New England to Wisconsin and South to Ohio and Pennsylvania Inclusive*. Boston and Cambridge, MA: James Munroe and Company.

Green Fire: Aldo Leopold and a Land Ethic for Our Time. 2011. Dir. Ann Dunsky, Steven Dunsky, and David Steinke. Bullfrog Films.

Greenwood, David A. 2017. Making Restoration History: Reconsidering Aldo Leopold's Arboretum Dedication Speeches. *Restoration Ecology* 25 (5): 681–688.

Haglund, Brent M., and Thomas W. Still. 2005. *Hands-On Environmentalism*. San Francisco, CA: Encounter Books.

Hall, Marcus. 2005. *Earth Repair: A Transatlantic History of Environmental Restoration*. Charlottesville, VA: University of Virginia Press.

Harvey, Mark T. 2005. *Wilderness Forever: Howard Zahniser and the Path to the Wilderness Act*. Seattle, WA and London: Weyerhauser Environmental Books/University of Washington Press.

Harvey, Mark T., ed. 2014. *The Wilderness Writings of Howard Zahniser*. Seattle, WA and London: Weyerhauser Environmental Books/University of Washington Press.

Hornaday, William Temple. 1913. *Our Vanishing Wild Life: Its Extermination and Preservation*. New York, NY: Charles Scribner's Sons.

Jepsen, Alicia. 2009. *Running Free: The Baraboo River Restoration Story*. Monona, WI: Sand County Foundation.

Jordan, William R. III, ed. 1984. *Our First 50 Years: The University of Wisconsin-Madison Arboretum 1934–1984*. Madison, WI: University of Wisconsin-Madison Arboretum.

Kozik, Liz Anna. 2017. *Stories in the Land: Tales of the University of Wisconsin Arboretum*. Masters Thesis. Madison, WI: University of Wisconsin-Madison.

Knight, Richard L., and Suzanne Riedel, eds. 2002. *Aldo Leopold and the Ecological Conscience*. New York, NY: Oxford University Press.

Lannoo, Michael J. 2010. *Leopold's Shack and Ricketts's Lab: The Emergence of Environmentalism*. Berkeley, CA: University of California Press.

Laubach, Stephen A. 2014. *Living a Land Ethic: A History of Cooperative Conservation on the Leopold Memorial Reserve*. Madison, WI: University of Wisconsin Press.

Leopold, Aldo. 1915. *Game and Fish Handbook*. Albuquerque, NM: USDA Forest Service District 3.

Leopold, Aldo. 1916. Game Conservation: A Warning, Also an Opportunity. *Arizona* 7 (1–2): 6.

Leopold, Aldo. 1918. The Popular Wilderness Fallacy: An Idea That Is Fast Exploding. *Outer's Book—Recreation* 58 (1): 43–46.

Leopold, Aldo. 1921. The Wilderness and Its Place in Forest Recreational Policy. *Journal of Forestry* 19 (7): 718–721.

Leopold, Aldo. 1923. *Watershed Handbook*. Albuquerque, NM: USDA Forest Service District 3.

Leopold, Aldo. 1925. A Plea for Wilderness Hunting Grounds. *Outdoor Life* 56 (5): 348–350.

Leopold, Aldo. 1927. Game Cropping in Southern Wisconsin. *Our Native Landscape* 1 (October).

Leopold, Aldo. 1931. *Report on a Game Survey of the North Central States*. Madison, WI: Sporting Arms and Ammunition Manufacturers' Institute.

Leopold, Aldo. 1933. The Conservation Ethic. *Journal of Forestry* 31 (6): 634–643.

Leopold, Aldo. 1934a. What is the University of Wisconsin Arboretum, Wild Life Refuge, and Forest Experiment Preserve? In *Our First 50 Years: The University of Wisconsin-Madison Arboretum 1934–1984*, ed. William R. Jordan III, 2–5. Madison, WI: University of Wisconsin-Madison Arboretum.

Leopold, Aldo. 1934b. The Arboretum and the University. *Parks & Recreation* 18 (2): 59–60.

Leopold, Aldo. 1935a. Coon Valley: An Adventure in Cooperative Conservation. *American Forests* 41 (5): 205–208.
Leopold, Aldo. 1935b. Why the Wilderness Society? *The Living Wilderness* 1 (September): 6.
Leopold, Aldo. 1939a. The Farmer as a Conservationist. *American Forests* 45 (6): 294–299, 316, 323.
Leopold, Aldo. 1939b. A Biotic View of the Land. *Journal of Forestry* 37 (9): 727–730.
Leopold, Aldo. 1940a. History of the Riley Game Cooperative, 1931–1939. *Journal of Wildlife Management* 4 (3): 291–302.
Leopold, Aldo. 1940b. Origins and Ideals of Wilderness Areas. *The Living Wilderness* 5 (July): 7.
Leopold, Aldo. 1947. The Ecological Conscience. *Bulletin of the Garden Club of America* (September): 45–53.
Leopold, Aldo. 1968 [1949]. *A Sand County Almanac, and Sketches Here and There*. New York, NY: Oxford University Press.
Leopold, Aldo. 1986 [1933]. *Game Management*. Madison, WI: University of Wisconsin Press.
Leopold, Aldo. 1991. *The River of the Mother of God and Other Essays by Aldo Leopold*, eds. Susan Flader and J. Baird Callicott. Madison, WI: University of Wisconsin Press.
Leopold, Aldo. 1993 [1953]. *Round River: From the Journals of Aldo Leopold*, ed. Luna B. Leopold. New York, NY: Oxford University Press.
Leopold, Aldo. 1999. *For the Health of the Land: Previously Unpublished Essays and Other Writings*, eds. J. Baird Callicott and Eric T. Freyfogle. Washington, DC: Shearwater Books/Island Press.
Leopold, Aldo. 2013. *A Sand County Almanac & Other Writings on Ecology and Conservation*, ed. Curt Meine. New York, NY: Library of America.
Leopold, Aldo, and Alfred E. Eynon. 1961. Avian Daybreak and Evening Song in Relation to Time and Light Intensity. *Condor* 63 (4): 269–293.
Leopold, Aldo, and Sara Elizabeth Jones. 1947. A Phenological Record for Sauk and Dane Counties, Wisconsin, 1935–1945. *Ecological Monographs* 17 (1): 81–122.
Leopold, Aldo, and Reuben Paulson. 1934. Helping Ourselves: Being the Adventures of a Farmer and a Sportsman Who Produced Their Own Shooting Ground. *Field and Stream* 39 (4): 32–33, 56.
Leopold, Estella B. 2016. *Stories from the Leopold Shack: Sand County Revisited*. New York, NY: Oxford University Press.

Liegel, Konrad. 1982. The Pre-European Settlement Vegetation of the Aldo Leopold Memorial Reserve. *Transactions of the Wisconsin Academy of Sciences, Arts, and Letters* 70: 13–26.

Liegel, Konrad. 1988. Land Use and Vegetational Change on the Aldo Leopold Memorial Reserve. *Transactions of the Wisconsin Academy of Sciences, Arts, and Letters* 76: 47–68.

Lorbiecki, Marybeth. 2016 [1996]. *A Fierce Green Fire: Aldo Leopold's Life and Legacy*. New Edition. New York, NY: Oxford University Press.

McCabe, Robert A. 1978. The Stoughton Faville Prairie Preserve: Some Historical Aspects. *Transactions of the Wisconsin Academy of Sciences, Arts, and Letters* 66: 25–49.

McCabe, Robert A. 1987. *Aldo Leopold: The Professor*. Madison, WI: Rusty Rock Press.

Meine, Curt D. 1987. Building 'The Land Ethic'. In *Companion to A Sand County Almanac: Interpretive & Critical Essays*, ed. Baird Callicott, 172–185. Madison, WI: University of Wisconsin Press.

Meine, Curt D. 1988. *Aldo Leopold: His Life and Work*. Madison, WI: University of Wisconsin Press.

Meine, Curt D. 2002. Moving Mountains: Aldo Leopold and *A Sand County Almanac*. In *Aldo Leopold and the Ecological Conscience*, eds. Richard L. Knight and Suzanne Riedel, 14–31. New York, NY: Oxford University Press.

Meine, Curt D. 2004. *Correction Lines: Essays on Land, Leopold, and Conservation*. Washington, DC: Island Press.

Meine, Curt D. 2017. Restoration and "Novel Ecosystems": Priority or Paradox? *Annals of the Missouri Botanical Garden* 102 (2): 217–226.

Meine, Curt D., and Keefe Keeley, eds. 2017. *The Driftless Reader*. Madison, WI: University of Wisconsin Press.

Meine, Curt D., and Richard L. Knight, eds. 1999. *The Essential Aldo Leopold: Quotations and Commentaries*. Madison, WI: University of Wisconsin Press.

Molvar, Erik, and Jan Nesset, eds. 2014. *50 Years of American Wilderness*. Tampa, FL: Faircourt Media Group with Wilderness50.

Muir, John. 2018 [1913]. *The Story of My Boyhood and Youth*. Edinburgh: Birlinn Origin.

Nash, Roderick Frazier. 2014. *Wilderness and the American Mind*, 5th ed. New Haven, CT and London: Yale University Press.

Primack, Richard B. 2014. *Walden Warming: Climate Change Comes to Thoreau's Woods*. Chicago, IL and London: University of Chicago Press.

Ribbens, Dennis. 1987. The Making of *A Sand County Almanac*. In *Companion to A Sand County Almanac: Interpretive & Critical Essays*, ed. J. Baird Callicott, 91–109. Madison, WI: University of Wisconsin Press.
Root, Robert. 2017. *Walking Home Ground: In the Footsteps of Muir, Leopold, and Derleth*. Madison, WI: Wisconsin Historical Society Press.
Ross, John, and Beth Ross. 1998. *Prairie Time: The Leopold Reserve Revisited*. Madison, WI: University of Wisconsin Press.
Sachse, Nancy D. 1965. *A Thousand Ages: The University of Wisconsin Arboretum*. Madison, WI: University of Wisconsin Arboretum.
Simpson, John Warfield. 2003. *Yearning for the Land: A Search for Homeland in Scotland and America*. New York, NY: Vintage Books.
Sutter, Paul S. 2005. *Driven Wild: How the Fight Against Automobiles Launched the Modern Wilderness Movement*. Seattle, WA and London: Weyerhauser Environmental Books/University of Washington Press.
Tanner, Thomas, ed. 1987. *Aldo Leopold: The Man and His Legacy*. Ankeny, IA: Soil Conservation Society of America.
Thoreau, Henry David. 2004 [1854]. *Walden*. 150th Anniversary Edition, ed. J. Lyndon Shanley. Princeton, NJ and Oxford: Princeton University Press.
Warren, Julianne Lutz. 2016 [2006]. *Aldo Leopold's Odyssey*, 10th Anniversary Edition. Washington, DC: Island Press.
The Wilderness Society. 2014. *Wilderness: Our Enduring American Legacy*. Washington, DC: The Wilderness Society.
Woodworth, Paddy. 2013. *Our Once and Future Planet: Restoring the World in the Climate Change Century*. Chicago, IL: University of Chicago Press.
Zedler, Joy B. 1999. Ecological Restoration: The Continuing Challenge of Restoration. In *The Essential Aldo Leopold: Quotations and Commentaries*, eds. Curt Meine and Richard L. Knight, 116–120. Madison, WI: University of Wisconsin Press.

5

'The Superb Monotony of Saw Grass Under the World of Air': Marjory Stoneman Douglas, the Everglades, and Friends of the Everglades

5.1 Rivers of America Book Series: #33 *The Everglades: River of Grass*

The natural history of Florida has long enthralled a litany of U.S. literary naturalists—William Bartram, John James Audubon, John Muir, Charles Torrey Simpson, John Kunkel Small, Edwin Way Teale, Rachel Carson, and Marjorie Harris Carr. Other writers, playwrights, and artists have been drawn to Florida, including Harriet Beecher Stowe, Mark Twain, Winslow Homer, Zora Neale Hurston, Marjorie Kinnan Rawlings, Ernest Hemingway, and Tennessee Williams. But arguably the greatest commentator on Florida's environmental history—and the staunchest Everglades guardian—is Marjory Stoneman Douglas (1890–1998).

This chapter begins with Marjory Stoneman Douglas's arrival in Florida in 1915 and her career in journalism at her father's *The Miami Herald*, before becoming a successful short story writer. While at the *Herald*, Douglas was introduced to two philosophies that would infiltrate all aspects of her life and work—social justice and regionalism. The chapter pauses at Douglas's first book, 1947's *The Everglades: River of Grass* (see Douglas 2017), to examine the foundation of Douglas's

L. Smith, *Ecological Restoration and the U.S. Nature and Environmental Writing Tradition*, https://doi.org/10.1007/978-3-030-86148-3_5

ecological restoration ethos. The chapter then hops back to the mid-1920s, to consider Douglas's place in the campaign to secure an Everglades National Park. The national park was eventually dedicated a month after the publication of Douglas's *The Everglades: River of Grass*.

But running alongside (and preceding and often overlapping) the story of Everglades National Park is the story of draining and dredging in the Everglades, both for land reclamation and for navigation across the peninsula. The chapter uses major Everglades reclamation legislation from the nineteenth and twentieth centuries to provide political context to Douglas's evolving conservation and restoration sensibility. When Douglas joined the conservation campaign to protest a jetport project in Big Cypress Swamp on the northern boundary of Everglades National Park in the late-1960s, her earlier social activism was recast around environmental concerns, and led her to form Friends of the Everglades (FOE) to agitate for Everglades restoration. This moment marked the start of Douglas's late-career entry as an Everglades restoration activist.

Two years after Douglas's death in May 1998 aged 108, President Bill Clinton authorized the Comprehensive Everglades Restoration Plan (CERP), a bold, multibillion dollar, multi-project program to restore historical sheet flows of freshwater from Lake Okeechobee through the South Florida ecosystem into Florida Bay. Two decades on, progress on CERP has been sporadic, disclosing the (political) intricacies, mutabilities, and challenges of 'Everglades restoration.'

The chapter closes by considering another Florida motif that has also inspired literary environmental activism—the hurricane. Poetry in particular offers a poignant (and often painful) commentary on the ecological and cultural devastation wrought by the hurricanes of the Gulf Coast.

5.1.1 Working at *The Miami Herald*

Marjory Stoneman Douglas moved to Miami, Florida, in September 1915. She was 25 years old—six years older than the city. Born in the Midwest, and after a childhood in New England with her mother, maternal grandparents, and aunt, Marjory Stoneman graduated from

Wellesley College in Massachusetts in 1912. A childhood passion for writing continued at Wellesley, and in her junior year, Stoneman was appointed as editor of the college annual, *The Legenda*, a portent of her future career in the newspaper and publishing trade. She also became involved in the women's rights movement, joining the Suffrage Club, founded by classmate Myra Morgan. When her mother Florence Lillian Trefethen died from breast cancer soon after graduation, and uncertain of what she wanted to do after college, Stoneman enrolled on a training program in Boston that would allow her to work as an instructor training salesclerks in department stores. She had posts in St. Louis, Missouri, and Newark, New Jersey, and it was while she was working in Newark that she met and married Kenneth Douglas, the church and social service editor of the *Newark Evening News*, and three decades her senior (Douglas 1987, see also Davis 2009). The couple would separate less than two years later and eventually divorce. It was this parting that led to a reunion between Douglas and her estranged father Frank Bryant Stoneman, and an invitation to move to Florida. Once in the Sunshine State, she could also file for divorce. Her father, editor-in-chief of *The Miami Herald*, had earlier openly opposed Governor Napoleon Bonaparte Broward's plan to drain the Everglades in the early-1900s, and he provided Douglas's earliest introduction to the Everglades drainage controversy.

Douglas had not been in Miami long before she joined the *Herald*—hired first on temporary assignment, then permanently, as the society editor. Her first piece was published on October 25, 1915. With limited Miami society stories to cover, particularly during the summer months, Douglas migrated to sharing general assignment duties, and this became her primer to the city. During her first year in Miami, Douglas also became involved in social activism. In February 1916, she established the Business and Professional Women's League in Miami. Later in 1916, she was enlisted by Mary Bryan to campaign for women's suffrage, and during much of 1917 and 1918, Douglas chaired the Florida Equal Suffrage Association's press committee. She also joined fellow lobbyists Annie Douglass Broward and May Mann Jennings (both married to former Florida governors), and Ivy Stranahan, to speak before the House Committee on Amendments in the state capital, Tallahassee, in

April 1917 in support of the women's suffrage amendment (Douglas 1967, 1987, 2002; cf. Davis 2001, 2003a, b, 2009; Holmes 2004; Poole 2015).[1] From its first introduction in Congress in 1923, Douglas would also be a lifelong supporter of the Equal Rights Amendment. In the summer of 1916, when her father and stepmother left for a month-long vacation traveling across the American West, Douglas was put in charge of the *Herald*'s editorial page.

Meanwhile, the First World War was escalating in Europe, and Douglas was tasked with following the first local woman to enlist in the Naval Reserve. On March 27, 1917, at the U.S. Navy tug *Peoria*, Douglas got her story, as she instead became the first Florida woman to enlist (Douglas 1987; cf. Davis 2009). So began her short hiatus from *The Miami Herald*. Douglas was appointed chief yeoman (although within three months she was downgraded to yeoman first class) and was stationed in the reserve headquarters on Elser Pier in downtown Miami, working as a secretary and typist. After a year in the Naval Reserve, Douglas requested her release from the Navy and joined the American Red Cross in 1918 and was sent on assignment to Paris, with the Civilian Relief department (Douglas 1987, see also 1961). After the war ended, Douglas remained in Europe with the American Red Cross and traveled through Italy and the Balkans reporting on the transition of Red Cross clinics and centers to local authorities for the *Red Cross Bulletin*, and the *Herald*. Back in Paris, Douglas's father sent a cable offering her the position of assistant editor and column writer at the *Herald*.

Douglas returned to Miami in January 1920 amidst the tumult of the city's real estate boom.[2] Back at the *Herald* as assistant editor, Douglas worked on the editorial page and wrote her own daily column—'The Galley Proof' (soon shortened to 'The Galley'), which ran between March 7, 1920, and July 31, 1923. It was here that Douglas quickly

---

[1] Congress passed the 19th Amendment on June 4, 1919, but 36 states had to vote in favor before it became law. Tennessee became the 36th state to ratify the amendment on August 18, 1920, recognizing women's right to vote. Women of color did not secure the right to vote until 1965, with passage of the Voting Rights Act. But it would not be until May 13, 1969, that Florida ratified the 19th Amendment.

[2] Douglas was a fleeting participant in the land boom in the 1920s, when she and friend Marion (Archie) Manley spent $880 on an eight-acre agricultural venture in Big Cypress Swamp (Davis 2009), but the two women defaulted on their loan when the boom collapsed.

carved out a niche in the opening years of the Jazz Age as an articulate commentator on women's rights, racial justice, and the cultural history of southern Florida—and began to explore regionalism motifs and investigate environmental concerns across South Florida. As she had earlier politicized the society page, so she also politicized 'The Galley.' But Douglas's environmentalist sensibility was still incubating—although inspired by the ecology of South Florida, she also accepted the popular rhetoric on the 'wealth,' control, or taming of nature, of engineering and drainage, and of agricultural and commercial development in the Everglades. The Everglades regularly featured in her columns, and in her poems and verse that opened her columns (in Douglas 2002). The closing passage of Douglas's final column on July 31, 1923, focused on the Miami River and the Everglades, a prophetic signaling to her future Everglades activism (in Douglas 2002). Douglas would reverse her Progressive conservationist position later in the twentieth century, as the biological, ecological, and hydrological importance of the Everglades became apparent.

Douglas also used her Galley column at the *Herald* to voice her commitment to social justice (inspired by her paternal Quaker ancestry and heritage)—opening up dialogue on education, public health, and child welfare. She also used her column as a platform to speak out against convict labor leasing in the South. Of all the columns she wrote for the *Herald*, Douglas was most proud of the column she wrote on April 20, 1923, a ballad entitled 'Martin Tabert of North Dakota' (in Douglas 2002, cf. Douglas 1987). It was written in response to the death of 22-year-old Martin Tabert, who was arrested for vagrancy in Tallahassee and sent to a labor camp where he was beaten and killed. The poem was read before the state legislature and contributed to the state abolishing beatings in labor camps in early-1924 (Douglas 1967, 1987, 2002). In 1955, Douglas became a charter member of the American Civil Liberties Union (ACLU) of Greater Miami, the first ACLU chapter in the South (see Davis, 2001, 2009).

With her father's encouragement, Douglas established the *Herald* Baby Milk Fund in 1922, the first charity in Miami unsupported by a church, to distribute milk to underprivileged families. But financial constraints at the newspaper meant the milk fund could not extend to African American families (Douglas 1967, 1987, 2002). In the late-1940s, Douglas joined the board of the Coconut Grove Citizens' Committee for Slum Clearance (later Coconut Grove Cares), set up by friend and neighbor Elizabeth Virrick with pastor Theodore Gibson, to lobby for an ordinance requiring all homes in Dade County to have indoor plumbing installed (Douglas 1967, 1987; also Davis and Frederickson 2003). The measure took two years to pass, and water mains and sewer lines were eventually extended to every street through the segregated African American district. Douglas would revisit issues of social and racial justice in her powerful young adult novel, *Freedom River: Florida 1845* (Douglas 1953), centered on the lives and relationships of three boys—an abolitionist white settler, a Black slave, and a Miccosukee Indian. Her social activism would later meld into her environmental activism.

It was during her three-year assistant editorship at the *Herald* that Douglas was first introduced to the Everglades (see Douglas 1967). To the cypress swamps, mangrove forests, and sawgrass marshes of the Everglades. But Douglas is quick to qualify what 'knowing' the Everglades requires:

> Let me say right away that knowing the Everglades does not necessarily mean spending long periods of time walking around out there. Unlike other wilderness areas, where the naturalist is a hiker, camper, and explorer, the naturalist in the Everglades must usually appreciate it from a distance. [...] I know it's out there and I know its importance. I suppose you could say the Everglades and I have the kind of friendship that doesn't depend on constant physical contact.
>
> —Douglas 1987: 135, 233

But in early-1924, growing increasingly exhausted under the pressure of her commitments to the newspaper, Douglas was diagnosed with nerve fatigue, and she resigned from the *Herald*. During her recovery, she found solace in crafting short stories. Douglas had first begun to experiment with one-line aphorisms—reworking popular literary quotations—and submitting them to magazines such as *Smart Set*, while still at the *Herald*, and progressed to short stories and novellas (Douglas

1987). At the suggestion of her agent in New York, Robert Thomas Hardy, Douglas began to compose short stories for *The Saturday Evening Post* magazine (and other magazines including the *Ladies' Home Journal* and the *Chicago Tribune* Sunday magazine). This shift marked Douglas's departure (and independence) from the *Herald*, and the start of her freelance writing career. Often, the stories were about Florida (18 of Douglas's short stories set in Florida for the *Post* are collected across Douglas 1990, 1998).

It was in 1924, too, that Douglas began to build her cottage on a long, narrow lot in Coconut Grove, just south of Miami. Her 943-square-foot cottage featured just one small bedroom, an equally small kitchen and bathroom, and was dominated by a large, airy 'workshop' filled with books and secondhand furniture (Douglas 1987). Completed in 1926, Douglas would reside at the 3744 Stewart Avenue property[3] until her death on May 14, 1998, aged 108. Douglas's cottage, a half-mile from Biscayne Bay, was finished just months after the September 1926 'Great Miami' hurricane abruptly halted Miami's real estate boom. Douglas was visiting family in Massachusetts when the hurricane struck, and the only damage to her almost-completed cottage was a high water mark on the exterior walls, broken window panes, some missing roof shingles, and seaweed and other debris inside (Douglas 1987).

Douglas's short story career was at its most prolific between 1926 and 1941. The editor of the *Post*, George Horace Lorimer, had a personal interest in environmental causes and encouraged Douglas to write about Everglades issues, including poaching, timber theft, and more. 'A Bird Dog in the Hand,' published on September 12, 1925, knits together Miami's real estate boom and land speculation with the draining and dredging of the Everglades. 'Goodness Gracious, Agnes,' published on October 17, 1925, features illegal wildcat fighting in the Everglades. 'Plumes,' published on June 14, 1930, speaks out against egret poachers and the feather trade. 'The Road to the Horizon,' Douglas's last short story for the *Post*, published on February 22, 1941, offers an intimate portrait of two major Everglades motifs—sawgrass and mosquitoes

[3] The Marjory Stoneman Douglas House was designated as a National Historic Landmark on April 22, 2015.

(included in Douglas 1990, 1998, and her first novel, *Road to the Sun*, expanded on the themes of 'A Bird Dog,' as Douglas 1952). The environment (and weather) of South Florida was a recurrent, supporting character. Dispatches from across the West Indies on volcanic eruptions, political instability, alongside recollections from her years in wartime Europe, further added to Douglas's magazine contributions. After her father passed away in February 1941, Douglas suffered another nervous breakdown. But with her small inheritance, Douglas was free of the magazine business and could, aged 51, pursue a long-dormant passion—writing a novel.

5.1.2 A Meeting with Hervey Allen, and the 'River of Grass' Idea Is Born

In the early-1940s, Hervey Allen, editor of Rinehart and Company's Rivers of America series, arrived at 3744 Stewart Avenue, Coconut Grove, with a proposition for Douglas. It was an invitation that would later lead to Douglas's recognition as the Grand Dame of the Everglades in the 1970s through 1990s (Davis 2001, 2009) and recipient of the Presidential Medal of Freedom in 1993. Allen's request was a simple one: would Douglas write a book about the Miami River? Douglas's response—that the Miami River was unremarkable and inconsequential (Douglas 1987)—soon saw the project expanded beyond the Miami River watershed, with Douglas arguing that the Miami River was part of the Everglades. Douglas openly confesses her bare, sparse knowledge of the Everglades at that moment: 'That it was there, that the birds were spectacular, that it should be a national park, and that it shouldn't be drained, that there were millions of acres of it. I'd been out in the Everglades no more than 20 times' (Douglas 1987: 190).

With her novel temporarily abandoned, Douglas signed her first book agreement on November 23, 1943, and embarked on her new assignment. Her first stop was *The Miami Herald* and her father's successor, John Pennekamp, who put her in contact with U.S. Geological Survey (USGS) hydrologist Garald G. Parker (see Parker et al. 1955 for the definitive, classic study of South Florida's hydrogeology, also Parker and

Cooke 1944). Douglas asked Parker whether the Everglades could be described as a river—and as a *river of grass*. Parker agreed, and with that confirmation, understanding of the Everglades ecosystem slowly shifted (on the evolution of the 'river of grass' idea, see Douglas 1987, also 1967 for an 'Everglades River' motif). Parker's influence echoes in the natural history theme of Douglas's book—and especially in its lengthy opening chapter, 'The Nature of the Everglades' (for other excellent volumes on the natural history of Florida and the Everglades, see McCally 2000; Davis and Arsenault 2005).

Douglas's *The Everglades: River of Grass* was published on November 6, 1947, with an initial print run of 7,500. By the end of the year, the book had sold out, and a further 5,000 copies were printed. The 33rd volume in the Rivers of America series, it is one of three books set in Florida in the 64-volume series. The publication of *The Everglades: River of Grass* (see Douglas 2017) came one month before the dedication of Everglades National Park on December 6, concluding a park boosterism campaign that had stretched across a quarter-century.

The opening to Douglas's (2017) *The Everglades: River of Grass* now stands as a classic of Everglades literature:

> There are no other Everglades in the world.
>
> They are, they have always been, one of the unique regions of the earth, remote, never wholly known. Nothing anywhere else is like them: their vast glittering openness, wider than the enormous visible round of the horizon, the racing free saltness and sweetness of their massive winds, under the dazzling blue heights of space.
>
> —Douglas 2017: 5

A few pages later, Douglas offers another description of the Everglades, and one that is perhaps even more iconic—'In the Everglades one is most aware of the superb monotony of saw grass under the world of air. But below that and before it, enclosing and causing it, is the water' (Douglas 2017: 14). Sawgrass and water, together, are for Douglas the two defining motifs of the Florida Everglades. Douglas iconized the Everglades as the River of Grass (echoing a Seminole translation of the Everglades as Pahayokee, or 'grassy water,' see Douglas 2017) (Fig. 5.1).

Fig. 5.1 Sawgrass prairie in Everglades National Park, Florida, in the mid-twentieth century. Courtesy of Florida Memory/State Archives of Florida

But Douglas's (2017) *The Everglades: River of Grass* is much more than a celebration of sawgrass and water. It tells of natural history, and ecology, hydrology, geology. A wilderness motif persists—and would later be politicized in Douglas's Everglades activism (also Douglas 1967, 1990, 2003). *The Everglades: River of Grass* also tells of Native Nations. European explorers and adventurers. Colonialism, slavery, military history, conflict, and war. Settlers and the frontier. South Florida's real estate boom, land speculation, and development and drainage. And finally, the promise and hope of restoration (a brief cultural history of the Everglades is also woven through Douglas 1967, 2003).

Revealed in the closing chapters of Douglas's (2017) *The Everglades: River of Grass* is a tentative formalizing of Douglas's early restoration sensibility. The final chapter, forebodingly entitled 'The Eleventh Hour,' opens with the sharp declaration, 'The Everglades were dying' (Douglas 2017: 349). The chapter sounds the alarm over pursuing urban planning and development at the expense of the Everglades. Three chapters earlier, Douglas uses the immense drainage infrastructure of the Everglades, much of it introduced during Broward's governorship, to set up the disruption to Everglades hydrology: 'to the intricate and subtle relation of soil, of fresh water and evaporation, and of runoff and saltwater intrusion, and all the consequences of disturbing the fine balance nature had set up in the past four thousand years—no one knew enough to look' (Douglas 2017: 286). For Douglas (2017), Everglades preservation had to be wholesale—not a hammock or marsh here, or a lake there, or a key somewhere else, but ecosystem-wide, watershed-wide (cf. critiques of the 'river of grass' motif for Everglades restoration reference models in McCally 2000; Grunwald 2006). 'The Eleventh Hour' is also a critical commentary on state-led reclamation programs (Douglas 2017). Douglas's (2017) *The Everglades: River of Grass* spoke to (and continues to speak to) public policy debates in the Everglades (see also McCally's 2000 ecological biography of the Everglades a half-century later).

5.2 Everglades National Park and the Marjory Stoneman Douglas Wilderness

More than a decade before she began work on *The Everglades: River of Grass*, Douglas was invited to join landscape architect and conservationist Ernest F. Coe's Tropic Everglades National Park Association, an advocacy group campaigning for an Everglades National Park. With her park association appointment in the late-1920s, Douglas's Everglades education began in earnest (on the history of, and Douglas's involvement with, Coe's national park campaign, see Douglas 1967, 1987, 2017; cf. Davis and Arsenault 2005; Grunwald 2006; Davis 2009). Douglas made several trips into the Everglades with Coe as the group developed and lobbied for its national park proposal, and later, with hydrogeologist Garald Parker and others, when researching *The Everglades: River of Grass* (Douglas 2017).

But before proceeding, a brief detour through the early history of key legislation in favor of reclaiming—that is, dredging and draining—the Everglades is necessary, to set the scene both for the establishment of Everglades National Park and for the metamorphosis from Douglas-as-writer to Douglas-as-environmental-writer-activist. Douglas's 108 years, spanning the last decade of the nineteenth century and all but 19-and-a-half months of the twentieth century, are impossible to untangle from a transformative century of Everglades reclamation and later restoration.

A cast of governors, railroad tycoons, land sales companies, plume hunters, conservationists, and others has irrevocably shaped the Everglades ecosystem. Not long after the slave state of Florida joined the Union in 1845, the state legislature instructed Congress to survey the Everglades for reclamation. The Florida legislature's early request to Congress led to the first scientific reconnaissance of the Everglades—Buckingham Smith's five-week survey of the swamp during the summer of 1847 (a century before Douglas's *The Everglades River of Grass* appeared), and his drainage report, *Report on the Everglades*, in June 1848 (in U.S. Congress, Senate 1848; cf. McCally 2000; Grunwald 2006; Davis 2009, 2017; Dunn 2019). Later followed the federal Swamp and Overflowed Land Grant Acts of 1849, 1850, and 1860, which promoted reclamation by granting Florida 20 million acres (equivalent to almost two-thirds of the state's land mass) of wetlands to use as it wanted (Grunwald 2006; Davis 2017). In the mid-nineteenth century, wetland was wasteland—the land needed to be reclaimed and redeemed (cf. Thoreau on redemption and swamps in Chapter 2). Land was bought 'by the gallon' (in Grunwald 2006). Reclamation, then, *was* conservation—it was the 'wise use' (as espoused by Gifford Pinchot, the architect of Progressive conservation, see Chapter 3) of Everglades lands.

The state soon began giving land away to be drained, dredged—and reclaimed. Through the rises and falls of the Internal Improvement Fund of the State of Florida (IIF), established in 1855, offering swamp and overflowed land grants to stimulate railroad construction. Through passage of the Riparian Act in 1856, allowing landowners to fill swampland—and, in 1921, tidal land. Through the vast drainage enterprises of Philadelphia industrialist and entrepreneur Hamilton Disston, and the agricultural precedent set by his 25-cents-an-acre, 4-million-acre Disston Land Purchase of February 1881. Through the Drainage of Counties Act of 1901. The railroad grants and dredging permits of Henry B. Plant and Henry M. Flagler. The plans, promises, and transactions of Disston's Atlantic and Gulf Coast Canal and Okeechobee Land Company, Flagler's Model Land Company, and other land sales companies (McCally 2000; Grunwald 2006; Davis 2017). Elsewhere, construction began of the

Tamiami Trail[4] (a Tampa-Miami highway, and the southernmost part of U.S. Highway 41) across the Big Cypress Swamp and northern Everglades. Everglades drainage laws (and a drainage tax) were approved in 1905, 1907, 1913, 1927. The Everglades Drainage District (EDD) was created in 1905.

In the 1870s and 1880s, Florida was a utopia for naturalists, for botanists, ornithologists, and entomologists. But by the mid-1880s, the feather trade was also booming in South Florida, supplying an insatiable millinery industry. As American—particularly New York—and European demand grew for more and more exotic plumes during the Gilded Age, plume hunters slaughtered huge colonies of wading birds, decimating rookeries along the shores of Lake Okeechobee, around Tampa Bay, and deep in the Everglades and across the Ten Thousand Islands. According to the American Ornithologists' Union (AOU), by 1886, five million birds a year were being slaughtered in the U.S. for the millinery trade (Allen 1886, on feather agents, hunters, and the lucrative dealing of plumes, see also McCally 2000; Grunwald 2006; McIver 2009; Punke 2016; Davis 2017). In a February 11, 1886, editorial manifesto in the popular hunting and fishing magazine, *Forest and Stream*, entitled 'The Audubon Society,' George Bird Grinnell announced the launch of a new national organization dedicated to the protection of wild birds—predominantly plume birds, but also songbirds and other nongame birds—and their nests and eggs (Grinnell 1886; cf. Merchant 2016; Punke 2016). Named for John James Audubon, the American ornithologist and artist, Grinnell would revisit and develop his Audubon Society proposal in subsequent issues of *Forest and Stream*, with other articles and letters also endorsing his proposal, and a year later, in February 1887, he launched *The Audubon Magazine*, devoted solely to birds and their protection. Grinnell's short-lived Audubon Society (1886–1889) was the

[4] Douglas had been in Florida for five years before she first visited the Everglades in 1920. At the time, construction of the Tamiami Trail was still underway in the northern Everglades (Douglas 1987), and Douglas and friends would go sightseeing, bird watching, and fishing from the end of the roadway (she also wrote at least two poems celebrating the highway, see Douglas 2002). This road through the Everglades, completed in 1928, would for many decades almost sever the flow of water from the Big Cypress Swamp north of U.S. 41 to the Everglades wetlands below it, until elevated on bridges and causeways.

first iteration of state Audubon societies and later a National Audubon Society.

A decade after Grinnell's first Audubon Society was established, the idea was resurrected by cousins Harriet Lawrence Hemenway and Minna B. Hall in Boston, Massachusetts, in 1896. The Massachusetts Audubon Society (or, Mass Audubon) was founded and led primarily by women, to campaign against the buying and wearing of feathered hats (cf. Mass Audubon and Walden restoration, in Chapter 2). Audubon membership often aligned closely with the membership of women's clubs. The establishment of Mass Audubon triggered a wave of similar endeavors, and by the close of the century, there were more than a dozen state Audubon societies, mostly located in the northeastern U.S. Grinnell's *The Audubon Magazine* was relaunched in 1889 as *Bird-Lore*, before reverting to *Audubon Magazine* in 1931, and *Audubon* in 1966. The Florida Audubon Society (hereafter, Audubon Florida) was founded in 1900 by nine women and six men (today, more than 40 chapters also operate in the state). Laura Norcross Marrs, also a member of Mass Audubon, was appointed chair of Audubon Florida's executive committee, a position she held until her death in 1926. Floridian conservation powerhouse Mary Barr Munroe had a reputation for confronting women wearing feathered hats in the street, and challenging and educating her patrons (on the history of women in Audubon Florida, see Poole 2015; also Davis 2003a, b, 2009; Davis and Frederickson 2003).

It was state Audubon chapters and the AOU that helped secure the first federal legislation for bird protection—the 1900 Lacey Act, prohibiting the interstate shipment and sale of birds killed in violation of state laws (on the history of the National Audubon Society, and its contributions to federal bird protection legislation, see especially Graham 1990; Merchant 2016; Punke 2016; Davis 2017). On the same day that President Theodore Roosevelt (hereafter, TR) wrote to Muir requesting him as the guide for his upcoming tour of Yosemite (see Chapter 3)—March 13, 1903—TR signed an executive order establishing the Pelican Island bird sanctuary outside Sebastian, 150 miles north of Miami—the first national wildlife refuge (NWR) in the U.S. Congress prohibited the import of wild bird feathers under the Tariff Act in 1913, and the Migratory Bird Treaty Act followed in 1918.

But the 1900 Lacey Act did not end plume hunting, and especially not in the lawless, frontier wilderness of the Everglades. In 1902, Guy Morrell Bradley was appointed as Audubon's—and Monroe County, Florida's—first game warden. Bradley's patrol included the Everglades, the Florida Keys, and the Ten Thousand Islands. Three years into his tenure, Bradley was killed in the line of duty on July 8, 1905, defending a rookery on the two small mangrove islands of Oyster Keys across the Florida Bay from his home in Flamingo from plume hunters (see McIver 2009 for an excellent biography of Bradley). The feather trade began to decline in the 1920s, reflecting changing tastes in middle- and upper-class fashion, but plume hunting continued into the 1940s. The Everglades would not be protected from plume (and alligator) hunters until it was designated as a national park, more than four decades after Bradley's death.

But a threat greater than plume hunting was emerging in the Everglades—drainage. Despite a checkerboard of early reclamation efforts across the Everglades, and James O. Wright's highly contentious *Report on Drainage of the Everglades* (in U.S. Congress, Senate 1912; Grunwald 2006), it was not until the summer of 1912 that calls came from land companies for a review and reassessment of the state's Everglades drainage program (see McCally 2000). Over the next 18 months, two reports produced by teams of hydraulic engineers were published which overhauled reclamation policy in the Everglades—the first report was commissioned by the land companies (Mead et al. 1912) and the second by the EDD for the state (in U.S. Congress, Senate 1914) (for further critique on the legislative context of reclaiming the Everglades, see Douglas 1967, 2017; McCally 2000; Grunwald 2006; Davis 2017; Dunn 2019). From this point on, reclamation efforts run alongside and overlap with Coe's bid for an Everglades National Park, and will be picked up again later in the chapter.

5.2.1 Everglades Protections I: Royal Palm State Park

Coe's Everglades National Park campaign was not the first attempt to protect the Everglades—it was preceded by a campaign to secure the smaller Royal Palm State Park. As with so much of Miami and South Florida history, Flagler and the Florida East Coast (FEC) Railroad feature in the backstory. In the nineteenth century, Royal Palm Hammock—or Paradise Key—in the southern Everglades, southwest of Homestead (and 50 miles south of Miami), was celebrated and revered by surveyors and scientists for its botanical diversity. In an effort to protect the island hammock from Flagler's railroad development down Florida's eastern seaboard, conservationists petitioned for the land to be set aside as a state park.

The fight to secure the protection of Paradise Key was largely orchestrated by women's groups. The Florida Federation of Women's Clubs (FFWC) first suggested the idea for an Everglades forest reserve in 1905, the same year Bradley was killed. Less than a year later, Edith Gifford and Mary Barr Munroe introduced a motion at a FFWC convention to protect Paradise Key. While the motion passed, nothing further happened (Davis 2009; Poole 2015). The adopted motion languished until May Mann Jennings, wife of former Florida governor William Sherman Jennings, became president of the federation in November 1914.

The proposed park fell on IIF land. But it sat adjacent to Model Land Company land. Mary Lily Kenan Flagler, third wife and widow of the railroad tycoon, and FFWC member, donated 960 acres as an endowment. Jennings enlisted her husband to draft a bill for Florida's first state park, and the federation began lobbying legislators. The FFWC bill was presented before and eventually passed the state legislature in June 1915. The state contributed another 960 acres. Years before women could vote, women's groups were often central in Progressive era social and political reform (Davis and Frederickson 2003; Poole 2015). Douglas moved to Miami in the fall of 1915 and was not actively involved in the state park campaign, but she supported increased Everglades protections, and used first her society page column and then her 'The Galley' column at the

Herald to report on the progress of later FFWC conservation campaigns (Douglas 2002; Davis 2009; Poole 2015) (Fig. 5.2).

On November 23, 1916, three months after the National Park Service (NPS) was established, the Royal Palm State Park was dedicated. The

Fig. 5.2 Royal Palm State Park, Homestead, Florida. Courtesy of Florida Memory/State Archives of Florida

FFWC was responsible for managing the state park—and it was the only women's club in the country to be doing so (Davis 2009). By 1921, an additional 2,080 acres had been donated to the state park, expanding its boundary to 4,000 acres (on the history of Royal Palm State Park, see Douglas 1987; Davis 2003a, 2009; Davis and Frederickson 2003; Grunwald 2006; Poole 2015). A commemorative plaque installed at the state park describes Royal Palm State Park as the 'nucleus' of Everglades National Park. The state park designation triggered the national park campaign—even though it protected just 0.1 per cent (according to Grunwald 2006) of the Everglades.

5.2.2 Everglades Protections II: Everglades National Park

By the early-1920s, support for the creation of a national park in the Everglades was growing among Florida's conservationists (see e.g. Bailey 1925; cf. Davis and Arsenault 2005; Grunwald 2006; Davis 2009). Similar conversations were happening 1,000 miles away in Washington, DC, in the Department of the Interior.

It was not long after Coe's arrival in Miami in 1925 that he launched his national park campaign. In his first three years in the city, Coe assembled the Tropic Everglades National Park Association, an advocacy group to lobby for the national park (with the *Tropic* motif politicized), persuaded Congress to study the national park idea, mapped out a tentative park boundary, and was chairing the Tropical Everglades National Park Commission, the state commission responsible for buying the land parcels (after Grunwald 2006). The office of *Herald* editor and drainage critic Frank Stoneman became a regular haunt of Coe, to share progress in his campaign (Douglas 1987). The work of Coe's Tropic Everglades National Park Association unfolded in the aftermath of the devastating 1926 and 1928 Miami hurricanes, and associated severe flooding from Lake Okeechobee, to argue against further development of the Everglades. Meanwhile, the state pushed ahead with greater control of Lake Okeechobee, further disconnecting the lake from the Everglades and cutting off the 'river of grass.'

Included within Coe's proposed Tropic Everglades National Park boundary were more than two million acres of the southern Everglades—from the Ten Thousand Islands, Florida Bay, and the upper Keys, and out into the coral reefs of the Atlantic, north to the Big Cypress Swamp and beyond, stretching 15 miles above the Tamiami Trail. In 1929, the year after Coe formed his Tropic Everglades National Park Association, the FFWC offered Royal Palm State Park for inclusion in Coe's proposed park, and this led to Douglas's invitation to join Coe's association (Davis 2003a, 2009). If Coe's proposal succeeded, it would create the second largest national park in the U.S. at that time, behind Yellowstone. The NPS supported Coe's proposal, but could do little until Florida deeded the land to the park service.

Coe was prepared for a long campaign, establishing himself as an Everglades park booster and lobbyist (as Muir and Johnson had done in Yosemite in the 1890s, see Chapter 3). His first test came when the NPS announced in January 1930 that a delegation would tour the Everglades. On February 11, Coe organized a flight over the Everglades in the Goodyear blimp for the national park delegation reviewing his proposal. Short on space in the cabin, Coe and Douglas spent the flight suspended in a small observation coop below. The flight was followed by a three-day bird-watching boat tour along Florida Bay to Cape Sable, the Shark River, and Tarpon Bay (Douglas 1967, 1987; Davis, 2003a, 2009; Grunwald 2006). This was Douglas's first major exposure to the Everglades, beyond excursions along the Tamiami Trail. In *Florida: The Long Frontier*, Douglas (1967) describes a boat tour on the Shark River with NPS rangers and members of Congress, exploring the Everglades and bird watching—and the powerlessness in departing, knowing that plume hunters were waiting nearby to enter the rookeries after the party. Her fears were confirmed the following evening by flares and echoes of gunshot. The tour also inspired many of her short stories in *The Saturday Evening Post*—including 'Plumes' in June 1930, 'Wings' in March 1931, and 'A Flight of Ibis' in December 1935 (see e.g. Douglas 1990, 1998; cf. Davis 2003a, 2009). A senatorial delegation also toured the Everglades later that year (Davis 2009).

As the origin story of Everglades National Park entered the halls of Congress in the early-1930s, literary activism again contributed to the conservation campaign. Coe, John Kunkel Small, and others testified during hearings. Douglas and USDA scientist Ralph Stoutamire (1932) co-wrote *The Parks and Playgrounds of Florida*, a 29-page Florida Department of Agriculture bulletin. Coe also distributed Small's (1929) *From Eden to Sahara: Florida's Tragedy*, and Douglas's (1923) Galley poem 'I Am the Mangrove'—and shipped coconuts—to members of Congress (Davis 2009). While the national park proposal continued to ebb and flow across the House and Senate, the NPS convened a special committee to review the Everglades bid. In 1931, landscape architect Frederick Law Olmsted Jr. and William P. Wharton of the American Forestry Association and the National Park Association spent 10 days exploring the Everglades with Coe (Davis 2009). Their report was submitted to the NPS before being presented to Congress in 1932 (Olmsted and Wharton 1932; U.S. Congress, Senate 1932; cf. Davis 2009). After further setbacks, the bill eventually passed the House, and President Franklin Delano Roosevelt authorized Everglades National Park on May 30, 1934 (although the *Tropic* was dropped in the park authorization bill).

But this is not the end of the national park's origin story. Disputes surrounding Coe's proposed boundary for the national park had to be resolved. The recommendations of D. Graham Copeland, chair of the state commission's boundaries committee, removed no more than 14 per cent of land from Coe's plan to variously appease game hunters, landowners and farmers, and politicians, centered on Big Cypress Swamp, Turner River, and Key Largo (after Grunwald 2006), but Coe's objections led to his removal as state commission chair in 1937 (Grunwald 2006; Davis 2009). With Coe's absence, the Tropical Everglades National Park Commission disbanded. In the early-1940s, the park boundary was renegotiated again and reduced to 1.3 million acres, removing more than three times as much land as Copeland's compromise, severing everything north of the Tamiami Trail, as well as Big Cypress Swamp, the Turner River area, the marshes of northeast Shark River Slough, Hole-in-the-Donut, and the upper Keys and coral reefs.[5]

[5] Many of these excluded areas would later host conservation battles to secure protections—with the establishment of John Pennekamp Coral Reef State Park in 1963, Big Cypress National

The heart of the southern Everglades was still protected, including most of the Ten Thousand Islands, Cape Sable, Florida Bay, Paradise Key, and most of Shark River Slough and Taylor Slough (Grunwald 2006; also Douglas 1987, 2003).

Before the national park could be designated, the state of Florida had to donate the land. When the Tropical Everglades National Park Commission was reinstated after the Second World War in April 1946, John Pennekamp, Frank Stoneman's successor at the *Herald*, was appointed legislative chairman and secured the release of funds to buy back the land (Grunwald 2006; Davis 2009; Douglas 2017). It took 13 years of negotiations, and a 'Declaration of Taking,' to acquire the lands for the park—with the land acquisition completed on June 20, 1947 (Davis 2009; Douglas 2017). In *The Everglades: River of Grass*, which was published just five months later, Douglas (2017) records this date as the birth of Everglades National Park.

America's 28th national park, Everglades National Park was formally dedicated by President Harry S. Truman on December 6, 1947, in a ceremony in Everglades City. Douglas and Coe both attended the ceremony, although Coe remained disappointed at being removed from the planning process, and at the smaller footprint of the national park than that recommended in his proposal. In early-December 1947, Everglades National Park encompassed just over 450,000 acres—roughly the same acreage as Lake Okeechobee. But the creation of the national park revisited the question of the dislocation and relocation of some Indigenous groups, including the Seminole and Miccosukee, to reservations outside the park boundary. Three years later, Everglades National Park was expanded to almost 1.28 million acres. By the end of the 1960s, a further 170,000 acres were added. Another 107,000 acres were added in 1989 under the Everglades National Park Protection and Expansion Act (Douglas 2003; Grunwald 2006; Davis 2009). The Everglades was recognized as the UNESCO Everglades and Dry Tortugas Biosphere Reserve in 1976, a United Nations World Heritage Site in 1979 (although it has twice appeared on the List of World Heritage in Danger, between

Preserve in 1974, Biscayne National Park in 1980, Ten Thousand Islands National Wildlife Refuge in 1996, and Hole-in-the-Donut and northeast Shark River Slough would also be added to Everglades National Park (see Grunwald 2006).

1993–2007, and 2010-present), and a Ramsar Wetland of International Importance in 1987 (Lowry 2009; Dunn 2019). Everglades National Park challenges and expands the national park idea—at its heart is a swamp, not the canyons, mountains, waterfalls, or geysers that defined earlier national parks like Yellowstone, Yosemite, or Grand Canyon. It was the first time a national park was designated to primarily protect flora and fauna, and the ecosystem that supports it.

Douglas's 1959 young adult novel, *Alligator Crossing* (see Douglas 2003), part of a national parks series, signals her burgeoning Everglades activism. The drama moves quickly from Miami to Everglades National Park, as 13-year-old protagonist Henry Albert Bunks pursues Arlie Dillon, the renegade alligator poacher who skinned an alligator Bunks had befriended and named George. The novel weaves together plume hunting, alligator baiting and poaching, Flagler's railroad, Bradley's murder, the fading frontier, the evolution of the national park, hurricanes, and more, to introduce the threats facing the Everglades. Beginning in 1972, and continuing for more than a dozen years, Douglas was invited to teach Everglades park rangers enrolled in the seasonal-training program (Davis 2009). Edward Abbey (Chapter 6) worked as a seasonal ranger at Everglades National Park between December 1965 and April 1966.

Coinciding with the 50th anniversary of the dedication of Everglades National Park in December 1997, and six months before her death, Congress designated 1.3 million acres—almost 86 per cent—of the national park as the Marjory Stoneman Douglas Wilderness Area (Davis 2009). Douglas is one of just 15 individuals honored in wilderness area designations (alongside e.g. Muir, TR, Leopold). Two weeks after her death, Douglas's ashes were scattered in Everglades National Park by park rangers.

5.3 Friends of the Everglades, and Its First Antagonists

Douglas's participation in Everglades battles did not end with the dedication of Everglades National Park. Indeed, it had barely begun. Two decades after the publication of *The Everglades: River of Grass* (Douglas

2017), and the national park dedication, and approaching 80 years old, Douglas emerged as an Everglades restoration iconoclast. In the 1920s, she had used her editorial column at *The Miami Herald* to report on (and tentatively lobby for protection of) the flora and fauna of the Everglades. In the late-1940s, she had written a defining text on the ecology, geology, and anthropology of the Everglades, changing the way America thought about the Everglades, and recasting the Everglades as a vibrant, slow-moving river rather than a fearsome swamp. As one Everglades activist notes, the 'river of grass' idea 'made it more of a living system. It really helped people understand what the Everglades were, and why they were worth saving.'[6] In the 1960s, she entered into the political quagmire of development tensions and frictions in the Everglades (Douglas's 1987 autobiography, *Voice of the River*, offers an excellent insight into the genesis of her restoration advocacy, cf. also Davis 2001 2003a, b, 2009; Holmes 2004; Poole 2015). The U.S. environmental movement was in its infancy in the 1960s, and by this time, the environmental ethos of Douglas's *The Everglades: River of Grass* (as Douglas 2017) had found philosophical allies in Aldo Leopold's 1949 *A Sand County Almanac* (as Leopold 1968), Rachel Carson's (1962) *Silent Spring*, and Interior Secretary Stewart Udall's (1963) *The Quiet Crisis*. All of these texts would mold U.S. environmentalism throughout the 1960s and 1970s and beyond.

5.3.1 In Defense of Big Cypress Swamp

In the early-1960s, two development projects were announced in Biscayne Bay—the SeaDade deep-water port, oil refinery, and industrial complex (to be sandwiched between Everglades National Park and the John Pennekamp Coral Reef State Park), as well as a causeway across Biscayne Bay to the islands of Elliott Key, and the city of Islandia project. A small coalition of conservationists formed to fight both projects and advocate for national park protections. After a five-year battle, the Tropical Audubon Society (TAS), the Mangrove chapter of the Izaak Walton League, the Safe Progress Association, and other lobbyists triumphed in stopping both projects. In the epilogue to *Florida: The Long Frontier*,

[6] FOE online interview, July 23, 2020.

Douglas (1967) celebrates the recent halting of the industrial seaport bid and hopes that a national monument bid already underway will succeed and replace it (Biscayne Bay had earlier been included in Coe's Tropic Everglades National Park proposal). Douglas's hope—and that of the TAS and other activists—was realized with President Lyndon B. Johnson's designation of Biscayne National Monument on October 18, 1968,[7] blocking both development projects and protecting mangrove forests, coral reefs, and islands in the bay (on the history of the SeaDade and Islandia struggles, the national monument bid, and a later controversy surrounding Florida Power and Light's proposed power plant at nearby Turkey Point and the thermal pollution problem, see Davis 2009).

But conservationists could not rest for long. As Douglas (1987: 224) argues, 'The most dangerous time is after some project like [Daniel K.] Ludwig's has been stopped. The opposition senses the lull, and uses the occasion to spawn numerous other idiocies that ought to be stopped. As soon as the Ludwig idea was killed, an equally ridiculous idea arrived to take its place.' The ridiculous idea? A jetport in the Everglades.

Six miles upstream of Everglades National Park in the Big Cypress Swamp, at the western boundary of Water Conservation Area 3 (WCA 3, which contains the upper Shark River Slough), construction had already begun of the Dade County Port Authority's 24,960-acre jetport following the ground-breaking on September 18, 1968. One runway was already complete, another was underway, and state permits and federal funding were confirmed. When completed, this six-runway jetport behemoth (quadruple the size of Miami International), plus a new city, would serve not just Miami, but Central and South Florida (including Orlando's Walt Disney World, which opened in October 1971). The jetport would also run a high-speed transit corridor through WCA 3 to connect the jetport to Miami and would disrupt the flow of freshwater into the northwestern Everglades National Park and the Ten Thousand Islands (Grunwald 2006; Davis 2009; Lowry 2009).

---

[7] The national monument was expanded in 1974 and again with the designation of Biscayne National Park on June 28, 1980.

TAS vice president and southeastern representative of the National Audubon Society, Joe Browder, began to mobilize conservationists. He organized the Everglades Coalition in April 1969, a partnership of 21 national (including the National Audubon Society, Sierra Club, Wilderness Society, Nature Conservancy, National Parks Association, and National Wildlife Federation) and one state (Audubon Florida) environmental groups opposed to the jetport, elevating the jetport issue to a national platform (for a detailed history of the jetport's legislative battle and its key actors, see e.g. Douglas 1987; Davis and Arsenault 2005; Grunwald 2006; Davis 2009; Lowry 2009; Dunn 2019). It is at this juncture that Douglas entered the fight to save the Big Cypress Swamp. As Douglas (1987) recalls in *Voice of the River*, it was a chance meeting with Browder's assistant and office manager, Judy Wilson, in a Coconut Grove grocery store one evening that set in motion her indoctrination as an Everglades activist. From there followed a phone call and a meeting with Browder at her Coconut Grove cottage the next day, before she called Browder a few weeks later to ask for a visit out to the jetport and industrial park construction site (also in Davis 2009). When Browder asked Douglas to issue a statement condemning the jetport, as a resident and respected authority with extensive experience of the Everglades and its history, she worried that the voice of one old lady would not be heard—only organizations had that political gravitas, political legitimacy. Browder suggested she start an organization (Douglas 1987).

While at the Fairchild Tropical Botanic Garden Ramble in Coral Gables in November 1969 to sign copies of her books, Douglas was musing on Browder's suggestion for a grassroots organization to advocate for Everglades conservation and restoration. Mentioning the idea of a 'Friends of the Everglades' organization to acquaintance Michael Chenoweth, he responded by handing her a dollar as his membership fee. With that transaction, Friends of the Everglades (FOE) had its first member—and, as Douglas (1987) jokingly describes it, an endowment.[8] By the following spring, FOE had a charter and a committee, with Douglas installed as president (Davis 2009; Poole 2015).

[8] Douglas also co-founded WATER! A Coalition of Citizens Concerned with Florida's Water Resources with Arthur R. Marshall in 1975 (Davis 2009).

So began Douglas's new career, age 79, as an environmental activist, and she headed FOE until her centenary year (Fig. 5.3).

Jack E. Davis (2003a, 2009) notes that a 92-year-old Douglas declared conservation a 'dead word' in a November 1982 speech to the Sarasota Wellesley Club, and she had long positioned FOE as a *restoration* organization. The mission statement of FOE is to 'preserve, protect, and restore the only Everglades in the world' (FOE 2021). In stump speeches and interviews across the state, and in articles, her argument never faltered or digressed—the biological and hydrological importance of the Everglades to the ecological health of the peninsula could no longer be ignored, and needed protection from the jetport. America could have a jetport in the Everglades, or a River of Grass, but not both (see Grunwald 2006; Davis 2009). For a long time, Douglas's relationship with the Everglades was a predominantly literary one. It was not until she joined the jetport fight (formalized through FOE) that she emerged as an Everglades activist.

Fig. 5.3 Marjory Stoneman Douglas (1890–1998) at a Friends of the Everglades event, early-1970s. Courtesy of Florida Memory/State Archives of Florida

The Everglades: River of Grass (Douglas 2017) was not written as a prescriptive, apostolic, or activist text, but as a (often sobering) journalistic record of natural and cultural history in the Everglades. But it became increasingly so in the decades that followed as the American environmental movement gained momentum. Douglas's own Hurricane House imprint (which ran for almost a decade, from the mid-1960s to the mid-1970s) reissued *The Everglades: River of Grass* in 1965, keeping it in print after Rinehart's final printing in 1959, at a time when renewed environmental attention was being focused on the Everglades.

On September 17, 1969, the Department of the Interior released its *Environmental Impact of the Big Cypress Swamp Jetport* (Leopold et al. 1969).[9] USGS hydrologist Luna B. Leopold, son of Aldo Leopold (Chapter 4), was the lead author. Leopold et al.'s (1969: 1) opening sentence was damning for the future of the jetport: 'Development of the proposed jetport and its attendant facilities will lead to land drainage and development for agriculture, industry, housing, transportation, and services in the Big Cypress Swamp which will inexorably destroy the south Florida ecosystem and thus the Everglades National Park.' Two more reports released in the days and weeks that followed consolidated the findings of the Leopold Report (explored in Davis 2009).

The death knell for the jetport came resoundingly on New Year's Day 1970. In his New Year address to the nation, President Richard M. Nixon announced that the 1970s would be the decade of the environment—before annulling the jetport in Big Cypress Swamp (Davis 2003a, Grunwald 2006; Lowry 2009; Dunn 2019).[10] The first Earth Day, organized by Wisconsin Senator Gaylord A. Nelson, was held on April 22, 1970. And a slew of new environmental legislation would be introduced before the end of the decade, including the 1970 National Environmental Policy Act, Reorganization Plan No. 3 of 1970, which

[9] Leopold et al.'s (1969) *Environmental Impact of the Big Cypress Swamp Jetport* was the first environmental impact statement (EIS) conducted in the U.S., and became federally mandated under Section 102 in Title I of the 1969 National Environmental Policy Act, signed into law on January 1, 1970.

[10] The 'Jetport Pact,' signed on January 16, 1970, agreed to the relocation of the jetport (Davis 2009). The 24,960-acre, one-runway airport on the eastern boundary of what is now Big Cypress National Preserve is today an aviation training facility—the Dade-Collier Training and Transition Airport.

established the Environmental Protection Agency (EPA), the 1970 Clean Air Act, the 1972 Clean Water Act, the 1973 Endangered Species Act, the 1976 National Forest Management Act, and the 1976 Federal Land Policy and Management Act. But the Big Cypress fight was not yet over. Parcels of the swamp were still privately owned and vulnerable to residential and commercial development. Browder, the Everglades Commission, and other conservationists, politicians, and political staffers and advisors continued to defend the swamp. President Nixon proposed the establishment of a Big Cypress National Fresh Water Reserve in February 1972, and on October 4, 1974, President Gerald Ford established the Big Cypress National Preserve (on the Browder-led fight for a Big Cypress preserve, see Grunwald 2006, Davis 2009, Dunn 2019) (Fig. 5.4).

With an increased commitment across the U.S. to environmental protection in the 1970s, there followed in Florida a flurry of 'battle books,' in defense of the subtropical environment (following in the literary environmental activism tradition of Stegner's (1955) *This Is Dinosaur: Echo Park Country and Its Magic Rivers*, see Chapter 6). Douglas's regionalism was quickly politicized across several activist volumes, including William Ross McCluney's (1971) *The Environmental Destruction of South Florida: A Handbook for Citizens*, and Robert B. Rackleff's (1972) *Close to Crisis: Florida's Environmental Problems*. A decade later, FOE published its own 'battle' volume—Leonard Pardue et al.'s (1982) *Who Knows the Rain?: Nature and Origin of Rainfall in South Florida*, for which Douglas wrote the Foreword—the first in its South Florida Regional Studies Series (after Davis 2009).

Fig. 5.4 Big Cypress National Preserve, Florida. Photograph by Robert M. Overton, 1992, courtesy of Florida Memory/State Archives of Florida

5.3.2 Restoring the Kissimmee, the Beginning

With the close of the jetport project,[11] Friends of the Everglades turned its attention to the water politics in and of the Everglades. To the draining, dredging, diking, plugging, and staunching of the Everglades. By the time Douglas became involved in Everglades conservation and restoration, the Everglades had accrued a long, checkered backstory of drainage and reclamation, overseen by state officials, land boosters and promoters, land sales companies, as well as agricultural pioneer-settlers, farmers, homesteaders, and commercial agriculture syndicates. Half the predrainage Everglades had vanished, lost to agriculture and development (Grunwald 2006; Lowry 2009; Dunn 2019).

In the early-1970s, Douglas joined with University of Miami ecologist, SeaDade-Islandia activist, and early FOE member Arthur R. Marshall, and Johnny Jones of the Florida Wildlife Federation, to form a formidable Everglades conservation trinity. Marshall had drafted an 18-part 'Marshall Plan' focused on restoring the historic slow sheet flow south through the Everglades to Florida Bay, beneath Alligator Alley (I-75) and the Tamiami Trail (U.S. 41), and onto the sloughs (Davis, 2003a, 2009, Grunwald 2006). And Marshall's plan began in the dechannelization of the Kissimmee River. Begun in the 1960s and completed in 1971, the U.S. Army Corps of Engineers (USACE) and the South Florida Water Management District (SFWMD, the largest of Florida's five WMDs) had supplanted the 103-mile-long, six-foot-deep river with the C-38 Canal—a 56-mile-long, 30-foot-deep canal (Grunwald 2006; Davis 2009; Lowry 2009; Dunn 2019). For Douglas, the

[11] But airport plans for the Everglades would not end with the Big Cypress jetport. A new airport controversy erupted in the 1990s at Homestead Air Force Base, between Everglades National Park and Biscayne National Park, closed in the aftermath of Hurricane Andrew in August 1992. The fight revealed rifts between the Everglades Coalition and the Clinton administration, as well as within the Everglades Coalition, pitting airport opponents such as Friends of the Everglades, Tropical Audubon, the Sierra Club, and the Natural Resources Defense Council, against the more moderate position of, e.g., National Audubon and Audubon Florida. And once more, the Department of the Interior had to defend Everglades National Park from another federal agency—this time, the Federal Aviation Administration (FAA). The controversy culminated, concluded in the days leading up to, and immediately after, the 2000 presidential election (see Grunwald 2006; Lowry 2009 for commentary on the Homestead airport conservation fight).

Kissimmee River was the lynchpin in the health of the Everglades: 'Until we get the Kissimmee River valley restored, we'll never get all the water in the Everglades we need' (Douglas 1987: 231). In 1976, the trio spoke before the Senate Appropriations Committee in Tallahassee, requesting funds to restore the Kissimmee River. Although the request was denied, they did secure passage of a bill (largely drafted by Jones and Marshall)—the Kissimmee River Restoration Act of 1976—requesting the USACE study the dismantling of the Kissimmee Canal (Douglas 1987; Davis 2003a, 2009) (Fig. 5.5).

A quote from Jones in an article in the February 9, 1981 issue of *Sports Illustrated*—the swimsuit issue featuring Christie Brinkley—about the fading environmental legacy of Florida's governor Daniel Robert Graham added to the Everglades debate (Grunwald 2006). The magazine used Brinkley's photoshoot on Captiva Island to segue to a scathing commentary on the poor ecological condition of South Florida, pointing

Fig. 5.5 Unobstructed Kissimmee River flowing into Lake Okeechobee, Florida. Photograph by John Henry Davis, 1943, courtesy of Florida Memory/State Archives of Florida

to vanishing mangrove swamps, the C-38 Canal controversy, Lake Okeechobee and bacteria-infested canals, contaminated drinking water, beach development, and more (Boyle and Mechem 1981). Part-way through the article was Jones's statement celebrating Graham's work as an environmental senator, only to then note, 'but as governor he has wandered away from us' (Jones 1981 in Boyle and Mechem 1981: 93; cf. Grunwald 2006). Not long after the issue appeared on newsstands, Jones received an invitation to meet with the governor. Accompanied by Marshall, the pair presented an abridged version of the Marshall Plan and encouraged the governor to support the dismantling of the C-38 Canal. They also handed Graham a copy of FOE's (1982) Marshall Plan pamphlet, *For the Future of Florida, Repair the Everglades* (see also Grunwald 2006; Davis 2009).

Two years later, in August 1983, Governor Graham announced his 'Save Our Everglades' program, a 17-year program to prioritize Everglades restoration into the new millennium (following on from his 'Save Our Coasts' program in 1982, and 'Save Our Rivers' program in 1981). Douglas was one of three experts invited by the governor to review the program before its unveiling (Davis 2009). Much of Graham's program came straight out of the Marshall Plan playbook. Including restoration of a small section of the old Kissimmee riverbed, beginning the process of dechannelization. Adjustment of water flow under Alligator Alley to reconnect the agricultural area to the conservation areas, and under the Tamiami Trail to reconnect the WCAs to Everglades National Park (Douglas had also long lobbied for dismantling the dikes surrounding the WCAs, see Douglas 1987). Restoration of flows through Shark River Slough and Taylor Slough in the national park (Grunwald 2006; Davis 2009; Lowry 2009; Dunn 2019). While many of Graham's 'Save Our Everglades' proposals were uncontroversial, the Kissimmee River restoration was still divisive among Florida's cattle and sugar syndicates.

To guarantee the rollout of his program, Graham needed support from outside the state. He required federal endorsement. But he was confronted with a Reagan administration unwilling to support land purchases, and a USACE that demanded an economic rationale for dismantling the C-38 Canal, insisting that environmental projects sat outside its purview (Grunwald 2006). Unwilling to wait for the federal government to catch up to his program, Graham set to work. Even after passage of the 1976 Kissimmee River Restoration Act, the USACE had continued to stall on beginning dechannelization. But in 1984, one year into 'Save Our Everglades,' Graham announced the Kissimmee River Demonstration Project and invited Douglas to join him in planting a cypress tree at the ceremony beside the C-38 Canal. A year later, the pair returned for a boat tour along one of the first oxbows to be restored (Douglas 1987, cf. Davis 2003a, 2009). In a bid to reelevate the Everglades restoration question to a national platform, Graham resurrected the Everglades Coalition in 1985, the alliance of local and national environmental groups that Browder had brought together to earlier lobby against the Big Cypress jetport project (with FOE joining Audubon Florida, the National Wildlife Federation, and others). When the Everglades Coalition hosted its first annual conference in 1986, the restoration of the Kissimmee River was the only issue on agenda.

A few months later, Graham was elected to the U.S. Senate, and he carried the Kissimmee River restoration issue to Capitol Hill, as his principal environmental legislative priority (meanwhile, in November 1986, Congress authorized the Emergency Wetland Resources Act, to protect and restore wetlands). In 1990, when a public works bill crossed his desk, he added a clause that expanded the mission of the USACE beyond economic projects, authorizing the USACE to undertake environmental projects. After the bill passed, Graham made sure that the Kissimmee River was the first stop on the USACE's new environmental journey (Grunwald 2006; Lowry 2009). Congress authorized the Kissimmee River Restoration Project in the 1992 Water Resources Development Act (see e.g. Bousquin 2014 for reviews and progress on the Kissimmee River Restoration Project).

5.3.3 Water Pollution, Restoration, and a Legislative Oxymoron: From the Marjory Stoneman Douglas Act to the Everglades Forever Act

Water contamination and pollution is a major disruptor to Everglades hydrology. Phosphorous pollution from sugarcane production is a leading contributor (on Big Sugar and Everglades water pollution, see Grunwald 2006; also Richardson 2008; Davis 2009; Lowry 2009; Lodge 2017; Dunn 2019). For Douglas (1987), pollution cleanup ranks just behind restoring the Kissimmee River as the leading environmental priorities in Everglades restoration. In the 1980s, Douglas emerged as a prominent anti-sugar activist (and supporter of Florida Rural Legal Services, which protected migrant farm laborer rights, see e.g. Davis 2001, 2003a), and FOE was involved in political dialogue of water quality legislation.[12]

When President Bill Clinton was inaugurated as the 42nd President of the U.S. on January 20, 1993, Everglades restoration activists were hopeful.[13] Not least because environmental politician Al Gore was now Vice President. And Bruce Babbitt, former president of the League of Conservation Voters, was appointed as Secretary of the Interior. There was still little to show from Graham's 'Save Our Everglades' program. For his first official trip, Secretary Babbitt traveled to Tallahassee, where he declared Everglades restoration as one of his leading legislative priorities. Babbitt formed the South Florida Ecosystem Restoration Task Force, to oversee the various federal agencies involved (Grunwald 2006; Lowry 2009). In Florida, new governor Lawton Chiles had a similar idea and created the Governor's Commission for a Sustainable South Florida, with a membership embracing environmentalists, water utilities companies, developers, farmers, bankers, and Native Nations groups, as well as government agencies.

But by July 1993, an announcement from Babbitt would unsettle and jar with the Everglades Coalition and other Everglades restoration activists. The 'Babbitt Agreement' set out a framework for settling the phosphorous lawsuits and had the political support of Florida's Senator Graham and Governor Chiles (Grunwald 2006). To Everglades activists, it was an awful compromise, prolonging cleanup deadlines from 2002 to 2006, and transferring the heavy cost to taxpayers, even if it included

[12] FOE online interview, July 23, 2020.

[13] While on the campaign trail in Miami the previous year, fellow Wellesley alumna Hillary Rodham Clinton had met with Douglas in her Coconut Grove cottage, and Douglas later gave her endorsement to the presidential candidate. Douglas met the President and First Lady later in 1993, when they again visited Miami (Davis 2009).

upward of 40,000 acres of filter marshes. George Barley, developer and Everglades Coalition activist, began petitioning for a penny-a-pound tax on sugar sales (Grunwald 2006). Penny-a-pound would descend into a bitter battle between environmentalists and sugar growers, the most acrimonious since the Big Cypress jetport. Babbitt was surprised by the backlash from Everglades activists to his deal securing the largest cleanup and restoration in U.S. history. In 1993, George Barley and Paul Tudor Jones II founded the Everglades Foundation, to advance scientific research on the Everglades. The following year, Barley also founded the political nonprofit Everglades Trust. The Everglades Foundation launched the Save Our Everglades Amendment to challenge the sugarcane industry (Grunwald 2006; Davis 2009; Lowry 2009).

On Babbitt's next trip to Florida, to meet with the Everglades Coalition in Miami in January 1994, the reception was cooler still. Douglas had been awarded the Presidential Medal of Freedom, the highest civilian honor, by President Clinton less than two months earlier, on November 30, 1993, but Babbitt declined to meet with her Friends of the Everglades organization on his trip. With settlement talks with sugar growers at an impasse, Chiles introduced a cleanup bill in February 1994—the Marjory Stoneman Douglas Act. But the 1994 Douglas Act was a legislative oxymoron—a bill heavily crafted by sugar interests and lobbyists, delaying final phosphorous standards until 2006 under the banner of an anti-sugar activist (see also Chiles's poor 1991 Marjory Stoneman Douglas Everglades Protection Act, in Davis 2009). A now 103-year-old Douglas did not waste any time in informing the governor of her disapproval and requested her name be removed immediately from the bill.[14] It was a political hiccough for Chiles, but the bill had the endorsement of the Clinton administration, and it passed overwhelmingly in the Florida legislature, under the amended title, the Everglades Forever Act (on the politics of the Everglades Forever Act, see Davis 2001, 2009; Grunwald 2006; Lowry 2009; Lodge 2017).[15]

[14] FOE online interview, July 23, 2020.

[15] The cleanup deadline has been repeatedly pushed back further by subsequent Florida governors—first to 2016 by Governor Jeb Bush in 2003, then to 2025 by Governor Rick Scott in 2012.

5.3.4 Friends of the Everglades After Douglas

In the decades since Douglas's death, the grassroots advocacy of FOE has continued. Today, FOE continues to fight toxic algae blooms in Lake Okeechobee, and toxic discharges into the Caloosahatchee and St. Lucie estuaries, the Everglades, and Florida Bay. FOE is lobbying for better management of Lake Okeechobee and has been involved in various legal actions to stop the SFWMD from backpumping runoff from sugarcane plantations into Lake Okeechobee, as well as to secure the treatment of agricultural chemicals discharged from the Everglades Agricultural Area (EAA) into the Everglades. FOE has commissioned a study into the SFWMD's proposal for an EAA Storage Reservoir 30 miles south of Lake Okeechobee (see Mitsch 2019),[16] and with the Sierra Club, and Center for Biological Diversity, submitted joint comments to the USACE's Final EIS for the reservoir (Umpierre et al. 2020).

Other campaigns and petitions include protecting water quality and preventing toxic algae across the state, halting oil drilling in Big Cypress National Preserve, tackling the public health threat from sugarcane burning, protecting the endangered Florida panther and its habitat by halting new tollways, and regulating runoff from sugar fields in the EAA to stop methyl-mercury pollution in the Everglades, and FOE is part of the Hold the Line Coalition, working to halt the expansion of Dade County's Urban Development Boundary (UDB) into the Everglades. In 2020, TAS, FOE, and 1,000 Friends of Florida were successful in blocking a toll road project outside of the UDB. In 2019, the Southern Alliance for Clean Energy, FOE, and TAS were plaintiffs in a lawsuit involving the Turkey Point nuclear plant and reached a settlement over its pollution of Biscayne Bay.[17]

FOE remains a small grassroots nonprofit, and in December 2019, the organization merged with the Bullsugar Alliance, the nonpolitical group of Bullsugar.org.[18] Bullsugar.org began in 2014 to call out the political power of Big Sugar in Florida and beyond, and works to stop

[16] FOE online interview, July 23, 2020.

[17] FOE online interview, July 23, 2020.

[18] FOE online interview, July 23, 2020.

toxic discharges into the Caloosahatchee and St. Lucie estuaries, and restore the flow of clean freshwater to Florida Bay. Bullsugar.org is a founding partner (with the Everglades Foundation, Everglades Trust, Patagonia, and others) of the Now or Neverglades Declaration (see Now or Neverglades Declaration 2018).

5.3.5 A Social Activism Coda: Douglas's Legacy and Never Again MSD

Seventy-one years after the publication of *The Everglades: River of Grass*, and 49 years after Douglas set up Friends of the Everglades, her legacy of courage, advocacy, and activism would be resurrected less than 50 miles north of her Coconut Grove home in a way—and for a cause—it should never have had to be. On February 14, 2018, a shooting at Marjory Stoneman Douglas High School in Parkland, Florida, killed 14 students and three faculty members (established in 1990, the high school was named to honor Douglas's centenary). A coalition of 20 student survivors—including Emma Gonzalez, David Hogg, and Cameron Kasky—formed Never Again MSD to advocate for stricter regulation and legislation to prevent gun violence. On March 24, Never Again MSD organized March For Our Lives, a national student-led walkout and protest in Washington, DC, demanding action on gun control—and the third largest protest in U.S. history, surpassed only by the Women's March in 2017 and 2018. And in the days and weeks after the tragedy—the ninth school campus shooting already that year in the U.S.—Douglas's words, and her philosophy, were politicized anew across walkouts and rallies:

> Be a nuisance where it counts; [...] Do your part to inform and stimulate the public to join your action. [...] Be depressed, discouraged, and disappointed at failure, and the disheartening effects of ignorance, greed, corruption, and bad politics—but never give up.
>
> —Douglas 1980: 13

Douglas's name is now forever entwined with the massacre that unfolded at the high school that bears her name—but, more significantly, with the

Fig. 5.6 Student demonstrators at a Never Again MSD rally in Florida's state capital, Tallahassee, on February 21, 2018, one week after the Parkland shooting. One of the signs displays the phrase 'Douglas Strong.' Photograph by Adam Watson, courtesy of Florida Memory/State Archives of Florida

surging wave of political activism against the National Rifle Association (NRA) that followed (on the Never Again and March For Our Lives movement, see especially Hogg and Hogg 2018; Cullen 2019; Lerner 2019) (Fig. 5.6).

5.4 Theorizing, Politicizing, and Enacting Restoration in the Everglades

The restoration history of the Everglades is entangled with, and emerges out of, the region's history of reclamation. Resuming the whistle-stop tour through the legislative landscape of Everglades reclamation, the journey pauses at another Everglades reclamation plan, to set the legislative context within which Douglas and FOE were thrust from the late-1960s onward.

After a hurricane, tropical storms, and severe flooding befell the Everglades in the summer and fall of 1947, and less than a year after the dedication of Everglades National Park, Congress authorized the Central and Southern Florida (C&SF) Project for Flood Control and Other Purposes as part of the Flood Control Act on June 30, 1948. Through the C&SF, the USACE was tasked with controlling, taming the Everglades once and for all (U.S. Congress, House 1948). The USACE was joined by various state and federal partners—the Central and Southern Florida Flood Control District,[19] the Soil Science Society of Florida, the USDA Soil Conservation Service, and the USGS—as co-contributors toward an emergent water control policy (Douglas 1967; McCally 2000; Grunwald 2006; Dunn 2019).

The $208-million C&SF program proposed an incredibly ambitious water control system—and the largest earth-moving project since the Panama Canal at the turn of the twentieth century—underwritten by upward of 2,000 miles of levees and canals, and hundreds of spillways, floodgates, pumps (Davis 2003a, 2009; Grunwald 2006). Scheduled for completion in 1960, the C&SF would dramatically exceed this timeframe (and budget), before triggering a 'Restudy' of the C&SF, and the Comprehensive Everglades Restoration Plan (CERP) (see McCally 2000; Grunwald 2006; Lodge 2017; Dunn 2019 on the history of the C&SF Project). Under the C&SF, the Everglades was divided into units—the northern Everglades was recast as farmland (with the Kissimmee Valley as a cattle empire, and the upper Everglades an agricultural empire), the central Everglades as massive reservoirs, and the eastern Everglades as farms and suburbs. The southern Everglades was protected as Everglades National Park (Grunwald 2006).

The USACE began a threefold infrastructural project in the Everglades. Through the installation of a 100-mile-long perimeter levee running north–south through the eastern Everglades, to protect towns and cities along the Gold Coast from direct flooding (but severing almost 20 per cent of the peripheral Everglades from the interior). Through the designation of the 700,000-acre EAA in the upper Everglades, at Lake

[19] With the retirement of the Everglades Drainage District in 1949, a new, expanded district was introduced—the Central and Southern Florida Flood Control District (later renamed the South Florida Water Management District).

Okeechobee's southern border (encompassing more than a quarter of the historic, predrainage Everglades). And through the creation of three giant water conservation areas (WCAs) between the EAA and Everglades National Park, the northernmost of which (WCA 1) became the Loxahatchee NWR in 1951, renamed the Arthur R. Marshall Loxahatchee NWR in 1988 to honor Marshall (Grunwald 2006; Davis 2009; Lodge 2017; Dunn 2019). Lake Okeechobee was further disconnected from the Greater Everglades ecosystem by upgrades to the dike running along its southern shore—realized in the Herbert Hoover Dike—and water from the lake was sent west and east to the Gulf Coast and Atlantic Ocean along canals connecting to the Caloosahatchee and St. Lucie rivers. A half-million acres south of the lake were drained for sugarcane plantations, vegetable fields, and cattle ranches. The sheet flow south through Everglades National Park to Florida Bay, and to the Gulf of Mexico and Biscayne Bay, stalled.

A restoration rhetoric also featured in the C&SF and had echoes of Graham's 'Save Our Everglades' program and the 'Marshall Plan,' as well as Coe's earlier plan for a Tropic Everglades National Park. Legislative attention was variously focused on expanding Big Cypress National Preserve, restoring the Turner River area and Lake Toho, protecting the coral reefs in the Florida Keys National Marine Sanctuary, and restoring Taylor Slough inside Everglades National Park (Grunwald 2006). In 1989, Graham helped secure congressional approval for an ambitious project to expand the park boundary by 107,000 acres and restore Shark River Slough (the Everglades National Park Protection and Expansion Act). The wetland marshes and swamps of northeast Shark River Slough had earlier fallen foul of the revised park boundary in the 1940s, later compounded by the C&SF diverting the slough west to protect eastern Everglades land. The project authorized the restoration of the slough and its relocation back east. Also included in the bill was a flood control program for the Eight-and-a-Half-Square-Mile Area, a low-lying east Everglades neighborhood at risk by the Shark River Slough restoration work (Grunwald 2006; Lowry 2009).[20] These restoration programs

[20] But the USACE flood control plan for the Eight-and-a-Half-Square-Mile Area neighborhood would hold up the restoration of Shark River Slough for several years (Grunwald 2006).

marked the first *environmental* projects undertaken by the USACE and were applauded by the Everglades Coalition, but they were still incredibly localized. FOE, partnering with Everglades National Park rangers, installed culverts and filled old canals, in an effort to increase water flow through the national park (Douglas 1987). Meanwhile, Everglades restoration activists resumed strategizing the restoration of the Everglades ecosystem, modeled after the Marshall Plan.

In the early-1990s, the Everglades Coalition drafted *Everglades in the 21st Century* (Everglades Coalition 1992) and began lobbying for a comprehensive 'Restudy' of the C&SF Project. To recast the C&SF as not just a flood control and water supply project, but an environmental project, too. The Restudy of the C&SF was commissioned by the 1992 Water Resources Development Act (with further congressional support for the Restudy coming with the 1996 Water Resources Development Act).

In October 1998, five months after Douglas's death, the USACE submitted its nine-volume draft plan to Congress to overhaul the C&SF (USACE 1998, cf. 1999a, b). The Restudy plan—the Comprehensive Everglades Restoration Plan (CERP)—was unveiled by Vice President Gore in West Palm Beach later that month (for further critical commentary on the history of the C&SF Restudy or CERP, see Voss 2000; Grunwald 2006; Davis 2009; Dunn 2019). On July 1, 1999, Vice President Gore presented the Restudy (or CERP) to Congress, with an accompanying pamphlet, *Rescuing an Endangered Ecosystem: The Plan to Restore America's Everglades* (USACE 1999b). As CERP worked its way through congressional hearings, Douglas's legacy and *The Everglades: River of Grass* (Douglas 2017) were invoked and quoted from several times (detailed in Grunwald 2006). FOE and Everglades National Park opposed the legislation (packed with urban and agricultural interests, and neglecting the national park), but other Everglades activists (including Audubon, the National Parks Conservation Association, the National Wildlife Federation, Defenders of Wildlife) rallied behind the legislation.

On December 11, 2000, five weeks after the presidential election, in a bipartisan ceremony in the Oval Office attended by Florida governor Jeb Bush and other invited politicians, President Clinton signed into law the Water Resources Development Act. Title IV of the Act authorized the

Comprehensive Everglades Restoration Plan (CERP). With its authorization, it became the largest ecological restoration project in the world at that time. It was a 'rescue mission' (Grunwald 2006) for the Everglades, and a *repair* project as much as a *restoration* project. To borrow again James Aronson et al.'s (1993) phrasing, it is restoration *sensu lato* that describes 'Everglades restoration.' Nine months afterward, the U.S. was left reeling by the events of 9/11—and the Iraq War followed less than two years later.

An ambitious, four-decade restoration program, featuring 68 project components, CERP had a projected cost of at least $7.8 billion. CERP is anchored in above- and belowground water storage reservoirs, and aquifer storage and recovery wells, as well as filter marshes, seepage controls, barrier removals, and water conservation and wastewater treatment plants (USACE 1999a, b; Voss 2000; Grunwald 2006; Lowry 2009; National Academy 2021). The restoration of Lake Okeechobee and the Greater Everglades is central, as is increasing water supply for the region. Now halfway through that timeframe, costs have already exceeded $20 billion (Dunn 2019). And not one of the 68 components has reached completion, although some are close.[21]

The restoration of the Everglades is 'long, slow work—incremental work.'[22] Even late in her life, after decades of activism, Douglas did not witness restoration of the Everglades. 'Everglades restoration' is progressing, but as CERP enters its fifth administration, the mutability of 'Everglades restoration' endures (see e.g. Davis and Ogden 1997; Weisskoff 2005; Grunwald 2006; Richardson 2008; Lowry 2009; Zweig and Kitchens 2010; Lodge 2017; Osborne and Wetzel 2017; National Academy of Sciences 2021). In its eighth biennial review of *Progress Toward Restoring the Everglades* in 2020, the National Academies of Sciences, Engineering, and Medicine repeated the recommendation from earlier progress reports in the series—more water from Lake Okeechobee needs to be sent south, through the Everglades (see National Academies 2021). Climate change and rising sea levels add another complication to Everglades restoration (Grunwald 2006; Davis 2009; Lowry 2009;

[21] FOE online interview, July 23, 2020.

[22] FOE online interview, July 23, 2020.

Lodge 2017). Much of the Everglades is no more than a few feet above sea level. Already, sea level rise is leading to the collapse of fragile peat moss in the coastal Everglades, with probable saltwater incursions into freshwater ecosystems.[23] There remains in reserve W. Hodding Carter's (2004) radical proposition—abandon CERP altogether, and designate the historic, predrainage Everglades ecosystem as a national park, from the Kissimmee Chain of Lakes into Lake Okeechobee, and the estuaries, and south through the Everglades to Florida Bay and the Keys.

5.5 After Katrina: Poems from the Gulf Coast as Environmental Advocacy, Justice

The Everglades is not the only font for Floridian literary environmental advocacy and activism. Over the last decade, poetry has become a powerful platform for writers to describe—and process and come to terms with—the destruction and devastation wrought by hurricanes along the Gulf Coast (cf. also Douglas 1958 for journalistic dispatches and scientific reporting on hurricanes from Florida, North Carolina, Jamaica, Martinique, and Cuba). I wrote this closing section of the chapter while Hurricane Dorian, a dangerous Category 5 storm, and the second-strongest Atlantic hurricane on record, slammed the Bahamas in late-August and early-September 2019. The timing was uncomfortable and humbling. But it reinforced the power of poetry and verse to give a voice to entanglements with natural disasters.

One of the earliest contributions to this Gulf Coast hurricane canon is Philip C. Kolin and Susan Swartwout's (2006) edited collection, *Hurricane Blues: Poems about Katrina and Rita.* Few, if any, other poetry collections had been born of natural disasters before Kolin and Swartwout's (2006) *Hurricane Blues* appeared. Featuring almost 100 poets and more than 100 poems, it is a testament to the hurricane and tells of just two of the more than two dozen hurricanes and tropical and subtropical storms that made up the 2005 Atlantic hurricane season. Hurricane

[23] FOE online interview, July 23, 2020.

Katrina, the twelfth storm of the season, struck in late-August, while Hurricane Rita, the season's eighteenth storm, struck in mid-September. Kolin and Swartwout's (2006) *Hurricane Blues* tracks the path of the storms along the Gulf Coast, up the Florida Panhandle and west across Louisiana, Texas, and beyond, pausing with a fierce intensity—much as the two storms did—on New Orleans. The sections of the anthology present vignettes of encounters with the pregnant, impending storms, of the storms making landfall, of flooding, of the immediate aftermath, of grief and mourning, and of resolutions and actions. The poems are often eyewitness accounts. Documentary. Personal and intimate. Inventories of sorrow and loss. Mundane. Painful. Sad. Resilient. Elegiac. Urgent. Political. Hopeful.[24]

Rebecca Dunham's (2017) *Cold Pastoral: Poems* is a poetic, lyric response to a triptych of recent U.S. natural and environmental disasters—placing Hurricane Katrina alongside the Deepwater Horizon explosion and oil spill in September 2010, and the Flint water crisis that began in April 2014—although Deepwater Horizon receives the greatest attention. Deepwater Horizon also provides the context for David Gessner's (2012) *The Tarball Chronicles: A Journey Beyond the Oiled Pelican and Into the Heart of the Gulf Oil Spill*, and Terry Tempest Williams's (2016) chapter on Gulf Islands National Seashore in *The Hour of Land: A Personal Topography of America's National Parks*, celebrating the NPS centennial. A different approach is taken by Gabriel Ojeda-Sagué in his 2019 bilingual poetry collection, *Losing Miami*, for he instead offers a speculative reimagining of a Miami lost to sea level rise and hurricanes—of the loss of Cuban-American culture, community, and language, above and beyond urban infrastructure (Ojeda-Sagué 2019).

[24] In the 2012 film, *Beasts of the Southern Wild*, magical realism is melded with a recasting of the events of Hurricane Katrina to tell the story of six-year-old Hushpuppy and her father, Wink, as they await the arrival of—and endure—a storm in their southern Louisiana bayou community known as the 'Bathtub.'

Works Cited

Allen, Joel Asaph. 1886. The Present Wholesale Destruction of Bird-Life in the United States. *Science* 7 (160): 191–195.

Aronson, J., C. Floret, E. LeFloc'h, C. Ovalle, and R. Pontanier. 1993. Restoration and Rehabilitation of Degraded Ecosystems in Arid and Semi-Arid Lands. I. A View from the South. *Restoration Ecology* 1 (1): 8–17.

Bailey, Harold H. 1925. *The Birds of Florida*. Baltimore, MD: Waverly Inc.

Beasts of the Southern Wild. 2012. Dir. Ben Zeitlin. Court 13 Films.

Bousquin, Stephen G., ed. 2014. Special Section: Kissimmee River Restoration Project. *Restoration Ecology* 22 (3): 345–434.

Boyle, Robert H., and Rose Mary Mechem. 1981. There's Trouble in Paradise. *Sports Illustrated* (February 9): 82–96.

Carson, Rachel. 1962. *Silent Spring*. Boston, MA: Houghton Mifflin.

Carter, W. Hodding. 2004. *Stolen Water: Saving the Everglades from Its Friends, Foes, and Florida*. New York, NY: Atria Books.

Cullen, Dave. 2019. *Parkland*. New York, NY: Harper.

Davis, Jack E. 2001. Green Awakening: Social Activism and the Evolution of Marjory Stoneman Douglas's Environmental Consciousness. *The Florida Historical Quarterly* 80 (1): 43–77.

Davis, Jack E. 2003a. "Conservation Is Now a Dead Word:" Marjory Stoneman Douglas and the Transformation of American Environmentalism. *Environmental History* 8 (1): 53–76.

Davis, Jack E. 2003b. Up from the Sawgrass: Marjory Stoneman Douglas and the Influence of Female Activism in Florida Conservation. In *Making Waves: Female Activists in Twentieth-Century Florida*, ed. Jack E. Davis and Kari Frederickson, 147–176. Gainesville, FL: University Press of Florida.

Davis, Jack E. 2009. *An Everglades Providence: Marjory Stoneman Douglas and the American Environmental Century*. Athens, GA: University of Georgia Press.

Davis, Jack E. 2017. *The Gulf: The Making of an American Sea*. New York, NY: Liveright/W.W. Norton & Company Inc.

Davis, Jack E., and Kari Frederickson, eds. 2003. *Making Waves: Female Activists in Twentieth-Century Florida*. Gainesville, FL: University Press of Florida.

Davis, Jack E., and Raymond Arsenault, eds. 2005. *Paradise Lost? The Environmental History of Florida*. Gainesville, FL: University Press of Florida.

Davis, Steven M., and John C. Ogden. eds. 1997. *Everglades: The Ecosystem and Its Restoration.* Boca Raton, FL: St. Lucie Press.
Douglas, Marjory Stoneman. 1952. *Road to the Sun.* New York, NY: Rinehart and Company.
Douglas, Marjory Stoneman. 1953. *Freedom River: Florida 1845.* New York, NY: Charles Scribner's Sons.
Douglas, Marjory Stoneman. 1958. *Hurricane.* New York, NY: Rinehart and Company.
Douglas, Marjory Stoneman. 1961. *The Key to Paris.* Pennsylvania, PA: J.B. Lippincott Company,
Douglas, Marjory Stoneman. 1967. *Florida: The Long Frontier.* New York, NY: Harper and Row.
Douglas, Marjory Stoneman. 1980. How You Can Protect the Environment. *GeoJourney: The Magazine of Florida's Natural Resources* 1 (2): 13.
Douglas, Marjory Stoneman. 1987. *Voice of the River: An Autobiography with John Rothchild.* Sarasota, FL: Pineapple Press Inc.
Douglas, Marjory Stoneman. 1990. *Nine Florida Stories by Marjory Stoneman Douglas.* Gainesville, FL: University Press of Florida.
Douglas, Marjory Stoneman. 1998. *A River in Flood, and Other Florida Stories by Marjory Stoneman Douglas.* Gainesville, FL: University Press of Florida.
Douglas, Marjory Stoneman. 2002. *The Wide Brim: Early Poems and Ponderings of Marjory Stoneman Douglas*, ed. Jack E. Davis. Gainesville, FL: University Press of Florida.
Douglas, Marjory Stoneman. 2003 [1959]. *Alligator Crossing.* Minneapolis, MN: Milkweed Editions.
Douglas, Marjory Stoneman. 2017 [1947]. *The Everglades: River of Grass*, 70th Anniversary. Sarasota, FL: Pineapple Press Inc.
Douglas, Marjory Stoneman, and Ralph Stoutamire. 1932. *The Parks and Playgrounds of Florida.* Tallahassee, FL: Florida Department of Agriculture.
Dunham, Rebecca. 2017. *Cold Pastoral: Poems.* Minneapolis, MN: Milkweed Editions.
Dunn, John M. 2019. *Drying Up: The Fresh Water Crisis in Florida.* Gainesville, FL: University Press of Florida.
Everglades Coalition. 1992. *Everglades in the 21st Century.* St. Petersburg, FL: Everglades Coalition.
Friends of the Everglades. 1982. *For the Future of Florida, Repair the Everglades.* Coconut Grove, FL: Friends of the Everglades.
Friends of the Everglades. 2021. About us. *Friends of the Everglades.* http://www.everglades.org/about.

Gessner, David. 2012. *The Tarball Chronicles: A Journey Beyond the Oiled Pelican and Into the Heart of the Gulf Oil Spill*. Minneapolis, MN: Milkweed Editions.
Graham, Frank Jr. 1990. *The Audubon Ark: A History of the National Audubon Society*. New York, NY: Alfred A. Knopf Inc.
Grinnell, George Bird. 1886. The Audubon Society. *Forest and Stream: A Weekly Journal of the Rod and Gun, Angling, Shooting, the Kennel, Practical Natural History, Fishculture, Protection of Game, and the Inculcation in Men and Women of a Healthy Interest in out-Door Recreation and Study* 26 (3): 41.
Grunwald, Michael. 2006. *The Swamp: The Everglades, Florida, and the Politics of Paradise*. New York, NY: Simon & Schuster.
Hogg, David, and Lauren Hogg. 2018. *#NeverAgain: A New Generation Draws the Line*. New York, NY: Random House.
Holmes, Madelyn. 2004. *American Women Conservationists: Twelve Profiles*. Jefferson, NC, and London: McFarland & Company, Inc. Publishers.
Kolin, Philip C., and Susan Swartwout, eds. 2006. *Hurricane Blues: Poems about Katrina and Rita*. Cape Girardeau, MO: Southeast Missouri State University Press.
Leopold, Aldo. 1968 [1949]. *A Sand County Almanac, and Sketches Here and There*. New York, NY: Oxford University Press.
Leopold, Luna B., Arthur R. Marshall Jr., and Manuel Morris. 1969. *Environmental Impact of the Big Cypress Swamp Jetport*. Washington, DC: Department of the Interior.
Lerner, Sarah, ed. 2019. *Parkland Speaks: Survivors from Marjory Stoneman Douglas Share Their Stories*. New York, NY: Crown Books for Young Readers/Random House Children's Books.
Lodge, Thomas E. 2017. *The Everglades Handbook: Understanding the Ecosystem*, 4th ed. Boca Raton, FL: CRC Press.
Lowry, William R. 2009. *Repairing Paradise: The Restoration of Nature in America's National Parks*. Washington, DC: Brookings Institution Press.
Marston, Anson, S.H. McCrory, and George B. Hills. 1927. *Report of the Everglades Engineering Board of Review to Commissioners of Everglades Drainage District*. Tallahassee, FL: T.J. Appleyard.
McCally, David. 2000. *The Everglades: An Environmental History*. Gainesville, FL: University Press of Florida.
McCluney, William Ross, ed. 1971. *The Environmental Destruction of South Florida: A Handbook for Citizens*. Coral Gables, FL: University of Miami Press.

McIver, Stuart B. 2009. *Death in the Everglades: The Murder of Guy Bradley, America's First Martyr to Environmentalism.* Gainesville, FL: University Press of Florida.
Mead, Daniel W., Allen Hazen, and Leonard Metcalf. 1912. *Report on the Drainage of the Everglades of Florida with Special Reference to the Lands of the Everglades Land Sales Company, Everglades Land Company, Everglades Sugar and Land Company in the Vicinity of Miami, Florida.* Chicago, IL: Board of Consulting Engineers.
Merchant, Carolyn. 2016. *Spare the Birds!: George Bird Grinnell and the First Audubon Society*. New Haven, CT and London: Yale University Press.
Mitsch, William J. 2019. Restoring the Florida Everglades: Comments on the Current Reservoir Plan for Solving Harmful Algal Blooms and Restoring the Florida Everglades. *Ecological Engineering* 138: 155–159.
National Academies of Sciences, Engineering, and Medicine. 2021. *Progress Toward Restoring the Everglades: The Eighth Biennial Review—2020*. Washington, DC: National Academies Press.
Now or Neverglades Declaration. 2018. Declaration. *Now or Neverglades Declaration.* http://www.noworneverglades.com.
Ojeda-Sagué, Gabriel. 2019. *Losing Miami*. Brooklyn, NY: The Accomplices.
Olmsted, Frederick Law Jr., and William P. Wharton. 1932. The Florida Everglades: Where the Mangrove Forests Meet the Storm Waves of a Thousand Miles of Water. *American Forests* 38 (142–147): 192.
Osborne, Todd, and Paul Wetzel, eds. 2017. Special Issue: Synthesis of Everglades Research and Ecosystem Services (SERES). *Restoration Ecology* 25.S1: S1–S106.
Pardue, Leonard, Jessie Freeling, L.J. Greenfield, and P.T. Gannon Jr. 1982. *Who Knows the Rain?: Nature and Origin of Rainfall in South Florida.* Coconut Grove, FL: Friends of the Everglades.
Parker, Garald G., and C. Wythe Cooke. 1944. Late Cenozoic Geology of Southern Florida, with a Discussion of the Groundwater. *Florida Geological Survey Bulletin* 27.
Parker, Garald G., G.E. Ferguson, S.K. Love, et al. 1955. *Water Resources of Southeastern Florida, with Special Reference to the Geology and Groundwater of the Miami Area.* Geological Survey Water-Supply Paper 1255. Washington, DC: U.S. Government Printing Office.
Poole, Leslie Kemp. 2015. *Saving Florida: Women's Fight for the Environment in the Twentieth Century*. Gainesville, FL: University Press of Florida.
Punke, Michael. 2016 [2007]. *Last Stand: George Bird Grinnell, the Battle to Save the Buffalo, and the Birth of the New West*. London: The Borough Press.

Rackleff, Robert B., ed. 1972. *Close to Crisis: Florida's Environmental Problems.* Tallahassee, FL: New Issues.

Richardson, Curtis J. 2008. *The Everglades Experiments: Lessons for Ecosystem Restoration.* New York, NY: Springer.

Small, John Kunkel. 1929. *From Eden to Sahara: Florida's Tragedy*. Lancaster, PA: The Science Press Printing Company.

U.S. Army Corps of Engineers. 1998. *Central and Southern Florida Project Comprehensive Review Study: Draft Integrated Feasibility Report and Programmatic Environmental Impact Statement*. Jacksonville, FL: U.S. Army Corps of Engineers.

U.S. Army Corps of Engineers. 1999a. *Central and Southern Florida Project Comprehensive Review Study: Final Integrated Feasibility Report and Programmatic Environmental Impact Statement*. Jacksonville, FL: U.S. Army Corps of Engineers.

U.S. Army Corps of Engineers. 1999b. *Rescuing an Endangered Ecosystem: The Plan to Restore America's Everglades.* Jacksonville, FL: U.S. Army Corps of Engineers.

U.S. Congress. House. 1928. *Letter from the Chief of Engineers, United States Army, Transmitting to the Chairman of the Committee on Commerce, United States Senate, in Response to a Letter Dated December 6, 1928, a Report on the Caloosahatchee River and Lake Okeechobee Drainage Areas, Florida, with a View to Improving for Navigation and the Control of Floods, April 9, 1928.* 70th Cong., 1st sess., H. Doc. 215.

U.S. Congress. House. 1948. *Comprehensive Report on Central and Southern Florida for Flood Control and Other Purposes.* 80th Cong., 2nd sess., H. Doc. 643.

U.S. Congress. Senate. 1848. *Report of Buckingham Smith, Esq.* 30th Cong., 1st sess., Ref. Com. 242.

U.S. Congress. Senate. 1912. *Everglades of Florida Acts, Reports, and Other Papers, State and National, Relating to the Everglades of the State of Florida and Their Reclamation.* 62nd Cong., 1st sess., S. Doc. 89.

U.S. Congress. Senate. 1914. *Florida Everglades: Report of the Everglades Engineering Commission to the Commissioners of the Everglades Drainage District and the Trustees of the Internal Improvement Fund, State of Florida, 1913.* 63rd Cong., 2nd sess., S. Doc. 379.

U.S. Congress. Senate. 1932. *The Proposed Everglades National Park: Report of a Special Committee of the National Parks Association Appointed to Study All the Features in Connection with the Proposed Everglades National Park in the State of Florida.* 72nd Cong., 1st sess., S. Doc. 54.

Udall, Stewart L. 1963. *The Quiet Crisis*. New York, NY: Holt, Rinehart and Winston.

Umpierre, Diana, Jaclyn Lopez, and Eve Samples. 2020. RE: Comments on the Final Environmental Impact Statement (EIS) for Central and Southern Florida, Everglades Agricultural Area (EAA). *Friends of the Everglades*. http://www.everglades.org/wp-content/uploads/2020/05/EAA-Storage-Reservoir_Final-EIS_Joint-Comments_Sierra-Club_Center-for-Biological-Diversity_Friends-of-the-Everglades_02–24–20–3.pdf.

Voss, Michael. 2000. The Central and Southern Florida Project Comprehensive Review Study: Restoring the Everglades. *Ecology Law Quarterly* 27 (3): 751–770.

Weisskoff, Richard. 2005. *The Economics of Everglades Restoration: Missing Pieces in the Future of South Florida*. Cheltenham: Edward Elgar Publishing Limited.

Williams, Terry Tempest. 2016. *The Hour of Land: A Personal Topography of America's National Parks*. New York, NY: Sarah Crichton Books/Farrar, Straus and Giroux.

Zweig, Christa L., and Wiley M. Kitchens. 2010. The Semiglades: The Collision of Restoration, Social Values, and the Ecosystem Concept. *Restoration Ecology* 18 (2): 138–142.

6

'The Canyonlands Did Have a Heart, a Living Heart:' Edward Abbey, Glen Canyon, and the Glen Canyon Institute

6.1 Park Ranger. Fire Lookout. Renegade Outdoorsman: The Environmental Politics of Edward Abbey

On March 21, 1981, during a protest rally at Glen Canyon Dam on the Colorado River, near Page, Arizona, five Earth First! (EF!) activists cascaded 300 feet of tapered black plastic down the center of the dam's façade, to symbolize a crack in the dam. The 'crack' was EF!'s first public act of civil disobedience and was inspired by the anarchic ambitions of Edward Abbey's (1927–1989) eponymous quartet of eco-saboteurs in his 1975 protest novel, *The Monkey Wrench Gang* (as Abbey 2004). Abbey also attended the rally and spoke from the back of a pickup truck (see Abbey 1981 in Philippon 2004). For a few hours, Abbey's anarchic environmentalism on the page sparked the brief and creative guerrilla performance of the plastic 'cracking' of the dam in Page.

This chapter begins by examining the influences behind Abbey's anarchic sensibility, and how this was molded into an anarchic, radical environmental philosophy that endorsed civil disobedience in defense of the American wilderness. The discussion journeys through Abbey's high

L. Smith, *Ecological Restoration and the U.S. Nature and Environmental Writing Tradition*, https://doi.org/10.1007/978-3-030-86148-3_6

school and college years and accompanying civil disobedience performances, to then consider how his employment as a seasonal park ranger and fire lookout across the West's national forests, national monuments, and national parks further helped to refine and consolidate his anarchic environmentalism. The desertscapes of the American West would swiftly emerge as significant anchors in Abbey's philosophy and writings. And for Abbey, there was no greater affront to the desert wilderness than the U.S. Bureau of Reclamation's (USBR) Glen Canyon Dam and Lake Powell, constructed in the late-1950s and early-1960s.

To understand Abbey's loathing of Glen Canyon Dam on the Colorado River, discussion broadens out to consider the Bureau of Reclamation's dam-building tradition in the Colorado River Basin. The episode that eventually condemned the Glen Canyon river corridor as a dam site unfolded upstream, near the confluence of the Green and Yampa rivers, when a coalition of national conservation organizations defeated a USBR dam project within the boundary of Dinosaur National Monument. The restoration of Glen Canyon became a major trope in Abbey's canon, nonfiction and fiction alike. He mounts a literary defense against the 'hydraulic empire' (after Worster 1985) in an effort to confront and upend political dialogue and (re)shape water policy. But where his famous fictional quartet in *The Monkey Wrench Gang* (Abbey 2004) failed in their endeavor to destroy the dam, nature almost succeeded, when massive snowmelt entered Lake Powell in June 1983. But inspired by Abbey's ideas, Glen Canyon restorationists are proposing programs for draining Lake Powell, led by Glen Canyon Institute's 'Fill Mead First.'

The chapter concludes by considering Abbey's legacy of desert anarchism beyond Glen Canyon. The Trump administration's massive downsizing of Utah's Bears Ears and Grand Staircase-Escalante national monuments in late-2017 revisited the quiet politics and gentle literary activism of the chapbook in defense of southern Utah's public lands.

6.1.1 On Abbey, Anarchism, and the Slickrock Desert of Utah, Arizona, and New Mexico

Abbey's first encounter with the American West was as a 17-year-old hitchhiker, leaving the family farm in the Allegheny Mountains in western Pennsylvania for three months in the summer of 1944 between his junior and senior year at high school, to embark on a personal pilgrimage west. This trip (recounted in Abbey 1984, 1991a, 2003) cemented the start of Abbey's decades-long preoccupation with the deserts of the American West. Following an eight-month deployment in Naples, Italy, at the end of the Second World War, where he served in the Military Police, Abbey returned to the United States—first to Pennsylvania, and then on to New Mexico.

It was while he was a student at the Indiana State Teacher's College in Pennsylvania (now Indiana University of Pennsylvania) that Abbey's anarchic sensibility first emerged and took root—and began to gain notoriety. An anti-draft notice (quoting Thoreau and civil disobedience) he posted on a public bulletin board drew the attention of the FBI, and Abbey would eventually amass an FBI dossier two inches thick, at more than 130 pages (Abbey 2003; cf. Bishop 1995; Calahan 2001). Abbey's action swiftly positioned him as a closer ally to the beatniks of the 1950s than to either Thoreau or Muir, although Abbey would remain on the periphery of the Beat Generation literary movement. It would not be until Abbey's penultimate novel, the pseudo-autobiographical *The Fool's Progress: An Honest Novel* (Abbey 1988b) that the Beat influence of Abbey's youth would be sharpest in his writing.

In 1948, Abbey moved to Albuquerque, and attended the University of New Mexico on the G. I. Bill, graduating in 1951 with a BA in Philosophy and Literature. He began writing while still an undergraduate student and was briefly the editor of the student literary publication, *The Thunderbird*. But he was replaced (and almost expelled) after another anarchic outburst, which saw the cover of the March 1951 issue boldly proclaim, 'Man will never be free until the last king is strangled with the entrails of the last priest,' sarcastically and gleefully misattributing eighteenth-century French philosopher Denis Diderot to Louisa May Alcott (Bishop 1995; Calahan 2001; cf. Abbey also plays with, riffs off this refrain in Abbey 1989, 2006). The issue also included Abbey's (1951) essay 'Some Implications of Anarchy.' Following this episode, the magazine was promptly suspended for a year. It was during his final year as an undergraduate that he began work on his first novel, *Jonathan Troy*. Following graduation, Abbey was awarded a Fulbright scholarship to study the works of Robert Burns[1] at the University of Edinburgh, but spent much of his year in Scotland revising *Jonathan Troy* (later published as Abbey 1954), and as a base for touring Europe.

From Edinburgh, Abbey returned once again to Albuquerque (interrupted briefly for work and then school back east, including a fleeting two-week stint enrolled on a graduate program at Yale in Connecticut), to re-enroll at the University of New Mexico for an MA in Philosophy, between 1954 and 1960.[2] Abbey defended his graduate thesis—*Anarchism and the Morality of Violence*—on August 5, 1959, and was awarded his degree the following year. Abbey's (1959) thesis on an ethics of political violence (although devoid of environmental concerns) stands as a prophetic signaling to Abbey's subsequent endeavors—and lasting legacy—in defense of public lands in the West (see also 'Theory of Anarchy,' in Abbey 1988a).

1956 would become a pivotal year in the evolution and early consolidation of Abbey's anarchic environmental philosophy. For not only was Abbey in the midst of researching his graduate thesis, but his second novel, *The Brave Cowboy: An Old Tale in a New Time* (as Abbey 1992), was published, which included the first appearance of his anachronistic and quixotic anarchist hero, Jack Burns (see also Abbey 1990a, 1991c, 2004).[3] But the summer of 1956 is also significant for it heralded the start of Abbey's itinerant employment as a park ranger and fire lookout, beginning in Arches National Monument near Moab, Utah (Arches became a national park on November 12, 1971). 1956 is also the year

[1] Burns's surname would be borrowed by Abbey and reassigned to one of his most legendary fictional characters—the renegade cowboy, Jack Burns.

[2] Between 1957 and 1958, Abbey joined the Stegner Fellowship program at Stanford, a two-year creative writing fellowship founded by Wallace Stegner. Although the fellowship brought Abbey to California and closer to the Beat Generation's San Francisco heart, Abbey remained on the periphery (Wendell Berry was a Stegner Fellow the year after Abbey).

[3] The novel was later adapted for cinema, released as *Lonely Are the Brave* in 1962.

construction began on Glen Canyon Dam, 150 miles away. Abbey would regularly return to the Desert Southwest, spending upward of seventeen seasons working as a park ranger and fire lookout across the West's national forests, national monuments, and national parks (Fig. 6.1).

Abbey's posting as a seasonal park ranger in Utah's Arches National Monument would prove monumental to the evolution and consolidation of the 29-year-old writer's environmentalist (and anarchist) sensibilities (evident throughout Abbey 1990b). His seasons at Arches were followed by posts across the American West and beyond, including in Casa Grande Ruins National Monument, Petrified Forest National Monument, Sunset Crater Volcano National Monument, Canyonlands National Park, Everglades National Park, Death Valley National Park, Lee's Ferry, and Organ Pipe Cactus National Monument (described in Abbey 1984, 1988b, 1991b, d, 1994, 2003, 2006). As a fire lookout and firefighter, Abbey spent further seasons in Gila National Forest (cf. Aldo Leopold's tenure in Chapter 4), the North Rim of Grand Canyon National Park, Lassen Volcanic National Park, Atascosa Peak in Coronado National Forest, Aravaipa Canyon Wilderness, Numa Ridge in Glacier National Park, and Aztec Peak in Tonto National Forest

Fig. 6.1 Balanced Rock in Arches National Park, Moab, Utah. Abbey's park service trailer was parked close to Balanced Rock when he was a seasonal ranger in 1956 and 1957. The original trailer site off Willow Flats Road is now part of the Arches work center. Photographs by the author, November 2012

(recounted in Abbey 1984, 1988b, 1989, 1991a, b, d, 1994, 2003, 2006, 2014).

But Abbey also wrote of his own clandestine, subversive, and localized efforts to impede the progress of industrial development across his beloved slickrock desert. His writings offer 'tips on desert etiquette' (Abbey 1991a; see also 1988a, 2003, 2006; cf. Calahan 2001; Loeffler 2002) and disclose instances of civil disobedience and mischief, or, 'eco-defense.' Of leveling and/or burning billboards (Abbey 1990b,1991a, d, 2006). Of pulling up survey stakes and ribbons marking out a new road in Arches while working as a seasonal ranger (Abbey 1990b, 1991b, 2006).[4] Of immobilizing bulldozer fuel tanks with sugar and sand (Abbey 1991b, d, 2006). Of discarding empty beer cans and bottles along the highway (Abbey 1988b, 1989, 1991a, 2006).

It is these two threads—of wilderness custodian *and* eco-saboteur, coalescing as a personal, anarchic, radical, guerrilla environmental philosophy—that set Abbey apart as a staunch activist and defender of public lands in the West, as a desert anarchist. 'The idea of wilderness,' Abbey exults at the start of 'Shadows from the Big Woods,' an essay in *The Journey Home* (Abbey 1991a: 223, also 1984, 1991d), 'needs no defense. It only needs more defenders.' In much the same way that the 'Thoreau Country' descriptor served both to align Henry David Thoreau with Walden Pond, and as a political tool in the defense of the Concord pond and woods and other New England landscapes (see Chapter 2), so too does the trope of 'Abbey's Country'—with the exception that Abbey himself originated and pro-actively employed the label, while Thoreau's only arose posthumously (Abbey 1990b, 1991b, 2003). Abbey's Country became synonymous with the redrock, slickrock canyonlands of southern Utah and northern Arizona. Even with his death from esophageal varices on March 14, 1989, at the age of 62, Abbey upheld (and prolonged) his desert anarchism. Following his

[4] Abbey first met fellow Glen Canyon activist Katie Lee in 1969 after Lee, who read *Desert Solitaire* (Abbey 1990b) with her third husband on their honeymoon in the southern Utah canyonlands, mailed Abbey a memento from one of their hikes—an uprooted survey stake and ribbon (with the remainder disposed of over a canyon rim)—and a copy of her album, *Folk Songs of the Colorado River* (see Lee 2004a, b; Lincoln 2020). The pair would become good friends, and separately two of the sharpest voices advocating the restoration of Glen Canyon and the Colorado River.

instructions, and in his final renegade, anarchic act, Abbey was wrapped in a sleeping bag and buried—illegally—in a remote, secret spot in the desert somewhere outside Tucson by friends Jack Loeffler, Doug Peacock, and his fifth wife Clarke Cartwright Abbey's father and brother-in-law (Calahan 2001; Loeffler 2002; Lincoln 2020).

6.1.2 When Abbey Saw Glen Canyon

Abbey's antagonism of Glen Canyon Dam and Lake Powell on the Colorado River is stark, abrasive. In a May 1, 1976 letter to *High Country News*, he declares, 'If I knew how to blow up Glen Canyon Dam, I'd be out there working on it tonight' (included in Abbey 2006: 63). Yet, his writings—both nonfiction and fiction—on the dam and reservoir, and the submerged canyons, draw attention to the controversy engulfing western water law and hydropower dam projects in the American West in the latter half of the twentieth century and question the ecological and social costs of these installations (cf. Spurgeon 2009 on the intersections between literature, western water policy, and public discourse). Abbey was a prominent and harsh, outspoken, and unflinching critic of the Bureau of Reclamation, part of the Department of the Interior, responsible for dam-building projects in the West (on the history and legacy of the USBR, see Worster 1985; Martin 1989; Reisner 1993; Fradkin 1996; Farmer 1999; Cassuto 2001; Wehr 2004; McCool 2012; Beard 2015).

Glen Canyon Dam and Lake Powell became mainstays in Abbey's writings, featuring in more than a dozen books and essays, including his pseudo-memoir *Desert Solitaire: A Season in the Wilderness* (Abbey 1990b), protest novel *The Monkey Wrench Gang* (Abbey 2004), the essays 'God's Plan for the State of Utah: A Revelation,' 'The Second Rape of the West,' and 'Down the River with Major Powell' in *The Journey Home* (Abbey 1991a), 'Down There in the Rocks' in *Abbey's Road* (Abbey 1991b), 'Down the River with Henry Thoreau,' 'Running the San Juan,' 'In the Canyon,' 'River Rats,' and 'Floating' in *Down the River* (Abbey 1991d), 'How It Was,' 'The Damnation of a Canyon,' and 'A Colorado River Journal' in *Beyond the Wall: Essays from the Outside* (Abbey 1984),

'Lake Powell by Houseboat' and 'River Solitaire: A Daybook' in *One Life at a Time, Please* (Abbey 1988a), collected epigrams, aphorisms, and poems in *A Voice Crying in the Wilderness (Vox Clamantis in Deserto)* (Abbey 1989) and *Earth Apples: The Poetry of Edward Abbey* (Abbey 1994), the sequel to *The Monkey Wrench Gang*, *Hayduke Lives!* (Abbey 1990a), as well as featuring in numerous journal entries (see Abbey 2003) and letters and correspondence (Abbey 2006).

The medium of the book becomes, for Abbey, an essential tool for environmental activism. He admits in the introduction to *Desert Solitaire* (Abbey 1990b) that the book is an elegy to the Arches National Monument he knew as a ranger more than a decade before. To an already disappearing, transformed, and transmuted landscape. But it is also, more importantly, a rallying cry, a call-to-arms: 'You're holding a tombstone in your hands. A bloody rock. Don't drop it on your foot—throw it at something big and glassy. What do you have to lose?' (Abbey 1990b: xiv). Elsewhere, he lightheartedly casts books as 'paper clubs' (e.g. Abbey 1988a, 2003). In *Down the River*'s introduction, Abbey (1991d) describes his writing as an antidote to despair.

Other writers have recorded the loss of the river canyons to the dam and reservoir, and spoken out against USBR policy and legislation, narrating and curating a collective memory of Glen Canyon (see e.g. Gross's 2003 anthology of Glen Canyon literature, which features accounts of the Glen before and after the dam). For Jared Farmer (1999), a Glen Canyon canon is 'the literature of the lost.' But none have approached the intensity of Abbey (or, indeed, Katie Lee) either in articulating a duty to protect the canyon, or in creating a vocabulary of a 'Glen Canyon restored.' For two decades, Abbey's writings, alongside invited lectures and speeches, offered polemics on the politics, policy, and governance of the dam.

Abbey would make only one float trip through Glen Canyon before it was lost beneath the waters of Lake Powell. Although he would later reluctantly and morosely return to the reservoir to explore the side canyons by motorboat (with these Lake Powell trips told in Abbey 1988a, 1991b). When Abbey and UNM friend Ralph Newcomb embarked on their river trip in the summer of 1959, the dam was already under construction. Abbey's journals from three years earlier, kept during his first season at Arches and coinciding with the early construction of the dam, see him already mourning the loss of Glen Canyon (e.g. May 16 and August 26, 1956, entries in Abbey 2003). It would not be until 1968's *Desert Solitaire*, his pseudo-autobiography as a ranger at Arches, that Abbey would recount his Glen Canyon encounter (Abbey 1990b). The longest chapter of *Desert Solitaire*, 'Down the River,' vividly documents his adventure, and Abbey (1990b) writes with awe and reverence of the soon-to-be-flooded canyon's beauty (see also Abbey 1991d, 2003, cf. Farmer 2003 on Abbey's literary memory of the Glen).[5]

Abbey and Newcomb were not the first to run the Colorado River through Glen Canyon. Nor the most celebrated. In 1869, Major John Wesley Powell, who fought in the Union Army during the Civil War and lost his right arm at the Battle of Shiloh, led the first Anglo-American expedition down the Green and Colorado rivers in a fleet of wooden dories, conducting a geological survey of the region. It was on this scientific expedition that Powell gave Glen Canyon its name, on August 3, 1869 (Powell 2002, and on Powell's 1869 and 1872 Colorado River expeditions and his legacy in the West, see Stegner 1992; Dolnick 2001; Worster 2001; Lago 2018; Ross 2018, also Vernon's 2002 fictional retelling). Powell was instrumental in mapping the West, becoming the second director of the U.S. Geological Survey (USGS) and was an early architect behind the idea of 'reclaiming' the American desert and its rivers. Powell died the same year that the Reclamation Service was established by President Theodore Roosevelt under the Newlands Reclamation Act—1902 (the Reclamation Service was renamed the Bureau of Reclamation in 1923). River running outfits through the Glen exploded in the postwar 1950s and 1960s.

Nor is Abbey the only writer to recount river trips through Glen Canyon. Powell's *The Exploration of the Colorado River and Its Canyons*, first published in 1895, is the inaugural literary and scientific text on the Glen (Powell 2002). Actress, folksinger, river runner, and activist Katie Lee's (2006) *Glen Canyon Betrayed: A Sensuous Elegy* is structured

[5] But in *Desert Solitaire*, Abbey (1990b) plays with, manipulates timescales—as he does elsewhere in his writings. Much like Thoreau's *A Week* or *Walden*, which both condense their timeframes, *Desert Solitaire* condenses Abbey's 1956 and 1957 seasons at Arches into a single 'season in the wilderness'—and absorbs outliers, such as the 1959 Glen run.

primarily around journal entries from various river running expeditions throughout 1954, 1955, 1956, and 1957, and concludes with infrequent later trips to the ever-swelling reservoir of the late-1950s and 1960s (but was not published until the close of the twentieth century, long after Abbey's books and essays—first as *All My Rivers are Gone: A Journey of Discovery Through Glen Canyon* in 1998,[6] and then as *Glen Canyon Betrayed* in 2006). But if decades passed before her journals appeared in print, her folk songs about the Colorado River, songs of protest, were immediate. In the 1950s, Lee sang in clubs across the U.S.—including Denver, Chicago, St. Louis, Tucson, Hollywood, Blanding, New York City—drawing attention to what was quietly unfolding in a canyon in southeast Utah (Lee 2004b, 2006).

Lee's three 'We Three' river trips through the Glen in the fall of 1955, 1956, and 1957 with friends boatman Frank Wright and photographer Tad Nichols underpin these accounts (Lee 2006; also Nichols 1999; *Drowning River* 2007; *DamNation* 2014; Lincoln 2020). Lee's journal from the first day of her first Glen Canyon trip in July 1954 records the rumor of a dam and reservoir in Glen Canyon, and another proposed dam project upstream on the Green River through Dinosaur National Monument (e.g. July 4, 1954 entry in Lee 2006). Her disbelief intensifies over the following years, even after the dam infrastructure begins to appear on canyon walls. Beginning with this 1954 trip, Lee ran the Glen every year for the next decade, until Colorado River waters began to rise behind the dam (Lee 2004b, 2006; *Drowning River* 2007). On the few occasions that Lee returned to Lake Powell, she often explored the reservoir on a small boat named SCREWD RIVER (Lee 2006; Lincoln 2020).

Abbey and Newcomb spent almost a fortnight floating down 150 miles of the Colorado River through Glen Canyon on a drugstore raft, starting on June 25 at Hite and ending at the abrupt, enforced take-out point at Gunsight Butte, near the Crossing of the Fathers, more than 10 miles upstream of the dam construction site at Lee's Ferry. Lee's river trips tell a very different story—of a very different Colorado River. On Lee's 1955 'We Three' trip through the Glen, her journals record her

[6] It was at Abbey's encouragement that Lee adapted a draft novel into the memoir *All My Rivers are Gone* (Lee 1998).

anger and sorrow as the trio pass through Mile 15, the confirmed site of the dam. Her journals from 1956 to 1957 mark the slow despoliation of the canyon as the appearance of tools and machinery, survey equipment and flags, and white lead paint along the canyon walls foretells of the impending dam construction. In 1956, the trio ignore the notice restricting travel on down to Lee's Ferry, but in 1957 have to disembark at Kane Creek, just upstream of the Crossing of the Fathers. The day after the trio's 1956 trip ended, on October 15, President Dwight D. Eisenhower authorized via telegraph the first of many explosive charges in the dam's construction (Lee 2006; also Farmer 1999; Fedarko 2013).

As Abbey and Newcomb near the end of their 1959 trip—and the dam construction site at Lee's Ferry—there is a growing disbelief between the two men that the dam and reservoir will submerge the canyon. But as the realization settles, they begin to fantasize scenarios that might yet befall the construction—a mass transfer of engineers to other projects, a dearth of cement or dynamite, and more. It is while daydreaming on the river that Abbey first vocalizes his fantasy of the dam's demise, musing,

> some unknown hero with a rucksack full of dynamite strapped to his back will descend into the bowels of the dam; there he will hide his high explosives where they'll do the most good, attach blasting caps to the lot and with angelic ingenuity link the caps to the official dam wiring system in such a way that when the time comes for the grand opening ceremony, when the President and the Secretary of the Interior and the governors of the Four-Corner states are all in full regalia assembled, the button which the President pushes will ignite the loveliest explosion ever seen by man, reducing the great dam to a heap of rubble in the path of the river.
>
> —Abbey 1990b: 165

And in this tableau, Cactus Ed's (Abbey's literary persona) radical, anarchic restoration pathology was born. Of a Glen Canyon restored.

If *Desert Solitaire*'s 'Down the River' chapter is an elegy to what is 'gone—or going under fast' (Abbey 1990b: xii, with the refrain repeated in 1990a, 1991a), then the essay 'The Damnation of a Canyon' (in Abbey 1984) stands out as Abbey's most sober, somber, and severe sermonizing of Glen Canyon Dam, Lake Powell, and the USBR. This later essay poignantly and evocatively describes the before-and-after scenarios in Glen Canyon (see also Abbey and Hyde 1971). The sterile, stagnant, barren reservoir is set against the noisy, chaotic, turbulent, unruly, fecund, fertile river. It is tamed set against untamed. Domesticated against wild. Silence against cacophony. The river and reservoir become, for Abbey, metaphors for life and death. The 'damnation' word play recurs across Abbey's canon (Abbey and Hyde 1971; Abbey 1984, 1988a, 2006; see also Wehr 2004; *DamNation* 2014). Abbey is especially scathing in his critique of Lake Powell. He often substitutes 'lake' for 'reservoir' in his writings (Abbey 1990b, as does Lee 2006; cf. Stegner 2017 and 'Glen Canyon Submersus' essay), and refers to Lake Powell as Lake Foul (in e.g. Abbey 1990a, 1991d, Abbey 1981 in Philippon 2004). Despite Abbey's intense antipathy toward the dam and reservoir, he worked as a seasonal ranger at Lake Powell between May and October 1967, first patrolling the Wahweap and Bullfrog marinas, before patrolling a 23-mile stretch of river below the dam at Lee's Ferry (Abbey 2003; cf. Calahan 2001).

Abbey regularly casts Glen Canyon and its environs as the 'living heart' of the canyon country (Abbey and Hyde 1971, Abbey 1984, 1990b, 1991b, d, also the Colorado River as the 'living river' in Abbey and Hyde 1971, Abbey 1984, 1990a, b). But he also struggles with this idea, noting too that the desert has no heart (Abbey 1990b, 1991b). Abbey's essay 'The Damnation of a Canyon' (Abbey 1984: 95) begins with a confession or admission that conclusively situates and underscores Glen Canyon at the center of the Colorado Plateau—and of Abbey's Country: 'There was a time when, in my search for essences, I concluded that the canyonland country has no heart. I was wrong. The canyonlands did have a heart, a living heart, and that heart was Glen Canyon and the golden, flowing Colorado River' (cf. Lee 2006 on the Glen's 'Wild Secret Heart'). It is this heartland, heart's-blood framing that would later be politicized and revisited in political dialogue and debates on proposals to decommission Glen Canyon Dam and drain Lake Powell.

Abbey's ecological restoration sensibility runs the gamut of the impatient, violent, sudden removal of the dam by explosive charges (Abbey 1990b, 2004; cf. Silko 1991; Meloy 1994; Lee 2006), to decommissioning and abandoning the hydropower plant and draining the reservoir (Abbey and Hyde 1971, Abbey 1984), to awaiting the patient, slow compounding double-act of evaporation paired with sediment accumulation (Abbey 1988a, 1991d; cf. Brower in McPhee 1977). The essay 'In the Canyon' (in Abbey 1991d) offers one of Abbey's most thorough edicts for the restoration of Glen Canyon, where he sagely advises,

> Take care of it. How? We can leave that to the engineers. They built it, they can unbuild it. With the dam gone, we can save and restore the natural riverine ecology of the [Grand] Canyon, bring back the great spring floods that used to flush out the channel every year, clear out the tamarisk, bring back the driftwood, rebuild the sand beaches (now being eroded away and not replaced) and make the river, as it used to be and should be, warm and violent and golden and full of catfish—*muy Colorado!* The red-brown god.
>
> Fantasy, you say. Perhaps. But if we don't do it, Nature will. In a few more centuries the dams will be filled with mud and sand, will become great waterfalls and then, as erosion does its work, will be reduced to polished stumps of concrete and re-bar, foaming rapids full of vee-waves and suckholes, a challenge to boat people, nothing more. Any river with the power to carve through the ancient limestones, granites, and schists of the Kaibab Plateau will have little trouble with the spongy cement deposited here, once upon a time, by some dimly remembered clan of ant-folk known as the Bureau of Reclamation.
>
> —Abbey 1991d: 147–148

Abbey offers restorationists a roadmap to, and an ethic for, restoring the Glen. Both Abbey (1991d, cf. Abbey 1981 in Philippon 2004) and Lee (2006) note the temporary ugliness of a would-be-drained Glen Canyon—with its bleached, bathtub-ring canyon walls, exposed mudflats and sediment stockpiles, waterlogged and rotting trees, and accumulated junk from decades as a watersport playground—but point to the raw power of the pre-dam river in flash flood to scour lateral (and central) canyons clear of obstructions, whether boulders, rocks, trees, driftwood, or cattle, deer, and other wildlife. With 'time and the river flowing' (after Leydet 1964, Lee 2006), desert varnish would once more streak canyon walls, and restoration would benefit upstream and downstream riparian ecosystems (also Brower 1997, Carothers and House 2000, Ingebretsen 2000). In Abbey's (1991d) 'In the Canyon' essay, as in 'The Damnation of a Canyon' (Abbey 1984), he draws attention to the ecological disruption downstream from the dam, through the Grand Canyon, caused by a river that no longer runs warm and laden with silt.

The closing chapter of Abbey's final (and posthumous) novel *Hayduke Lives!* (Abbey 1990a) is titled simply 'Resurrection,' an echo of the first chapter, 'Burial'—and although these refer to the misadventures of an old desert tortoise (a symbol of the desert), together they might also speak to Abbey's deeper restoration pathology. Equally for Lee, the promise of ecological restoration is clear in her framing of Glen Canyon as 'the once-and-future Glen Canyon' (as Lee 2004b, 2006). The restoration of Glen Canyon is, for Lee, a question of *when* rather than *if* (Lee 2006).

6.2 Two Dams, a Cautionary Tale

The origin story of Glen Canyon Dam began in the early-1950s in another proposed dam project upstream. In the dam project tagged for Whirlpool Canyon below Echo Park and Split Mountain Canyon on the Green River in Dinosaur National Monument (hereafter consolidated as just Echo Park). Both dams have a shared history of literary activism. But the intent of that environmental writing-as-advocacy diverges dramatically.

Before proceeding, a brief exposition of western water law and water rights along the Colorado River is necessary. The vast drainage basin of the Colorado River extends across seven western U.S. states, with seven competing sets of rights-claims (as well as two Mexican states). With the signing of the seven-state Colorado River Compact (CRC) agreement on November 24, 1922, in Santa Fe, New Mexico, the waters of the river were regulated between Upper Basin (Colorado, New Mexico, Utah, Wyoming) and Lower Basin (Arizona, California, Nevada) states, with the artificial boundary between the Upper Basin and Lower Basin set at Lee's Ferry, immediately downstream of Glen Canyon, on the Utah-Arizona border (the complicated history of the CRC and water allocation agreements, policy, governance, and reform is explored in Worster 1985; Martin 1989; Reisner 1993; Berger 1994; Fradkin 1996; Cassuto 2001;

Grossman 2002; Wehr 2004; Adler 2007; Powell 2008; Fleck 2016; Owen 2017).

The Compact[7] accords the upstream and downstream states a guaranteed flow of 75 million acre-feet (maf) every decade (or, 7.5 maf each year).[8] There are, however, at least two major flaws in this formalized allocation, with the repercussions continuing into contemporary water resource management. First, is that the USGS data underwriting the 1922 Compact is based on an anomalously wet period, with waters quickly overallocated, and states in deficit. Second, the Compact neglected the rights-claims to Colorado River waters beyond the U.S.-Mexico border toward the Sea of Cortés. Mexico was not granted legal rights to the waters of the Colorado River until 22 years after the CRC, when the 1944 U.S.-Mexico Water Treaty allocated 1.5 maf a year to Mexico. Even so, water barely reaches the Colorado River Delta and the Sea of Cortés, and that that does is plagued by high salinity, after flowing through dams and diversions (see Adler 2007). The rights-claims of Native Nations—principally Navajo and Hopi—remain excluded from the Compact. Nevertheless, the Law of the River is still enforced.

The 1922 Colorado River Compact heralded a number of major legislative actions throughout the Upper Basin and Lower Basin states to secure state water rights. The 1928 Boulder Canyon Project Act authorized the 726-foot-high Boulder Dam (renamed Hoover Dam for President Herbert Hoover in 1947) and Lake Mead in Black Canyon on the Nevada-Arizona border, constructed through the Depression and Dust Bowl years of 1931 and 1935, beginning a program of development of the lower Colorado River for use by the Lower Basin states. At the time of its construction, it was the largest dam in the world. Beneath Lake Mead is the take-out point of Powell's celebrated 1869 expedition down the Colorado River through Glen and Grand Canyon (Fig. 6.2).

It would be more than a quarter-century before an equivalent program was rolled out by the Upper Basin states. In 1956, the same year Abbey started as a seasonal ranger at Arches National Monument and *The Brave*

---

[7] Lee (2006) describes her own Compact with the canyons of the Colorado River, a compact of remembrance and finding solace in what remains.

[8] An acre-foot corresponds to water covering an acre of land at a depth of one foot—or, 325,851 gallons.

Fig. 6.2 Hoover Dam and Lake Mead on the Colorado River, and the Colorado River (and the Mike O'Callaghan-Pat Tillman Memorial Bridge/U.S. Route 93) downstream of the dam, on the Nevada-Arizona state line. Photographs by the author, November 2012

Cowboy was published, the Colorado River Storage Project Act authorized the construction of the massive Colorado River Storage Project (CRSP). The CRSP sanctioned the development of the upper Colorado River to meet the 1922 Compact entitlements of the Lower Basin states (the institutional and political context surrounding the CRSP will not be further rehearsed here, but see Reisner's 1993 classic on water history and policy in the American West, and other volumes on western water law, politics, and policy, including McPhee 1977, Worster 1985, Martin 1989, Fradkin 1996, Farmer 1999, Cassuto 2001, Grossman 2002, Wehr 2004, Adler 2007, Powell 2008, McCool 2012, Beard 2015, Fleck 2016, Owen 2017). The CRSP provided water security through the long-term regulatory storage and delivery of water, and approved the construction of four major dams and 11 minor dams and irrigation projects on the upper stretch of the Colorado River and its tributaries. Two of the four major dams proposed under the CRSP were Echo Park Dam on the Green River on the Colorado-Utah border, and Glen Canyon Dam on the Colorado River on the Utah-Arizona border.

6.2.1 The Fight for Wilderness in Echo Park, and *This Is Dinosaur*

Back to Echo Park in Dinosaur National Monument. A conservation battle in Dinosaur in the early-1950s was the first major challenge the

USBR had faced to its dam-building inventory in its almost five-decade history—and it would not be the last. The question of dams in Dinosaur National Monument, established in 1915 (and expanded in 1938 from 80 acres to 210,000 acres to include the canyonlands upstream), replayed earlier national park preservation arguments by John Muir, the Sierra Club, and others in Yosemite's Hetch Hetchy Valley in the 1900s and 1910s (see Chapter 3). But the Dinosaur battle also predated major U.S. environmental and conservation legislation that would be used in later dam fights—for example, the 1969 National Environmental Policy Act, and the 1973 Endangered Species Act. Amidst the political theater and conservation politics surrounding the dam project in Dinosaur National Monument emerged a tale of literary activism in defense of public lands, national park units, and the idea of wilderness.

A literary activism narrative began less than a month after Secretary of the Interior Oscar Chapman announced his approval of two dams in Dinosaur—in Whirlpool Canyon below Echo Park and in Split Mountain Canyon—in late-June 1950, and continued for five years. In an article in *The Saturday Evening Post* on July 22, 1950, entitled 'Shall We Let Them Ruin Our National Parks?,' Bernard DeVoto challenged the USBR's intention to build two dams within Dinosaur National Monument (as DeVoto 2005). Other articles in the national press kept attention focused on Dinosaur and the parks system.

But as the battle for Echo Park became increasingly embroiled in a larger congressional debate on the CRSP, conservationists sought to wrest back celebration of the wilderness of Dinosaur National Monument (on the battle for Echo Park and its legacy in U.S. conservation rhetoric, see Harvey 2000, 2005; also Martin 1989; Reisner 1993; Berger 1994; Fradkin 1996; Farmer 1999; Calahan 2001; Cassuto 2001; Wehr 2004; Brower 2005; DeVoto 2005; Powell 2008; McCool 2012; Fedarko 2013; *DamNation* 2014; Nash 2014). At the same time as conservation organizations and scientists were testifying at hearings on the Dinosaur dams, filming documentaries, organizing river trips, and more, another literary project worked to elevate the dam issue in public and political rhetoric.

The Sierra Club was a leader in the Dinosaur conservation campaign, along with the National Parks Association, Wilderness Society, National

Wildlife Federation, Izaak Walton League, and others. At the encouragement of David R. Brower (the Sierra Club's first executive director, 1952–1969), the Sierra Club published the first in what would become a series of Exhibit Format landscape photography books—or 'battle books' (Trimble in Sumner 2002a). With Wallace Stegner recruited as editor, together with photographer Philip Hyde, 1955's *This Is Dinosaur: Echo Park Country and Its Magic Rivers* documents what would be sacrificed to the dam construction. *This Is Dinosaur* (Stegner 1955) helped shape conservation advocacy and public discourse at Echo Park. The book was the first of its kind to publicize a unit of the parks system and support a conservation campaign. Early copies were sent to Congress to lobby against the dam (Martin 1989; Brower 1997; Miller 2000; Powell 2008; Spurgeon 2009). The book stands as the archetype of of-the-moment, reactionary literary environmental activism, and Dinosaur National Monument persists as the homeland of literary environmental activism and advocacy in the West.

With the attention of conservation groups focused on an enclave in Dinosaur National Monument, the river corridor of Glen Canyon went unheeded. In a conservation campaign built upon a threat to the parks system, outliers did not register on the collective conservation radar at that time. Glen Canyon was not included in the national monuments and parks inventory, although the whole canyon—and, indeed, much of the canyon country of southern Utah—had been considered for protection under Secretary of the Interior Harold L. Ickes's 1934 proposal for an Escalante National Monument of almost 4.5 million acres (Martin 1989; Reisner 1993; Berger 1994; Harvey 2000). The only objection to the Glen Canyon dam site from conservationists concerned the possible inundation of the adjacent 160-acre Rainbow Bridge National Monument by the encroaching waters of Lake Powell (Reisner 1993; Fradkin 1996; Farmer 1999; Harvey 2000; DeVoto 2005; Lee 2006; Powell 2008, see also Abbey 2004). It was not until 1972 that Glen Canyon was designated as the 1.25-million-acre Glen Canyon National Recreation Area (Glen Canyon Institute today supports converting Glen Canyon

NRA into a national park,[9] cf. Lee's 2006 call for stronger protections as a national park).

The Sierra Club withdrew its opposition to a dam in Glen Canyon in the mid-1950s, to protect Echo Park and Dinosaur National Monument. The Dinosaur dam project was eventually canceled in November 1955 and removed from the CRSP bill. The dam project was relocated upstream on the Green River (and outside the national monument), to Flaming Gorge, Utah. The defeat of the dam would stand as an important episode in the run-up to the 1964 Wilderness Act (cf. Chapter 4).

6.2.2 Losing Glen Canyon, and the Eulogy of *The Place No One Knew*

The Colorado River Storage Project Act was passed on April 11, 1956. Its signing authorized the Curecanti unit on the Gunnison River, Flaming Gorge Dam on the Green River, Navajo Dam on the San Juan River, and Glen Canyon Dam on the Colorado River. Conservationists secured language in the Act to 'preclude impairment' of Rainbow Bridge National Monument from the new reservoir behind Glen Canyon Dam, although this was later overturned (Reisner 1993; Fradkin 1996; Farmer 1999; Harvey 2000, 2005; Miller 2000; Lee 2006; Powell 2008; Lincoln 2020). Other legislation followed soon after, including the 1968 Colorado River Basin Project Act, which authorized the Central Utah Project, the Central Arizona Project, and more.

Construction of the 710-foot-high Glen Canyon Dam (the fourth highest in the U.S.) began within months of the approval of the CRSP and was completed in 1963 (for a history of the dam's construction, see Martin 1989; Reisner 1993; Farmer 1999; Powell 2008; *DamNation* 2014). The dam cost close to $300 million. At its base, it measures 300 feet thick, narrowing to 25 feet at the crest, which spans 1,560 feet between the canyon walls. 4.9 million cubic yards of concrete was placed. Eighteen construction workers died, and 348 were seriously injured (Fradkin 1996; Miller 2000; Grossman 2002; Waterman 2010;

[9] GCI telephone interview, February. 6, 2013, GCI online interview, August 14, 2020.

Fedarko 2013). On January 21, 1963, the diversion tunnels closed, and the Colorado River was halted behind the dam. As well as bookending the Grand Canyon, the Hoover and Glen Canyon dams bookended the USBR's golden age of dam building.

Glen Canyon Dam is a 'cash-register' dam, built for the water storage capacity of Lake Powell (ensuring the Upper Basin states meet the downstream delivery obligations of the CRC), and subsequent hydropower generation (Reisner 1993; Farmer 1999; Grossman 2002; Powell 2008; McCool 2012). With more than 120 arterial canyons branching off the central Glen Canyon, it took two decades for Lake Powell to fill to capacity—storing 27 maf of Colorado River water. But sedimentation is reducing storage capacity by 37,000 af a year and could begin impairing dam operations within 100 to 150 years (Fradkin 1996; Carothers and House 2000; GCI 2000; Harvey 2000, Ingebretsen 2000; Waterman 2010; Fedarko 2013). Lake Powell, the second largest reservoir in the country, extends 155 miles upstream through Glen Canyon, 30 miles through Cataract Canyon, and 75 miles along the San Juan River and other tributaries (after Carothers and House 2000), creating almost 2,000 miles of shoreline (longer than the Pacific coastline from northern Washington to southern California). 15 miles of Glen Canyon remain, downstream of the dam (Fig. 6.3).

It was as the reservoir waters began to rise that a version of Dinosaur's literary activism began to unfold in Glen Canyon. Meanwhile, Lee's folk songs kept the Glen Canyon dam issue in the public imagination. For Glen Canyon restorationists, it is environmental writers who have 'kept the flame alive—the vision alive—of restoring Glen Canyon. [...] They weren't thinking about the "practical" side of things, they were thinking about the beauty and the vision and the solitude and the ethics of the whole thing. They didn't care whether it was politically feasible or not, that wasn't really the point.'[10] Photographer Eliot Porter's (1963) *The Place No One Knew: Glen Canyon on the Colorado* is another of the Sierra Club's Exhibit Format photography books, following Stegner's (1955) *This Is Dinosaur*. But if Stegner's (1955) *This Is Dinosaur* is a manifesto on keeping dams and reservoirs out of national monuments and national

[10] GCI telephone interview, February 6, 2013.

Fig. 6.3 Glen Canyon Dam and Lake Powell on the Colorado River, and the Colorado River (and the Glen Canyon Dam Bridge/U.S. Route 89) downstream of the dam, from the Carl Hayden Visitor Center at Glen Canyon Dam, Page, Arizona. Photographs by the author, November 2012

parks, defending a swath of canyon country, Porter's (1963) *The Place No One Knew* is an obituary, eulogizing a disappearing canyon. But the evocative title is misleading—the Glen was less 'the place no one knew,' and more the place, as Farmer (1999) and Lee (2002) separately argue, that Brower and the Sierra Club did not know until it was too late. As Brower reflects in his foreword to *The Place No One Knew* (Porter 1963: 7), 'Glen Canyon died in 1963 and I was partly responsible for its needless death. So were you. Neither you nor I, nor anyone else, knew it well enough to insist that at all costs it should endure. When we began to find out it was too late.' In the following decades, he recast his 'regret to restoration' (Brower 1997) in defense of the Glen.

Other volumes also combine landscape photography with prose essays to record the river and canyons lost beneath Lake Powell (see Hyde 1979; Inskip 1995; Nichols 1999; Crampton 2009). Tad Nichols's (1999) volume features photographs from his 'We Three' trips with Lee and Wright. Eleanor Inskip's (1995) volume includes some of Lee's photographs from these trips. But it is not just the Glen's inundation that has featured in photographic collections. Annette McGivney's (2009) volume, *Resurrection: Glen Canyon and a New Vision for the American West*, with photographs by James Kay, and Rebecca Solnit's

(2017) *Drowned River: The Death and Rebirth of Glen Canyon on the Colorado*, with photographs by Mark Klett and Byron Wolfe, document the reemergence of the Glen's lateral canyons as Lake Powell's waters recede.

Two years after the Sierra Club published Porter's (1963) *The Place No One Knew*, the Bureau of Reclamation responded with its own pamphlet, *Lake Powell: Jewel of the Colorado* (Bureau of Reclamation 1965). Written by Floyd Dominy, Bureau of Reclamation Commissioner between 1959 and 1969, the pamphlet throws shade on the language and tone of Porter's (1963) volume, for Dominy also frames his descriptions of Lake Powell in terms of discovery, exploration, and adventure and celebrates the reservoir's unspoiled, unexploited, wild beauty and solitude (Bureau of Reclamation 1965; *DamNation* 2014, for critiques of the 'jewel' trope, see Abbey 1984, 1988a; Berger 1994; Farmer 1999; Lee 2006; Adler 2007; Waterman 2010).

6.2.3 Grand Canyon Imperiled, and *Time and the River Flowing*

On January 21, 1963—the same day the diversion tunnels were sealed at Glen Canyon Dam—Interior Secretary Stewart Udall announced plans for two more cash-register dam projects on the Colorado River. With the announcement emerged a new conservation battleground—the Grand Canyon in Arizona (on the Grand Canyon dams fight, see especially Pearson 2002; also McPhee 1977; Martin 1989; Reisner 1993; Zakin 1993; Fradkin 1996; Lee 2006; Powell 2008; McCool 2012; Fedarko 2013). The two dams, part of the Pacific Southwest Water Plan,[11] were scheduled for Marble Canyon and Bridge Canyon.

The story of the Grand Canyon dams fight is again the story of Sierra Club activism—and literary activism. What began in Echo Park (and Glen Canyon and Rainbow Bridge) in the 1950s was relocated to and recast for the Grand Canyon in the 1960s. The Sierra Club had

[11] Another component of the PSWP—the proposed Hooker Dam on the Gila River in New Mexico—jeopardized the Gila National Forest and neighboring Gila Wilderness, part of Aldo Leopold's conservation legacy (Zakin 1993).

briefly supported the Bridge Canyon dam proposal in 1949, but the proposal was blocked the next year. A decade-and-a-half on, after the events in Echo Park and Glen Canyon, the Sierra Club position had shifted. In 1963, Brower began building a constituency for conservation in Grand Canyon. He established a Grand Canyon Task Force to coordinate the conservation campaign, that included Luna Leopold, a USGS chief hydrologist and the son of Aldo Leopold (Chapter 4), who had earlier consulted on the battle for Echo Park. Porter's (1963) *The Place No One Knew* was published in mid-1963, within months of Udall's Grand Canyon announcement. The volume serves not only as a custodian of the lost Glen Canyon upstream (cf. Spurgeon 2009; also Meloy 1994), but its timing, amidst the Grand Canyon dam furor, meant that the 'lost canyon' motif was politicized by conservationists in defense of another exposed canyon. Glen Canyon, with its incumbent dam and reservoir, was recast, as Farmer (1999) argues, as an 'anti-symbol' of western conservation rhetoric, that succumbed to folly, conceit, and arrogance. One year later, the Sierra Club published another book in its Exhibit Format series—about the Grand Canyon: François Leydet's (1964) *Time and the River Flowing: Grand Canyon*, featuring photography by Philip Hyde. When Brower and colleagues again testified before Congress, as they had during the Echo Park fight, he entered Porter's (1963) *The Place No One Knew* into the *Congressional Record* (Zakin 1993).

In a series of full-page advertisements beginning on June 9, 1966, in *The New York Times*, *The Washington Post*, the *Los Angeles Times*, the *San Francisco Chronicle*, and other national newspapers, Brower launched what became his 'Grand Canyon Battle Ads' campaign. 1966 also marked the 50th anniversary of the National Park Service (NPS) Organic Act. A July ad in *The New York Times* questioned, SHOULD WE ALSO FLOOD THE SISTINE CHAPEL SO TOURISTS CAN GET NEARER THE CEILING? (see e.g. Reisner 1993; Zakin 1993; Miller 2000; Powell 2008), with the language reminiscent of Muir's theological arguments in defense of Hetch Hetchy six decades earlier (e.g. Muir 2003). The newspaper canvass was another powerful advocacy tool in Brower's Grand Canyon campaign arsenal, together with Porter's (1963) *The Place No One Knew*, Leydet's (1964) *Time and the River Flowing*, and pamphlets, bulletins, testimonials, letter-writing campaigns to Congress, documentary movies, and more.

Abbey's (1990b) *Desert Solitaire* was published in 1968, during the Grand Canyon dams debate. *Desert Solitaire* (Abbey 1990b), like *The Place No One Knew* (Porter 1963) before it, sits across and speaks to both dam episodes. The 'Down the River' chapter of *Desert Solitaire* recounts Abbey's only float trip through Glen Canyon in the late-1950s as dam construction began, and in this account is an echo of Brower's Sistine Chapel motif, when Abbey draws parallels between inundating Glen Canyon and burying the Taj Mahal and Chartres Cathedral in mud (Abbey 1990b). But in *Desert Solitaire*, Abbey (1990b) also briefly notes the scenario unfolding on the periphery of the Grand Canyon. The popularity of *Desert Solitaire* (Abbey 1990b) also fed into the designation of Arches as a national park in November 1971.

But Brower's 'Battle Ads' drew the attention of the Internal Revenue Service (IRS), which ruled the ads amounted to lobbying to influence legislation and revoked the Sierra Club's tax-deductible status for donations and contributions (McPhee 1977; Reisner 1993; Fradkin 1996; Powell 2008; Fedarko 2013). Brower's tenure as Sierra Club executive director ended in 1969 after his Grand Canyon campaign (McPhee 1977; Martin 1989). But both dam projects were later suspended in 1968 and brought the USBR's dam-building golden age to a close. The Wild and Scenic Rivers Act was signed within days of the cancelation. In 1969, President Lyndon B. Johnson established the Marble Canyon National Monument, retiring the canyon as a possible dam site (six years later, the national monument was absorbed into Grand Canyon National Park under the Grand Canyon National Park Enlargement Act). Within four years, three major federal environmental protections that would frame later dam fights were also passed—the National Environmental Policy Act, the Clean Water Act, and the Endangered Species Act.

6.3 Introducing the Monkey Wrench Gang

> "The dam?"
> "Yes sir."
> "Not the dam."
> "Yes sir, we have reason to think so."
> "*Not* Glen Canyon *Dam!*"
> "I know it sounds crazy. But that's what they're after."
>
> —Abbey 2004: 6–7

Abbey's fictional quartet of eco-saboteurs made their debut on August 1, 1975, in the pages of *The Monkey Wrench Gang*, his fifth novel. In surgeon Doc Sarvis, polygamist river runner Joseph 'Seldom Seen' Smith, former Green Beret and Vietnam veteran George Washington Hayduke, and feminist revolutionary and Doc's receptionist Bonnie Abbzug, Abbey created a lead cast of allies embarking on a mission to reclaim and wrest public lands in the American West from the grip of corporate, bureaucratic development projects. The endgame for the gang—perhaps unsurprisingly—was the obliteration of Glen Canyon Dam.

Abbey's (2004) *The Monkey Wrench Gang* is in part an outrageous, exaggerated, hyperbolical denouement of Abbey's own anarchical attempts at—and hopes for—rebelling against the loss of wilderness, and imagines and anticipates how a more coordinated rebellious maneuvering could play out across the canyons, mesas, and plateaus of southern Utah. *The Monkey Wrench Gang* is cathartic, purgative. It is wickedly comic and satirical. But more seriously too, it also speaks back to what is at stake on the West's public lands.

Abbey's fever dream struck a chord with environmentalists. The novel galvanized a new type of environmental activism, fusing anarchism with environmentalism. (The 10th anniversary edition of the novel, illustrated by countercultural cartoonist R. Crumb, brought the symbolic aggression of the novel into sharp relief.) *The Monkey Wrench Gang* (Abbey 2004) was embraced by (and found political purchase amongst) environmentalists as an instruction manual or training manual almost for direct action and eco-sabotage (or, ecotage)—and inspired the establishment of radical environmental advocacy group Earth First! in the early-1980s (as well as the much more controversial Earth Liberation Front, with members soon classified as eco-terrorists by the FBI). I consider it an ecological restoration novel, albeit of guerrilla restoration. It is revolutionary in all senses of the word. Abbey subtly weaves technical details about the dam, reservoir, and adjacent highway suspension bridge, and descriptions of the canyon and river, into the novel's plotlines.

The destruction of Glen Canyon Dam remains tantalizingly out of reach of the gang throughout the novel. But the anticipation of the dam's demise buoys the gang as they disrupt numerous projects across the American West. The gang swiftly compile a longlist of possible targets—from billboards and bulldozers to power lines, pipelines, highway projects, railroads, bridges, to power plants, mines, dams. No industrial intrusion into the desert is spared. Some ecotage projects are spontaneous and impulsive, others are painstakingly recced, planned, and strategized. The gang bury caches of tools and food supplies across the Four Corners region of Utah, Colorado, Arizona, and New Mexico, to support their endeavors. The gang's targets are both large-scale and small-scale. Some targets are fictitious, others have real-life counterparts (most notably Glen Canyon Dam and Glen Canyon Dam Bridge, as well as the Peabody Coal Company, and the Black Mesa and Lake Powell Railroad), further embedding the novel in the canyonlands of southern Utah and northern Arizona. The Glen Canyon Dam Bridge succumbs to the gang's explosives. And time and again, Abbey's own civil disobedience and ecotage sensibilities, pathologies—and moralities—infiltrate the novel.

6.3.1 Damning the Dam

Abbey's protest novel is littered with fantasies that variously envisage bringing about the obliteration of the dam and the restoration of Glen Canyon on the Colorado River. Some are plausible (and, indeed, have historical precedent), others are pipedreams, the rest fall somewhere in between. But all serve to question the temporality and necessity of the dam.

One scenario turns to faith, and a confidence in divine intervention to restore the canyon. There is Seldom's prayer in the middle of the Glen Canyon Dam Bridge early in the novel for a 'little old *precision-type* earthquake' to compromise the dam (Abbey 2004; cf. 'God's Plan' in Abbey 1991a, Abbey 1981 in Philippon 2004). Seldom's prayer becomes a recurring motif in the novel. In essays, Abbey tells of his own playful attempts with friends to encourage along the downfall of the dam—whether through a collective prayer for localized vulcanism (Abbey 1988b, 1991a; cf. Calahan 2001; also Loeffler 2002 at Glen Canyon Dam Bridge with Abbey), or trying to levitate the dam through concentration and visualization (Abbey 1991d; cf. Meloy 1994; Prentiss 2015).

Other scenarios anticipate a sudden and violent intervention at the dam. There is Hayduke's suggestion to ambush the dam's control center and set down a backpack packed with dynamite (Abbey 2004), a return to Abbey's earlier daydream as he floated through the Glen with Newcomb in 1959 (Abbey 1990b). There is also Seldom's fantasy of houseboats packed with ammonium nitrate, diesel fuel, and a detonator charge, often embellished with the addition of 'trained dolphins,' to tow the fleet out beyond the boom and toward the rear of the dam (Abbey 2004, cf. the myth, rumor of dam sabotage in Abbey 1991c). More than one western writer has entertained the confluence of a hypothetical terminal medical diagnosis and dynamite at the dam (Brower in McPhee 1977; Lee in Berger 1994; Meloy 1994). In her song 'Muddy River,' Lee (e.g. 2004a, 2006) sings of her wish to 'bust out' Hoover Dam, and also ponders a storm ripping out dams down the Colorado River (Lee 2006; cf. storm motif in Abbey 1988b). Bruce Berger (1994) recounts how he hosted Lee's birthday celebrations in 1975, when Lee triumphantly trampled a chocolate birthday cake modeled after Glen Canyon Dam, sending a deluge of frosting, sponge, and wafer across his living room carpet. Dynamite has been used by eco-saboteurs on various dam and irrigation projects across the U.S., including at the Ament Dam near Grants Pass on the Rogue River in 1912, the Alabama Gates, part of the Los Angeles Aqueduct, outside Lone Pine in California in 1924, and the No Name Siphon aqueduct, also part of the Los Angeles Aqueduct, in 1927 (see Reisner 1993; Bishop 1995; Cassuto 2001; Adler 2007).

When the gang return in *Hayduke Lives!* (Abbey 1990a), published the year after Abbey's death, the acts of ecotage resume with a renewed fervor. The dam still lingers as the gang's eventual endgame—'maybe next year' (Abbey 1990a: 115)—but this focus is interrupted by a new threat to the canyonlands of southern Utah and northern Arizona: 'GOLIATH the Super-G.E.M. of Arizona, the Giant Earth Mover' slowly advancing toward the Neck, in the Island in the Sky district of Canyonlands National Park. The gang itself also has a more muted presence in the novel, supplanted instead by a growing band of EF! protesters. The closest the gang come to realizing their original fantasy is in Hayduke's 'special project,' a solo flight in a stolen Cessna when he drops a cargo of milkjugs filled with black latex paint onto the face of the dam, inscribing a giant, unavoidable X-marks-the-spot upon the dam's façade (Abbey 1990a).

6.3.2 Politicizing and Mobilizing the Monkey Wrench as an Ecotage and Restoration Motif

The political nuance of the monkey wrench escalates across Abbey's two protest novels. The monkey wrench begins as a subtle motif within *The Monkey Wrench Gang* (Abbey 2004), that by *Hayduke Lives!* (Abbey 1990a) has become a much more radical and political statement on anarchy, ecotage, eco-defense. In *The Monkey Wrench Gang* (Abbey 2004), there are very few allusions to the tool. A lone monkey wrench slides across the tarmac as the Glen Canyon Dam Bridge collapses. Direct mention of the Monkey Wrench Gang by name is rare. Although Abbey was an early pioneer and adopter of the monkeywrenching construct (Loeffler 2002), specific use of terms such as 'monkey wrench business,' 'monkeywrenching,' 'monkey business,' 'monkeying around' are equally sparse, even though the novel hinges on such action.

But this all changes in *Hayduke Lives!* (Abbey 1990a). There is still the suggestive, symbolic imagery—of wielding flying monkey wrenches, of throwing a monkey wrench in the works. But now, the tool is mobilized in the narrative as both noun and verb much more openly. In the later novel, the monkey wrench—and monkeywrenching—is explicitly a symbol of anarchy and eco-defense. A 4-foot antique cast iron monkey wrench is carried by one of the EF! protesters. The monkey wrench features prominently on flags displayed at demonstrations and rallies. The monkey wrench appears in subtler guises in the novel, too: an 'Art Goodwrench' leads a workshop on diesel mechanics (i.e. bulldozer sabotage) at the Earth First! Rendezvous on the north rim of the Grand Canyon.

There is also a philosophical collision between the monkey wrench and restoration in Abbey's two protest novels. As Abbey's friend (and inspiration for Seldom Seen) Ken Sleight attests, Abbey imagined the monkey wrench not as a symbol of destruction, but of *restoration* (in Meloy 1994; Williams 2006; Prentiss 2015; Lincoln 2020). I like Terry Tempest Williams's (2002: 7) framing of 'metaphorical monkey wrenches and shovels' to describe the creative guerrilla restoration activism begun in Abbey's literary productions. A 2014 documentary on Abbey's legacy by M. L. Lincoln, *Wrenched*, and an accompanying book, *Wrenched from the Land: Activists Inspired by Edward Abbey* (Lincoln 2020), continue the monkey wrench—restoration entanglements (also Miller 2018 on the cultural, political, and labor histories of the monkey wrench). Katie Lee, Ken Sleight, Doug Peacock (the inspiration for Hayduke), Dave Foreman (cofounder of EF! and the Rewilding Institute), Terry Tempest Williams, and many other activists feature across the documentary and book.

The monkey wrench as a symbol of ecotage was enthusiastically adopted by Earth First!, the radical environmental advocacy group, established in 1980. EF! was inspired by *The Monkey Wrench Gang*, and the affinity of the monkey wrench with ecotage was confirmed in the EF! manual *Ecodefense: A Field Guide to Monkeywrenching* (Foreman and Haywood 1993; cf. 'Eco-Defense' in Abbey 1988a; also Zakin 1993). Abbey wrote the 'Forward!' to *Ecodefense* (Abbey 1993), and honored the group (and many of its founding members) in *Hayduke Lives!* (Abbey 1990a).

The symbolic 'cracking' of dams—specifically of marginal and/or abandoned dams, or 'deadbeat dams' (after Beard 2015)—flourished among EF!ers and other environmental activists. It emerged as a popular creative, artistic rebuttal to the continued presence of controversial dams on several other western U.S. rivers. Six years after EF!'s first guerrilla 'crack' at Glen Canyon Dam in March 1981, two more clandestine 'nighttime beautification' (after Abbey 1990a, 2004) performances were staged by EF! activists (on the guerrilla theatricality of 'cracking' Glen Canyon Dam, see Standing 2012; also Zakin 1993; Barber 1996). The face of the O'Shaughnessy Dam on the Tuolumne River in Yosemite's Hetch Hetchy Valley (cf. Chapter 3) hosted a painted jagged crack and the war-cry 'FREE THE RIVER!---J. MUIR,' while the Glines Canyon Dam on the Elwha River in Olympic National Park, Washington, was emblazoned with another crack and 'ELWHA BE FREE' (Grossman 2002; *DamNation* 2014). In 2011, a giant pair of scissors cutting down a dotted line appeared overnight on the retired Matilija Dam on Matilija Creek near Ojai, California (*DamNation* 2014).

As guerrilla art installations, these 'cracks' embody Abbey's declaration that, 'One brave deed is worth a thousand books. Sentiment without action is the ruin of the soul' (Abbey 1984: xvi; 1989: 40; 2006: 126, 224, cf. *DamNation* 2014). The 'cracks' contribute to an 'iconography of sabotage' (Lindholdt 2015), but also question the necessity of many dams. Of the 'cracked' dams, Glines Canyon Dam is the first (and so far only) dam to have been removed—together with the Elwha Dam, it was demolished as part of the Elwha River restoration project between 2011 and 2014 (see McCool 2012 on the 1992 Elwha River Ecosystem and Fisheries Restoration Act, also *DamNation* 2014; Beard 2015). A different scenario entirely unfolded following the EF! 'cracking' of the O'Shaughnessy Dam—this is the only installation to have been blocked out with white paint. And so the legacy of Abbey, desert anarchist, eco-defender, wilderness warrior, endures. So potent and compelling is the entanglement of Abbey's legacy in the West that, 'It is true that some of my fiction was based on actual events. But the events took place after the fiction was written' (Abbey 1989: 58, see also Abbey's 2004 wry epigraph).

6.4 June 1983

Less than a decade after Abbey's (2004) *The Monkey Wrench Gang* was published, Glen Canyon Dam was confronted, in June 1983, with high inflow into the recently filled Lake Powell after the highest spring runoff in the Colorado River in a quarter-century. Life came very close to imitating art as this massive inflow nearly caused the dam to fail (cf. Hannon 1997). Snowmelt had outstripped fictional earthquakes, TNT, and rigged houseboats. If Abbey's cautioning in essays that natural forces could and probably would overwhelm the dam (e.g. Abbey 1991d) struck a little too close for comfort that summer, then the closing verse of Lee's protest song 'Pore Colly Raddy' from the 1950s feels uncomfortably astute. Although the song anticipates the river weakening the concrete foundation rather than breaching the rim of the dam, the dam's demise is forecast to unfold in June (Lee 2004a, 2006).

When Reclamation Commissioner Floyd Dominy led David Brower and writer John McPhee on an all-access tour through Glen Canyon Dam in the late-1960s, Dominy included a stop at the very bottom of the dam, below the original river bed, where the dam was 300-foot thick—and had cracked. Colorado River water was pouring through into passageways, corridors, and workspaces. Unalarmed, Dominy reasoned the concrete was still curing, and there was an emergency repair program in place if needed (in McPhee 1977, also Fradkin 1996).

But nothing prepared the USBR for what unfolded in June 1983. Kevin Fedarko's (2013) *The Emerald Mile: The Epic Story of the Fastest Ride in History Through the Heart of the Grand Canyon* comprehensively pieces together what happened at the dam early that summer, as USBR operators, engineers, and technicians raced to stabilize the reservoir level (further history on the dam's near-failure is also included in Martin 1989, Reisner 1993, Grossman 2002, Powell 2008, Waterman 2010, Owen 2017). At the start of June, the reservoir level was less than three inches from the top of the spillway tunnel gates. Lake Powell was filling to capacity, with no allowance for the storage of spring runoff. As the reservoir approached full capacity, a deluge of snowmelt floodwater was entering the reservoir almost twice as fast as it could be drained through the dam's penstocks into the turbines, and through the four river outlet pipes. Glen Canyon Dam was not the only Upper Basin dam approaching full capacity that summer, but it was the only one to suffer a severe compromise to its infrastructure.

Glen Canyon Dam had an emergency bypass system in place to mitigate high inflow events. Twin spillway tunnels each with a diameter of more than 40 feet, bored through the Navajo sandstone either side of the dam, each capable of channeling upward of 100,000 cubic feet per second (cfs) of water around the dam and into the Colorado River (Fedarko 2013). But the spillway tunnels had never been run at full capacity, and soon after the first tunnel—the east tunnel—was opened in early-June, releasing 20,000 cfs downstream, the sustained water pressure eroded the concrete lining in the tunnel, exposing the sandstone of the canyon cliff wall. Huge chunks of concrete, rebar, and sandstone were launched into the river. The west spillway was opened to allow inspection of the east spillway, but mostly water releases were directed through the compromised tunnel, with the west spillway often held in reserve. The reservoir discharge increased throughout June, and on June 28 peaked at 92,000 cfs, far exceeding what the dam had ever had to handle before—even during a test run in 1980 (Powell 2008; Waterman 2010; Fedarko 2013).

Meanwhile, the power plant was kept online continuously, with turbines and generators running at full capacity, to further draw down the reservoir. Still more water was released directly into the Colorado River, bypassing the power plant, through four river outlet pipes. The water pressure in the outlet pipes was severe enough to jettison manhole covers in the adjacent parking lot, and the outlets had to be shut down temporarily when leaks developed (Fedarko 2013). One early, improvised solution to the ever-rising waters lay in an incredibly low-tech enterprise—requiring little more than angle iron and marine-grade plywood. Dam maintenance crews of welders and carpenters constructed a watertight wall of flashboards across the tops of the spillway gates, which served to expand the capacity of the reservoir by four feet, instantly creating an additional 645,000 af of storage (Powell 2008; Waterman 2010; Fedarko 2013). As the snowmelt inflow began to drop at the start of July, the temporary flashboards were upgraded, and by mid-July were replaced by steel structures, doubling the height of the barrier.

The reservoir level had come within feet of the crest of the dam. Later in July, with the new flashboards installed and as the floodwaters abated, the spillway gates were closed to allow repair crews to enter the tunnels and inspect the damage. Both tunnels had gaping cavities near the 'elbow' (where the spillway tunnel intersected the old diversion tunnel), although the damage was most severe in the east tunnel. Work to repair the cavities began in late-August, and cost more than $32 million. 2,300 cubic yards of concrete was needed to plug the holes, the deepest of which extended 50 feet back into the sandstone (Powell 2008; Waterman 2010; Fedarko 2013).

6.5 Glen Canyon Institute and Fill Mead First

Since it was established in Salt Lake City, Utah, in 1996, the Glen Canyon Institute (GCI) has been a leading nonprofit organization advocating the restoration of Glen Canyon and the Colorado River, by draining Lake Powell and bypassing the dam.

In October 1996, the Glen Canyon Institute hosted a meeting at the University of Utah in Salt Lake City to begin exploring strategies for decommissioning Glen Canyon Dam. This was GCI's first public proclamation on the restoration of Glen Canyon. Panelists included GCI's president and founder Richard Ingebretsen, former Sierra Club executive director David Brower, who soon after joined GCI's board of trustees, river runner and activist Katie Lee, and two former Bureau of Reclamation personnel, Commissioner Daniel P. Beard and scientist David Wegner (Brower 1997; Wegner 2000). Both Lee and Brower, together with Abbey, writer Bruce Berger, and river runner Kent Frost, would be recognized by GCI as 'Glen Canyon Legends' (GCI 2018b).

At that early Salt Lake City meeting, the USBR revealed that a substantial volume of Lake Powell's water is lost to evaporation and bank storage each year—almost 1 maf—and with time, sediment will impede not just the reservoir's water storage capacity, but the dam's capacity for power generation (Brower 1997; Ingebretsen 2000; Wegner 2000; Grossman 2002; Waterman 2010). Less than a fortnight after this meeting, David Brower persuaded the Sierra Club board of directors to advocate the draining of Lake Powell. Four decades earlier, during the Echo Park fight, the board had overruled Brower's opposition to Glen Canyon Dam. On November 16, 1996, the board unanimously supported his motion for draining Lake Powell (Brower 1997; Ingebretsen 2000; Miller 2000). Brower again made his case in the March/April 1997 issue of *Sierra* magazine, in an article entitled, 'Let the River Run Through It' (Brower 1997). Congress, meanwhile, trivialized the GCI and Sierra Club proposal.

With the support of the Sierra Club and other conservation organizations, GCI launched a citizen-led environmental assessment (CEA) in 1996 to investigate decommissioning the dam (Ingebretsen 2000). 1996 also marks the year the USBR began an intermittent series of high flow experimental (HFE) releases from Glen Canyon Dam—artificial floods—to simulate the pre-dam flooding of the Colorado River through the Grand Canyon.[12] In March 1996, the USBR released 120 million gallons of water (equivalent to 370 af) from Lake Powell with the sole purpose of encouraging the restoration of beaches and sandbanks downstream through Lee's Ferry and the Grand Canyon (Grossman 2002; Adler 2007). The following year, the House Committee on Natural Resources (HCNR) held an oversight hearing on the Sierra Club's proposal. In holding the hearing, the HCNR Subcommittee on National Parks and Public Lands gave legitimacy to political dialogue and debate on the future decommissioning of Glen Canyon Dam (see Miller 2000; Wegner 2000; Beard 2015).

Released in December 2000, GCI's *Citizens' Environmental Assessment (CEA) on the Decommissioning of Glen Canyon Dam* sets out the effects of dam operations, and the potential impacts of decommissioning the dam and draining the reservoir (GCI 2000, the CEA is also explored in Ingebretsen 2000; Adler 2007). In the quarter-century since it was founded, GCI has also completed a number of scientific studies on the impacts of the dam, won a lawsuit requiring the USBR to reevaluate the impact of dam operations on endangered species, launched the Glen Canyon GIS Mapping Project, and Glen Canyon Living Atlas Project, and much

[12] YNP online interview, August 3, 2020.

more (GCI 2018a). GCI is also campaigning to stop the Lake Powell Pipeline and is building a constituency for a Glen Canyon national park proposal.

For GCI, the restoration of Glen Canyon requires a Colorado River Basin approach. As a Glen Canyon restorationist notes, 'you can't pick out one part of the basin by itself and ignore the rest [...] it can't just be this one part of the river.'[13] Mobilizing this Colorado Basin approach, GCI's proposition for restoring Glen Canyon revolves around consolidating the waters of Lake Powell and Lake Mead in Lake Mead (also Brower 1997). Rather than having two reservoirs standing at most at half-capacity, with double the evaporation and double the seepage, GCI favors moving to one reservoir at near-full capacity. Thus: Fill Mead First (hereafter FMF)—an expansion of Brower and the Sierra Club's proposal. In May 2021, water levels in Lake Powell registered at 34 per cent capacity, and Lake Mead registered at 37 per cent capacity—its lowest capacity since it was filled in the 1930s (GCI 2021; James 2021). Lake Powell is unlikely to fill to capacity again (Lee 2006; Powell 2008; Beard 2015; GCI 2018c; Lincoln 2020; cf. Fleck 2016 on Lake Mead).[14] In the history of the reservoir, 'full pool' has only been reached three times since the June 1983 bypass event, with only one of those requiring a major release, running at 50,000 cfs (Powell 2008; Waterman 2010). In 2005, Lake Powell was just three years away from reaching its 'dead pool' level (the lowest exit for reservoir releases, at the elevation of the dam's four river outlet pipes), but the following spring the reservoir slowly began to rise (Powell 2008; Spurgeon 2009). Just two decades separate Glen Canyon's historic flood and drought episodes.

If authorized, FMF would assign Lake Mead as the 'primary water storage and distribution facility for the upper and lower Colorado River basins' (GCI 2012, see also 2018c). GCI's Fill Mead First proposal seeks to modify the operation of Glen Canyon Dam through a three-part program—draw down and stabilize the reservoir to the lowest 'power pool' level of 3,490 feet (limited hydropower generation), then to the 'dead pool' level of 3,374 feet, before drilling new diversion tunnels to

[13] GCI telephone interview, February 6, 2013.

[14] GCI telephone interview, February 6, 2013, GCI online interview, August 14, 2020.

bypass the dam (GCI 2012, 2018c).[15] Abbey also made a similar proposition in a March 26, 1973 letter to Utah Senator Frank E. Moss (in Lee 2006). GCI renegotiates Abbey's polemic for outright dam removal toward draining the reservoir by bypassing the dam. 'Glen Canyon restoration' as promoted by GCI, aligns with James Aronson et al.'s (1993) restoration *sensu lato* goal—it is about ecosystem health.

GCI first submitted its FMF proposal to the Bureau of Reclamation in 2012 as an option on a Colorado River Basin study, and again in 2016 as part of a review of the operating plan for Glen Canyon Dam. But the proposal was rejected on both occasions (see GCI 2012, 2018c; cf. Bureau of Reclamation 2012, 2016). Despite its rejection, FMF is gaining legitimacy and political gravitas in dialogue on water policy reform: 'it is taken seriously by even the people who don't like it and don't want to do it.'[16] The FMF plan challenges the historic precedent of how and where to store water in the Colorado River system (Schmidt et al. 2016). For Utah writer Terry Tempest Williams (2006), GCI has transmuted Abbey's monkey wrench into a microphone, a bullhorn.

The call to decommission Glen Canyon Dam, drain Lake Powell, and 'Fill Mead First' has a surprising ally in former Reclamation Commissioner Daniel P. Beard, who served between 1993 and 1995. Beard's vision, formally laid down in his 2015 book, *Deadbeat Dams: Why We Should Abolish the U.S. Bureau of Reclamation and Tear Down Glen Canyon Dam*, vibrates against the USBR's position on the dam. In *Deadbeat Dams*, Beard (2015) speaks out on unnecessary and environmentally-damaging dams and traces the genesis of political conversations—of committee meetings and hearings—on the issue of dam removal in Glen Canyon. But he also considers the history of water rights legislation on the Colorado River, and the implications and complications of the decades-long drought plaguing the American West. While Beard (2015) cautiously praises Abbey's essays and protest novels for celebrating and drawing attention to the beauty of the former river canyon (his praise of Powell's 2002 travelogue is more forthcoming), he notes the double-edged sword that is the Abbey canon—how Abbey's

[15] GCI telephone interview, February 6, 2013.

[16] GCI telephone interview, February 6, 2013.

anarchical fictional quartet (as Abbey 1990a, 2004) became so ingrained with dam removal that it also often quarantined talk of dam removal amongst water wonks. Beard (2015) realigns Abbey's polemic for ecological restoration around a reappraisal of western water policy, and water reform. Much like GCI's (2012, 2018c) FMF plan, Beard's (2015) argument revolves around redundancy, and the temporality of dams, exemplified in a chapter title in *Deadbeat Dams* that asserts, 'Dams Aren't Forever.'

These two responses to Abbey's invitation and provocation for restor(y)ing Glen Canyon coalesce along an unexpected trajectory. Such is the institutionalized, politicized gravitas of Abbey's literary imaginary that both restorationists (in GCI) and reclamationists (in former Reclamation Commissioner Beard) find in Abbey's writings—however, indirectly or mediated and renegotiated—an argument for the restoration of Glen Canyon. As Sara L. Spurgeon (2009: 751) argues, 'While to date no legislation specifically tied to Abbey's writings has been passed, the continuing controversy surrounding Glen Canyon Dam and Lake Powell may change that.'

Regardless, Glen Canyon Dam is far from being considered for decommissioning. It is an incredibly incendiary issue (consider the hydrological, biological, technological, socioeconomic, political, and legal issues outlined in Farmer 1999; Carothers and House 2000; Miller 2000; Wegner 2000; Grossman 2002; Adler 2007). To dismantle the dam, or bypass the dam and divert the river around it, would require the reexamination and rewriting of historic, complex water law, and the renegotiation of contested rights-claims to the waters of the Colorado River, as well as adherence to federal environmental legislation including the 1969 National Environmental Policy Act, 1972 Clean Water Act, and 1973 Endangered Species Act.

But the decades-long drought engulfing swathes of the American West is already affecting Lake Powell. Reduced snowmelt, together with evaporation, is exposing previously submerged side canyons. The front-page story in *The Salt Lake Tribune* on June 20, 2021, announced the 'Return of Glen Canyon' (Maffly 2021). There are also many hanging canyons back in the side canyons that never got flooded, where vegetation and wildlife is beginning to recover and repopulate lower reaches of the canyons. Confirming Abbey's (e.g. 1991d) and Lee's (2006) belief that the ecological 'ugliness' that followed a drawn-down Lake Powell would be transitory, restorationists working in Glen Canyon describe hiking in side canyons where, 'you would not even know that [...] the reservoir had been there,' and note the resiliency of canyon ecosystems, 'that silt will wash away, and the willows and other native plants will come back.'[17] Restoration of the central canyon will be slower closer to the rear of the dam—where the canyon has been underwater for longer, and the sediment stockpile is greater (see Ingebretsen 2000 on metamorphosis; also McGivney 2009; Beard 2015; Solnit 2017). But some of the marinas farthest from the dam have been abandoned as the reservoir level has dropped.

The release of the USBR's latest operating plan for Glen Canyon Dam late in 2016 reveals a federal commitment to at least another two decades of operations at the dam (Bureau of Reclamation 2016). The operation of the dam and effects of water releases on the Grand Canyon are also informed by the 1992 Grand Canyon Protection Act. But while the restoration of Glen Canyon upstream of the dam remains uncertain, the river and canyons immediately downstream tell another pseudo-restoration story. For almost a quarter-century, the USBR has conducted a series of HFE releases from Glen Canyon Dam. These periodic releases are intended to support the restoration of sandbars, beaches, and backwater habitats affected by the operation of Glen Canyon Dam (Fradkin 1996; Wegner 2000; Cassuto 2001; Grossman 2002; Lee 2006; Adler 2007; *DamNation* 2014; see also Wegner 1991 on the Glen Canyon Environmental Studies). HFE releases began in March 1996 and have been repeated in 2004, 2008, 2012, 2013, 2016, and 2018 (Bureau of Reclamation 2019; also Farmer 1999; Ingebretsen 2000; Miller 2000; Grossman 2002; Adler 2007; Powell 2008).

But the flood flow experiments are not the canyon restoration's silver bullet. Not even temporarily. Rare historic seasonal flows could reach upward of 200,000 cfs through Glen Canyon and Grand Canyon, sometimes approaching 500,000 cfs, and at other times could almost disappear. Since the appearance of dams and diversions along the Colorado

[17] GCI telephone interview, February 6, 2013.

River, variable annual and seasonal flows have been stabilized, and at Glen Canyon Dam are controlled at between 8,000 and 25,000 cfs (Fradkin 1996; Carothers and House 2000; Miller 2000; Wegner 2000; Powell 2008; Fedarko 2013). Historic variations in water temperature and sediment load are also now highly regulated. The Colorado River water released from behind the dam is no longer warm and silty—'too thick to drink, too thin to plow' (Fradkin 1996; GCI 2000; Adler 2007)—but cold and clear, drawn from far below the reservoir surface. These cold water releases impair and kill downstream riparian ecosystems and species. For Glen Canyon restorationists, the HFE releases are 'really just a Band-Aid on an unsustainable system.'[18]

The Colorado River is then impounded once again in Lake Mead, behind Hoover Dam. But following the 2012 signing of the U.S.-Mexico Minute 319 agreement, an eight-week pulse flow release from the Morelos Dam (separated from Hoover Dam by at least a half-dozen other dams) on the Arizona-Mexico border in late-March 2014 briefly restored Colorado River water to the final stretch of the river—the Colorado River Delta and the Sea of Cortés (on the Minute 319 pulse flow and ecological restoration, see e.g. Gerlak et al. 2013; Bark et al. 2016; Fleck 2016; Glenn et al. 2017; also Adler 2007 on restoring water flows and ecosystems in Mexico).

6.6 Literature, Advocacy, and Two Utah National Monuments

Literature has long provided a creative platform for environmental advocacy and activism in the slickrock canyons of the Four Corners region. The Sierra Club's 'battle books,' and Abbey's nonfiction and fiction writings, are a large part of this powerful canon, but the last two decades have seen the emergence of a new model of writing-as-advocacy in southern Utah, as the voices of writers assembled and joined the fight to secure national monument designations in the redrock canyon country.

[18] GCI online interview, August 14, 2020.

6.6.1 The Quiet Politics of the Public Lands Chapbook

In the summer of 1995, and just a few weeks after sending an invitation letter for contributions, Utah writers and conservationists Stephen Trimble and Terry Tempest Williams's edited chapbook, *Testimony: Writers of the West Speak on Behalf of Utah Wilderness*, was in print (cf. the trade edition, as Trimble and Williams 1996, the default hereafter). The template was Stegner's (1955) *This Is Dinosaur*, and his 1960 'Wilderness Letter' (in Stegner 2017).[19]

The slim volume of essays and poems was a creative intervention responding to the ongoing hearings in Utah and Washington, DC surrounding the Utah Public Lands Management Act of 1995, a bill that would open up wilderness areas to development. Williams had earlier testified on conservation panels before hearings in Utah and Washington, DC, but was disappointed by the disinterest displayed by committee members to conservation concerns each time (Sumner 2002b; Williams 2012). Her solace was that, at least in the Washington, DC hearings, her testimony and that of fellow conservationists was entered into the *Congressional Record*. *Testimony* (Trimble and Williams 1996) drew together sympathetic voices from across the American West in support and celebration of Utah wilderness (two interviews with Trimble and Williams—Sumner 2002a, b—share *Testimony*'s origin story, see also Trimble 2008; Williams 2012). Contributors included Margaret E. Murie, Ann Zwinger, John McPhee, N. Scott Momaday, Barry Lopez, Ellen Meloy, Gary Paul Nabhan, and Rick Bass. In their opening salvo, Trimble and Williams (1996) firmly situate *Testimony* as an act of faith by the contributing authors, and an exercise in bearing witness, and applaud the power of story and storytelling to cut through partisan politics.

In late-September 1995, and with the support of the Southern Utah Wilderness Alliance (SUWA), Trimble and Williams traveled to Washington, DC for a press conference outside the Capitol, sent out a press release, and distributed copies of *Testimony* to congressional offices.[20]

[19] Environmental writer online interview, July 22, 2020.

[20] Environmental writer online interview, July 22, 2020.

Supporters of *Testimony* in the House and Senate wrote cover letters, and the chapbook was distributed to every member of Congress. Over the following months, Trimble and Williams began work on a trade book with Milkweed Editions. The expanded trade edition features an Appendix on America's Red Rock Wilderness Act (Trimble and Williams 1996, see also the Appendix in Williams 2002).

When the Utah Public Lands Management Act arrived on the floor of the U.S. Senate in late-March 1996, the Senate swiftly entered into a filibuster, led by New Jersey Senator Bill Bradley and Democrat allies (Sumner 2002a, b; Williams 2012; Trimble 2017). During the course of the March 26 filibuster against the Utah (anti-)wilderness bill, Trimble and Williams's (1996) *Testimony* chapbook was read aloud on the Senate floor—and, in a mirroring of Williams's own experience the year before of testifying on conservation panels before hearings in Utah and Washington, *Testimony* too entered into the *Congressional Record*. Eventually defeated, the Utah Public Lands Management Act did not move out of the Senate.

But Trimble and Williams's (1996) *Testimony* also had another life to live (after Thoreau 2004), beyond the anti-wilderness bill. On September 18, 1996, from a desk overlooking the South Rim of the Grand Canyon, President Bill Clinton established the 1.9-million-acre Grand Staircase-Escalante National Monument in southern Utah (Proclamation No. 6920, 1996, on the history of Grand Staircase-Escalante National Monument, see Trimble 2008).[21] Grand Staircase-Escalante was the first 'landscape-scale' national monument.[22] Its size was unprecedented. Williams spoke at the event, and President Clinton told her that *Testimony* had informed his administration's decision on the national monument (Sumner 2002b; Williams 2012; Trimble 2017). For environmental writer-activists, 'When we did the book, we weren't thinking of the monument so much as stopping this terrible wilderness bill. But then the book became part of the campaign.'[23]

[21] President Clinton signed the proclamation across the border in Arizona, a not-so-subtle overture pointing to the lack of support from the Utah congressional delegation for the new national monument.

[22] GSEP online interview, August 5, 2020.

[23] Environmental writer online interview, July 22, 2020.

In the fall of 1996, Milkweed Editions organized a joint reading of *Testimony* for the writers at Robert Redford's Sundance Mountain Resort in Utah (Sumner 2002a). The weekend gathering to talk about the release of the trade edition came just a few days after President Clinton signed the Grand Staircase-Escalante proclamation. The gathering hosted an outdoor group reading, and roundtable discussions on conservation, and writing-as-advocacy, and became 'a celebration of community and the power of words.'[24]

Trimble and Williams's (1996) *Testimony* continues the legacy of literary environmental advocacy and activism in the slickrock desert of Utah begun by Stegner's (1955)*This Is Dinosaur*, and Porter's (1963) *The Place No One Knew*, and later by Abbey's (1990a, 2004) protest novels *The Monkey Wrench Gang*, and *Hayduke Lives!* But *Testimony* has enjoyed something approaching a cult status of its own, with an ecological and political legacy that extends beyond southern Utah. It has inspired other collections, anthologies, and chapbooks across the U.S., featuring writers speaking out in defense of Alaska's Arctic National Wildlife Refuge (ANWR), and the Tongass Rain Forest, the Bolsa Chica Ecological Preserve in California, the Boundary Waters Canoe Area Wilderness in Minnesota, Montana's Yaak Valley, Petroglyph National Monument in New Mexico, and the Wasatch Mountains in Utah (collated in Sumner 2002a, b).[25] But *Testimony*'s legacy would go full circle and return to Utah two decades later.

Across the early-2010s, first Utah Diné Bikéyah (UDB), a Navajo nonprofit organization working to protect ancestral Native lands, then the Bears Ears Inter-Tribal Coalition (BEITC), a group of five sovereign tribal nations with strong ties to the Bears Ears region (Hopi, Navajo, Uintah and Ouray Ute, Ute Mountain Ute, and Zuni), petitioned the Obama administration to establish a Bears Ears National Monument protecting the twin Bears Ears buttes that preside over the Cedar Mesa in southern Utah. The BEITC's Bears Ears National Monument proposal was gaining momentum over (and in some ways, also morphed out of) a

---

[24] Environmental writer online interview, July 22, 2020.

[25] Environmental writer online interview, July 22, 2020.

proposal for a Greater Canyonlands National Monument.[26] At the same time, another anti-wilderness bill was being pushed by the Utah delegation—the Utah Public Lands Initiative (PLI). In the spring of 2016, as the BEITC national monument proposal vibrated against the PLI, a new literary advocacy project emerged in support of the proposed monument. SUWA brought together a regional community of conservationists, writers, and others, to collaborate and respond to the new threat to Utah wilderness.

Stephen Trimble, co-editor of *Testimony*, worked with Utah-based nonprofit publisher Torrey House Press (THP) to organize and edit a chapbook of essays and poems celebrating the Bears Ears. In a nod to its literary predecessor, the volume is titled, *Red Rock Testimony: Three Generations of Writers Speak on Behalf of Utah's Public Lands* (Trimble 2016). Contributors included activists, scientists, historians, former elected officials, and Native American tribal leaders. Pieces by Bruce Babbitt, Mark Udall, Brooke Williams, Shonto Begay, Lauret Savoy, David Gessner, George Handley, Amy Irvine, Jacqueline Keeler, Regina Lopez-Whiteskunk, Michelle Nijhuis, and Alastair Botsoi join new pieces by some earlier *Testimony* alum.

Working to a similarly tight timeframe, *Red Rock Testimony* (Trimble 2016) was also compiled in less than two months, and with the support of SUWA, was again distributed to Congress—and to the Bureau of Land Management (BLM), the NPS, and the U.S. Forest Service (USFS)—in June 2016. Former Interior Secretary Bruce Babbitt wrote the letter of sponsorship that went with the chapbook to every member of Congress.[27] An expanded trade edition was published by THP in July 2017 as *Red Rock Stories: Three Generations of Writers Speak on Behalf of Utah's Public Lands* (see Trimble 2017). And as with the trade edition of *Testimony* (Trimble and Williams 1996), *Red Rock Stories* (Trimble 2017) also includes an Appendix that traces the origin story of the chapbook—and the political theater of public lands issues that the trade

[26] THP online interview, July 28, 2020, GSEP online interview, August 5, 2020, GCI online interview, August 14, 2020.

[27] Environmental writer online interview, July 22, 2020.

edition entered into following the 2016 U.S. presidential election and inauguration of President Donald J. Trump in January 2017.

But *Red Rock Stories* (Trimble 2017) was not THP's only piece of Bears Ears literary activism. In late-June 2016, THP launched a Kickstarter campaign to fund the publication of *Edge of Morning: Native Voices Speak for the Bears Ears* (Keeler 2017), published as a trade book in June 2017. This anthology of Native writers, edited by Navajo/Yankton Dakota Sioux writer Jacqueline Keeler, features interviews with representatives from the BEITC, including Willie Grayeyes, Jonah Yellowman, Regina Lopez-Whiteskunk, and Alastair Bitsoi, and essays and poems from Indigenous leaders, activists (including members of UDB), writers, and scholars. Where *Red Rock Stories* (Trimble 2017) builds up a Bears Ears politics of place, *Edge of Morning* (Keeler 2017) turns to the stories and oral histories, and landscapes, passions, and activisms told by Native voices.[28]

Seven weeks after the 2016 presidential election, and with just three-and-a-half weeks of his presidency remaining, President Barack Obama established the 1.35-million-acre Bears Ears National Monument (550,000 acres smaller than the BEITC proposal) on December 28, 2016 (Proclamation No. 9558, 2016). Bears Ears National Monument is the first monument to be proposed by Native Nations, rather than conservation groups. It is also the first monument to be jointly administered by tribes and federal agencies—by the BEITC, BLM, and USFS (Keeler 2017; Trimble 2017; Burrillo 2020; Gessner 2020).[29] *Edge of Morning* (Keeler 2017) and *Red Rock Stories* (Trimble 2017) were published less than eight months after the proclamation, during the Trump administration. During his two terms in office, President Obama created or expanded more national monuments than any other U.S. president, protecting more than a half-billion acres of public lands and waters across 34 national monuments (for further discussion on the history and evolution of Bears Ears National Monument, see Keeler 2017; Trimble 2017; Robinson 2018; Strom 2018; Burrillo 2020;

[28] THP online interview, July 28, 2020.

[29] Environmental writer online interview, July 22, 2020, GSEP online interview, August 5, 2020.

Gessner 2020; Roberts 2021). But less than two months later, the Bears Ears National Monument designation would be under threat from the Trump administration, as the future of the 1906 Antiquities Act and the permanence of national monument designations was cast in doubt.

However, literature does not have the monopoly on creative activist platforms for defending Utah's public lands. The power of art-as-advocacy is revealed in the political cartoons of Pat Bagley, editorial cartoonist at *The Salt Lake Tribune*. His four-decade tenure at the *Tribune* has seen him respond to numerous conservation and public lands issues—and none more so than the friction surrounding Bears Ears National Monument. Whereas *Red Rock Stories* (Trimble 2017) and *Edge of Morning* (Keeler 2017) offer static, fixed commentaries, captured in the months pre- (and immediately post-) designation (although the online community storytelling platform Red Rock Stories does mitigate this, at: www.redrockstories.org), Bagley's cartoons have an immediacy in responding and reacting to state and federal decisions as they arise.

Across two dozen cartoons spanning a three-year period, Bagley succeeds in abridging the story of Bears Ears National Monument, distilling and cataloging its history to key moments and critical junctures, as they have played out across the transition from the Obama administration to the Trump administration. The quiet politics of Bagley's cartoons succeeds in renewing and reinvigorating public (and political) rhetoric and dialogue surrounding the monument.[30] Bagley

---

[30] The story of Bears Ears National Monument can be traced through Bagley's cartoons including: 'Bears Ears Monument' (April 5, 2016), 'The Bears Ears Inquisition' (April 28, 2016), 'Bad Badger Infestation' (Jun. 17, 2016), 'When is a Bishop a Pawn' (July 13, 2016), 'Bucket of Deplorables' (December 16, 2016), 'Bears Ears National Monument' (December 28, 2016), 'This Land Was Our Land' (January 27, 2017), 'Utah's Worst Idea' (February 18, 2017), 'Sometimes Bear Gets You' (February 23, 2017), 'This Was the Place' (March 16, 2017), 'This Land is Trumpland' (April 28, 2017), 'Zinke's Ears' (May 5, 2017), 'Bears Ears Bits' (June 12, 2017), 'Honey, I Zinkied the Monuments' (August 24, 2017), 'Beauty of Bears Ears' (November 30, 2017), 'Monumental Scam' (December 5, 2017), 'The Lost Tribes' (December 6, 2017), 'Bargain Basement Bears Ears' (December 15, 2017), 'Whose Land Is This?' (December 31, 2017), 'The Never Ending Story' (January 10, 2018), 'Image Problem' (January 26, 2018), 'GOP Talking Points' (March 6, 2018), 'Monumental Bull' (July 26, 2018), 'Wilderness Trafficking' (July 31, 2018), 'This Land Is Trump's Land' (August 19, 2018), 'Changes in San Juan County' (January 10, 2019), 'Colorado River Compact' (January 22, 2019), 'Attack on Bears Ears' (April 24, 2019), at: www.sltrib.com/opinion/bagley (see also Matt Wuerker's cartoons on the national monument review in *Politico*, at: www.politico.com/news/matt-wuerker).

marked the designation of Bears Ears National Monument in a simple yet powerful cartoon depicting a resolute President Obama staring down a queue of bulldozers already en route to the Bears Ears buttes (Bagley 2016).

6.6.2 After E.O. 13792: National Monuments, and the 1906 Antiquities Act in the Trump Administration

April 26, 2017. E.O. 13792. Presidential Executive Order on the Review of Designations Under the Antiquities Act.

Three months into the Trump administration, and Congress and the Utah GOP were already targeting the U.S.'s national monuments. Bears Ears National Monument was singled out in the Executive Order (Exec. Order No. 13792, 2017), requiring Interior Secretary Ryan Zinke to comment directly on the designation in his interim report. The Executive Order followed on from three resolutions in the Utah Senate by Utah's GOP to hobble the Grand Staircase-Escalante and Bears Ears national monuments. Utah's congressional delegation had earlier objected to President Clinton's designation of Grand Staircase-Escalante, and with the election of President Trump, and President Obama's designation of Bears Ears, old tensions became inflammatory again. Pat Bagley's (2017a) cartoon two days later on April 28, 'This Land is Trumpland,' covered the announcement of the review, depicting the President teeing off for a round of golf with the Bears Ears buttes in the distance, an oil rig already installed atop one, a golf flag waving from atop the other.

Southern Utah was to quickly become the flashpoint in a nationwide debate on public lands protections, the national parks and monuments system, and the legacy of the Antiquities Act. And the literary activism of the *Testimony* (Trimble and Williams 1996) and *Red Rock Testimony* (Trimble 2016) chapbooks would also be politicized anew—and imbued with a renewed, potent urgency—in defense of the canyon country. The trade books *Red Rock Stories* (Trimble 2017) and *Edge of Morning* (Keeler 2017) were published a month before Secretary Zinke submitted his final report. The unfolding national monument fight meant that for *Red Rock*

Stories, 'we thought it would land as a book of celebration, and instead, it landed as a battle book'[31]—'what began as a book of advocacy became a book of defense.'[32] *Red Rock Stories* and *Edge of Morning*, like *Testimony* before them, have many more lives unfolding.

Utah's Bears Ears and Grand Staircase-Escalante, and Maine's Katahdin Woods and Waters (cf. Chapter 2), were three of 27 national monuments placed under review. Secretary Zinke's four-day 'listening tour' to Bears Ears and Grand Staircase-Escalante in May 2017 prioritized meetings with members of the Utah delegation and San Juan County Commissioners, all long opposed to the monument. Although Zinke met with BEITC, and conservation nonprofit Friends of Cedar Mesa, other groups including Grand Staircase Escalante Partners (GSEP), the Sierra Club, SUWA, and UDB, as well as Chambers of Commerce, and Native community organizations were sidelined (see Bagley 2017b and 'Zinke's Ears'). A leaked memorandum on Secretary Zinke's report in mid-September recommended modifications to 10 national monuments, including a severe curtailing of the boundaries of both Bears Ears and Grand Staircase-Escalante (reported by Eilperin 2017a). On December 4, 2017, President Donald Trump traveled to Utah to confirm his administration's decision to shrink both Bears Ears and Grand Staircase-Escalante national monuments. In the days prior to the president's visit, documents obtained by *The Washington Post* revealed that Bears Ears was expected to be reduced by 85 per cent, and Grand Staircase-Escalante by almost 50 per cent (Eilperin 2017b).

With the signings, the Trump administration cut more than two million acres from the two monuments, replacing them with five smaller, disconnected monuments, and revoked the tribal co-management of the Bears Ears monument (Proclamation No. 9681, 9682, 2017). The 1.35-million-acre Bears Ears National Monument, designated just eleven months earlier, was reduced to 201,876 acres, becoming the Shash Jáa and Indian Creek national monuments, of 129,980 acres and 71,896 acres, respectively (Shash Jáa is the Navajo name for Bears Ears, but this renaming of the monument mutes the many names other sovereign tribes

[31] THP online interview, July 28, 2020.

[32] Environmental writer online interview, July 22, 2020.

also have for the area). The 1.9-million-acre Grand Staircase-Escalante National Monument, designated over two decades ago, was reduced to 1 million acres and three units—Grand Staircase at 209,933 acres, Kaiparowits at 552,034 acres, and Escalante Canyon at 242,836 acres. President Trump's political maneuvering signaled the largest rollback of protected public lands in American history and set an uncomfortable precedent that potentially exposed the entire national parks and monuments system to a similar fate. Bagley (2017c) invokes President Theodore Roosevelt, the original signatory of the Antiquities Act, to comment on the modifications to the monuments, depicting TR ecstatic at the creation of national monuments, and irate at their reduction.

Against the backdrop of this uproar, Glen Canyon Institute (2018d) cast the Bears Ears and Grand Staircase-Escalante regions as part of a 'Greater Glen Canyon' ecosystem—and any threat to the monuments as a threat to the health, integrity, and cohesion of the Glen Canyon and Colorado River watersheds. Glen Canyon (and the Glen Canyon NRA) is sandwiched between Grand Staircase-Escalante National Monument to the west, and Bears Ears National Monument to the east. The idea behind expanding and upgrading the Glen Canyon NRA to a national park designation is to protect watershed areas not included in the Bears Ears and Grand Staircase-Escalante boundaries, including areas along the Dirty Devil and Escalante river watersheds, and the Christmas Mountains roadless area.[33] Glen Canyon becomes inscribed anew, and politicized anew, in defense of the two southern Utah monuments.

6.6.3 Restoration Lawsuits Bruin at Bears Ears

Within hours of the signing ceremony, the first lawsuits were filed, arguing that the reduction of the boundaries was an unconstitutional use of the Antiquities Act. And within a matter of weeks, there were five separate lawsuits filed in the DC courts—three lawsuits suing the Trump administration over changes to Bears Ears, and two lawsuits suing over changes to Grand Staircase-Escalante.

[33] GCI online interview, August 14, 2020.

Alongside environmental writer-activists, Native Nations, and conservation advocacy groups, outdoor gear companies also spoke out in defense of the two Utah national monuments. On the day of the signing, Patagonia, a long-time defender of Bears Ears (and collaborator on *DamNation* 2014), which had launched its first-ever TV ad in March 2017 to support the Bears Ears monument, replaced its homepage with a provocative, stark white-on-black message that read simply: 'The President Stole Your Land' (Patagonia 2017). While Patagonia's response was no-holds-barred, REI's (2017) message on its website was more subtle, declaring: 'We ♡ Our Public Lands,' as was the lead banner on The North Face's (2017) website: 'Protect Bears Ears,' which also linked to a Kickstarter campaign by Friends of Cedar Mesa to fund a Bears Ears Education Center. And the Wilderness Society's parody website, Landmark Estates (2017), edged a little closer to reality, especially when its homepage boasted the update: 'Two more sold—parks are going fast! Make an offer today!'

Representative Rob Bishop, chairman of the HCNR, openly attacked Patagonia after the company joined as co-plaintiff on a lawsuit with Utah Diné Bikéyah, Friends of Cedar Mesa, Archaeology Southwest, Conservation Lands Foundation, Access Fund, National Trust for Historic Preservation, and Society of Vertebrate Paleontology, filed after the December 4 announcement. Bishop sent a letter to founder and CEO of Patagonia, Yvon Chouinard, inviting him to testify before the Committee on the Antiquities Act, national monuments, and federal land management (Bishop 2017). Chouinard (2017) on behalf of Patagonia declined, rejecting the 'disingenuous' invite, and the issue faded. Only to be resurrected in early-January 2018, when Bishop followed up with another letter to Chouinard (Bishop 2018), mocking the company for living within a 'limited ideological bubble' and accusing the company of misleading the public on public lands concerns in Utah. But the HCNR also took to Twitter, and in a poor riff on Patagonia's earlier homepage message, charged first that 'Patagonia Is Lying To You,' and later, that 'Patagonia Is ~~Lying~~ Hiding.'[34] The statements were regarded by conservationists and politicians alike as petty and unconstructive, and the HCNR was criticized for using a government social media account to target a private business. When attempts by the Utah GOP and other parties to shrink the Bears Ears monument in order to open up the vast oil and natural gas deposits in the region for exploration and extraction were exposed, Patagonia revisited its monochromatic campaign declaring, 'The President Stole Your Land and You Were Lied To' (Sheehy 2018).

In early-February 2018, Federal Judge Tanya S. Chutkan announced the consolidation of the five lawsuits into two cases, with one case dealing with the former Bears Ears National Monument, and the other, the former Grand Staircase-Escalante National Monument. Almost eight months later, in late-September 2018, came the confirmation that the two lawsuits would remain in Washington, DC, after the Trump administration had requested the lawsuits be moved to Utah. Litigation around the boundaries continued through 2019 and 2020, and with the inauguration of Joseph R. Biden Jr as the 46th President of the United States on January 20, 2021, the new administration immediately ordered an Interior Department review of the Trump administration's national monument rollbacks (in Exec. Order No. 13990, 2021). The Biden administration, in consultation with Native Nations, is exploring restoring the Grand Staircase-Escalante National Monument boundary, and restoring and possibly expanding the Bears Ears National Monument boundary. Following a request from the Biden administration, the U.S. District Court for the District of Columbia has paused its decision on the lawsuits while Interior completes its review. At the time of writing, a decision on the lawsuits is still pending.

[34] @NatResources. '@Patagonia doesn't want #MonumentsForAll, they just want your money #BearsEars.' *Twitter*. Twitter, December. 8, 2017, 20.54. twitter.com/NatResources/status/939236821971734530. @NatResources. 'Unfortunately, @Patagonia refused to appear publicly to defend their lies. Letter → bit.ly/2IZpBKp.' *Twitter*. Twitter, January 5, 2018, 21.52. twitter.com/NatResources/status/949398337521897472.

Works Cited

Abbey, Edward. 1951. Some Implications of Anarchy. *The Thunderbird* VI (3): 3–19.
Abbey, Edward. 1954. *Jonathan Troy*. New York, NY: Dodd, Mead & Company.
Abbey, Edward, and Philip Hyde. 1971. *Slickrock: Endangered Canyons of the Southwest*. San Francisco, CA: Sierra Club Books/Scribner Library.
Abbey, Edward. 1959. *Anarchism and the Morality of Violence*. Unpublished Master of Arts in Philosophy Thesis. Albuquerque, NM: University of New Mexico. http://digitalrepository.unm.edu/phil_etds/17/.
Abbey, Edward. 1984. *Beyond the Wall: Essays from the Outside*. New York, NY: Henry Holt and Company, LLC.
Abbey, Edward. 1988a. *One Life at a Time, Please*. New York, NY: Henry Holt and Company, LLC.
Abbey, Edward. 1988b. *The Fool's Progress (An Honest Novel)*. New York, NY: Henry Holt and Company, Inc.
Abbey, Edward. 1989. *A Voice Crying in the Wilderness (Vox Clamantis in Deserto): Notes from a Secret Journal*. New York, NY: St. Martin's Press.
Abbey, Edward. 1990a. *Hayduke Lives!* New York, NY: Back Bay Books/Little, Brown, and Company.
Abbey, Edward. 1990b [1968]. *Desert Solitaire: A Season in the Wilderness*. New York, NY: Touchstone.
Abbey, Edward. 1991a [1977].*The Journey Home: Some Words in Defense of the American West*. New York, NY: Plume.
Abbey, Edward. 1991b [1979]. *Abbey's Road*. New York, NY: Plume.
Abbey, Edward. 1991c [1980]. *Good News*. New York, NY: Plume.
Abbey, Edward. 1991d [1982]. *Down the River*. New York, NY: Plume.
Abbey, Edward. 1992 [1956]. *The Brave Cowboy: An Old Tale in a New Time*. New York, NY: Avon Books, Inc.
Abbey, Edward. 1993. Forward! In *Ecodefense: A Field Guide to Monkey-wrenching*, 3rd ed., eds. Dave Foreman and Bill Haywood, 3. Chicago, CA: Abbzug Press.
Abbey, Edward. 1994. *Earth Apples: The Poetry of Edward Abbey*, ed. David Petersen. New York, NY: St. Martin's Press.
Abbey, Edward. 2003 [1994]. *Confessions of a Barbarian: Selections from the Journals of Edward Abbey*, ed. David Peterson. Boulder, CO: Johnson Books.

Abbey, Edward. 2004 [1975]. *The Monkey Wrench Gang*. London: Penguin Books Ltd.

Abbey, Edward. 2006. *Postcards from Ed: Dispatches and Salvos from an American Iconoclast*, ed. David Petersen. Minneapolis, MN: Milkweed Editions.

Abbey, Edward. 2014 [1971]. *Black Sun: A Novel*. New York, NY: Harper Perennial Modern Classics.

Adler, Robert W. 2007. *Restoring Colorado River Ecosystems: A Troubled Sense of Immensity*. Washington, DC: Island Press.

Aronson, J., C. Floret, E. LeFloc'h, C. Ovalle, and R. Pontanier. 1993. Restoration and Rehabilitation of Degraded Ecosystems in Arid and Semi-Arid Lands. I. A View from the South. *Restoration Ecology* 1 (1): 8–17.

Bagley, Pat. 2016. Bagley Cartoon: Bears Ears National Monument. *The Salt Lake Tribune*. http://archive.sltrib.com/article.php?id=4758978&itype=CMSID.

Bagley, Pat. 2017a. Bagley Cartoon: This Land is Trumpland. *The Salt Lake Tribune*. http://archive.sltrib.com/article.php?id=5231650&itype=CMSID.

Bagley, Pat. 2017b. Bagley Cartoon: Zinke's Ears. *The Salt Lake Tribune*. http://archive.sltrib.com/article.php?id=5258409&itype=CMSID.

Bagley, Pat. 2017c. Bagley Cartoon: Monumental Scam. *The Salt Lake Tribune*. http://sltrib.com/opinion/bagley/2017/12/04/bagley-cartoon-monumental-scam/.

Barber, Katrine E. 1996. Wisecracking Glen Canyon Dam: Revisioning Environmental Mythology. In *Change in the American West: Exploring the Human Dimension*, ed. Stephen Tchudi, 127–143. Reno and Las Vegas, NV: University of Nevada Press.

Bark, Rosalind H., Catherine J. Robinson, and Karl W. Flessa. 2016. Tracking Cultural Ecosystem Services: Water Chasing the Colorado River Restoration Pulse Flow. *Ecological Economics* 127: 165–172.

Beard, Daniel P. 2015. *Deadbeat Dams: Why We Should Abolish the U.S. Bureau of Reclamation and Tear Down Glen Canyon Dam*. Boulder, CO: Johnson Books.

Berger, Bruce. 1994. *There Was a River: Essays on the Southwest*. Tucson, AZ and London: University of Arizona Press.

Bishop, James, Jr. 1995. *Epitaph for a Desert Anarchist: The Life and Legacy of Edward Abbey*. New York, NY: Touchstone.

Bishop, Rob. 2017. Letter to Mr. Yvon Chouinard, December 15, 2017. *House Committee on Natural Resources*. http://naturalresources.house.gov/uploadedfiles/bishop_to_chouinard_letter.pdf.

Bishop, Rob. 2018. Letter to Mr. Yvon Chouinard, January 5, 2018. *House Committee on Natural Resources*. http://naturalresources.house.gov/uploadedfiles/bishop_response_ltr_to_yvon.pdf.

Brower, David R. 1997. Let the River Run Through It. *Sierra Magazine* 82 (2): 42–45.

Brower, David. 2005. Dinosaurs, Parks, and Dams. In *A Green River Reader*, ed. Alan Blackstock, 179–186. Salt Lake City, UT: University of Utah Press.

Bureau of Reclamation. 1965. *Lake Powell: Jewel of the Colorado*. Washington, DC: Bureau of Reclamation.

Bureau of Reclamation. 2012. *Colorado River Basin Water Supply and Demand Study: Study Report*. Washington, DC: Bureau of Reclamation.

Bureau of Reclamation. 2016. Glen Canyon Dam Long-Term Experimental and Management Plan FEIS. *Bureau of Reclamation*. http://ltempeis.anl.gov/documents/final-eis/.

Bureau of Reclamation. 2019. Glen Canyon Dam High Flow Experimental Release. *Bureau of Reclamation*. http://usbr.gov/uc/rm/gcdHFE/.

Burrillo, R.E. 2020. *Behind Bears Ears: Exploring the Cultural and Natural Histories of a Sacred Landscape*. Salt Lake City, UT: Torrey House Press.

Calahan, James M. 2001. *Edward Abbey: A Life*. Tucson, AZ: University of Arizona Press.

Carothers, Steven W., and Dorothy A. House. 2000. Decommissioning Glen Canyon Dam: The Key to Colorado River Ecosystem Restoration and Recovery of Endangered Species? *Arizona Law Review* 42: 215–238.

Cassuto, David N. 2001. *Dripping Dry: Literature, Politics, and Water in the Desert Southwest*. Ann Arbor, MI: University of Michigan Press.

Chouinard, Yvon. 2017. Response to the House Committee on Natural Resources. *Patagonia*. http://patagonia.com/blog/2017/12/response-to-the-house-committee-on-natural-resources/.

Crampton, C. Gregory. 2009 [1994]. *Ghosts of Glen Canyon: History Beneath Lake Powell*. Salt Lake City, UT: Bonneville Books.

DamNation. 2014. Dir. Ben Knight and Travis Rummel. Patagonia Presents a Stoecker Ecological & Felt Soul Media Production.

DeVoto, Bernard. 2005. Shall We Let Them Ruin Our National Parks? In *A Green River Reader*, ed. Alan Blackstock, 170–178. Salt Lake City, UT: University of Utah Press.

Dolnick, Edward. 2001. *Down the Great Unknown: John Wesley Powell's 1869 Journey of Discovery and Tragedy Through the Grand Canyon*. New York, NY: HarperCollins Publishers.

Drowning River. 2007. Dir. M.L. Lincoln. M.L. Lincoln Films LLC.

Eilperin, Juliet. 2017a. Shrink at least 4 national monuments and modify a half-dozen others, Zinke tells Trump. *The Washington Post*. http://washingtonpost.com/national/health-science/shrink-at-least-4-national-monuments-and-modify-a-half-dozen-others-zinke-tells-trump/2017/09/17/a0df45cc-9b48-11e7-82e4-f1076f6d6152_story.html?utm_term=.c1f5c4f2540d.

Eilperin, Juliet. 2017b. Trump to cut Bears Ears National Monument by 85 percent, Grand Staircase-Escalante by half, documents show. *The Washington Post*. http://washingtonpost.com/news/energy-environment/wp/2017/11/30/trump-to-cut-bears-ears-national-monument-by-85-percent-grand-staircase-escalante-by-half-documents-show/?noredirect=on&utm_term=.dc2b78fc6ee0.

Exec. Order No. 13792, 82 Fed. Reg. 20429 (April 26, 2017).

Exec. Order No. 13990, 86 Fed. Reg. 7037 (January 20, 2021).

Farmer, Jared. 1999. *Glen Canyon Dammed: Inventing Lake Powell & the Canyon Country*. Tucson, AZ: University of Arizona Press.

Farmer, Jared. 2003. *Desert Solitaire* and the Literary Memory of an Imagined Place. *Western American Literature* 38 (2): 155–170.

Fedarko, Kevin. 2013. *The Emerald Mile: The Epic Story of the Fastest Ride in History Through the Heart of the Grand Canyon*. New York, NY: Scribner.

Fleck, John. 2016. *Water Is for Fighting Over, and Other Myths about Water in the West*. Washington, DC: Island Press.

Foreman, Dave, and Bill Haywood, eds. 1993. *Ecodefense: A Field Guide to Monkeywrenching*, 3rd ed. Chico, CA: Abbzug Press.

Fradkin, Philip L. 1996. *A River No More: The Colorado River and the West*. Berkeley and Los Angeles, CA: University of California Press.

Gerlak, Andrea K., Francisco Zamora-Arroyo, and Hannah P. Kahler. 2013. A Delta in Repair: Restoration, Binational Cooperation, and the Future of the Colorado River Delta. *Environment: Science and Policy for Sustainable Development* 55 (3): 29–40.

Gessner, David. 2020. *Leave It As It Is: A Journey Through Theodore Roosevelt's American Wilderness*. New York, NY: Simon & Schuster Inc.

Glen Canyon Institute. 2000. *Citizens' Environmental Assessment (CEA) on the Decommissioning of Glen Canyon Dam. Report on Initial Studies, December 2000*. Salt Lake City, UT: Glen Canyon Institute.

Glen Canyon Institute. 2012. Option Submittal Form: Fill Mead First. *Bureau of Reclamation*. http://usbr.gov/lc/region/programs/crb.study/60_Fill_Mead_First.pdf.

Glen Canyon Institute. 2018a. Who We Are. *Glen Canyon Institute*. http://glencanyon.org/who-we-are/.

Glen Canyon Institute. 2018b. Legends. *Glen Canyon Institute*. http://glencanyon.org/legends/.

Glen Canyon Institute. 2018c. Fill Mead First. *Glen Canyon Institute*. http://glencanyon.org/fill-mead-first/.

Glen Canyon Institute. 2018d. Greater Glen Canyon: Bears Ears and Grand Staircase-Escalante. *Glen Canyon Institute*. http://glencanyon.org/greater-glen-canyon-bears-ears-and-grand-staircase-escalante/.

Glen Canyon Institute. 2021. *Colorado River Lowdown: Newsletter of Glen Canyon Institute* 20 (5).

Glenn, Edward, Karl Flessa, Eloise Kendy, Patrick B. Shafroth, Jorge Ramírez-Hernández, Martha Gomez-Sapiens, Pamela L. Nagler (eds.). 2017. Environmental Flows for the Colorado River Delta: Results of an Experimental Pulse Release from the U.S. to Mexico. *Ecological Engineering* 106 (B): 629–808.

Gross, Mathew Barrett, ed. 2003. *The Glen Canyon Reader*. Tucson, AZ: University of Arizona Press.

Grossman, Elizabeth. 2002. *Watershed: The Undamming of America*. New York, NY: Counterpoint.

Hannon, Steven. 1997. *Glen Canyon: A Novel*. Denver, CO: Kokopelli Books.

Harvey, Mark T. 2000 [1994]. *A Symbol of Wilderness: Echo Park and the American Conservation Movement*. Seattle, WA and London: University of Washington Press.

Harvey, Mark T. 2005. *Wilderness Forever: Howard Zahniser and the Path to the Wilderness Act*. Seattle, WA and London: Weyerhauser Environmental Books/University of Washington Press.

Hyde, Phillip. 1979. *A Glen Canyon Portfolio*. Flagstaff, AZ: Northland Press.

Ingebretsen, Richard J. 2000. Foreword. *Stanford Environmental Law Journal* 19 (1): xi–xiv.

Inskip, Eleanor, ed. 1995. *The Colorado River Through Glen Canyon Before Lake Powell: Historic Photo Journal 1872 to 1964*. Moab, UT: Inskip Ink.

James, Ian. 2021. 'Red alert:' Lake Mead falls to record-low level, a milestone in Colorado River's crisis. *The Arizona Republic*. http://eu.azcentral.com/story/news/local/arizona-environment/2021/06/10/lake-mead-declines-new-low-colorado-river-crisis-deepens-arizona-drought/7621138002/.

Keeler, Jacqueline, ed. 2017. *Edge of Morning: Native Voices Speak for the Bears Ears*. Salt Lake City, UT: Torrey House Press.

Lago, Don. 2018. *The Powell Expedition: New Discoveries about John Wesley Powell's 1869 River Journey*. Reno and Las Vegas, NV: University of Nevada Press.

Landmark Estates. 2017. Landmark Estates. *The Wilderness Society*. http://parksforsale.com.

Lee, Katie. 1998. *All My Rivers are Gone: A Journey of Discovery Through Glen Canyon*. Boulder, CO: Johnson Books.

Lee, Katie. 2002. Glen Canyon—The Place We Knew. In *Land That We Love: Americans Talk About America's Public Lands*, ed. Barry Scholl, 57–59. Washington, DC: Department of Agriculture.

Lee, Katie. 2004a [1964]. *Folk Songs of the Colorado River*. Smithsonian Folkways Recordings.

Lee, Katie. 2004b. *Sandstone Seduction: Rivers and Lovers, Canyons and Friends*. Boulder, CO: Johnson Books.

Lee, Katie. 2006. *Glen Canyon Betrayed: A Sensuous Elegy*. Flagstaff, AZ: Fretwater Press.

Leydet, François. 1964. *Time and the River Flowing: Grand Canyon*. San Francisco, CA: Sierra Club.

Lincoln, M.L. 2020. *Wrenched from the Land: Activists Inspired by Edward Abbey*. Albuquerque, NM: University of New Mexico Press.

Lindholdt, Paul. 2015. *Explorations in Ecocriticism: Advocacy, Bioregionalism, and Visual Design*. Lanham, MD: Lexington Books.

Loeffler, Jack. 2002. *Adventures with Ed: A Portrait of Abbey*. Albuquerque, NM: University of New Mexico Press.

Maffly, Brian. 2021. Lost Glen Canyon wonders reemerge as Lake Powell dries up. *The Salt Lake Tribune*. http://sltrib.com/news/environment/2021/06/20/lost-glen-canyon-wonders/.

Martin, Russell. 1989. *A Story That Stands Like a Dam: Glen Canyon and the Struggle for the Soul of the West*. New York, NY: Henry Holt and Company, Inc.

McCool, Daniel. 2012. *River Republic: The Fall and Rise of America's Rivers*. New York, NY: Columbia University Press.

McGivney, Annette. 2009. *Resurrection: Glen Canyon and a New Vision for the American West*. Seattle, WA: Braided River.

McPhee, John. 1977 [1971]. *Encounters with the Archdruid*. New York, NY: Farrar, Straus, and Giroux.

Meloy, Ellen. 1994. *Raven's Exile: A Season on the Green River*. Tucson, AZ: University of Arizona Press.

Miller, Daegan. 2018. On Possibility; or, The Monkey Wrench. In *Future Remains: A Cabinet of Curiosities for the Anthropocene*, eds. Gregg Mitman, Marco Armiero, and Robert Emmett, 143–148. Chicago, IL: University of Chicago Press.

Miller, Scott K. 2000. Undamming Glen Canyon: Lunacy, Rationality, or Prophecy? *Stanford Environmental Law Journal* 19 (1): 121–207.

Muir, John. 2003 [1912]. *The Yosemite*. New York, NY: Modern Library.

Nash, Roderick Frazier. 2014. *Wilderness and the American Mind*, 5th ed. New Haven, CT: Yale University Press.

Nichols, Tad. 1999. *Glen Canyon: Images of a Lost World*. Santa Fe, NM: Museum of New Mexico Press.

Owen, David. 2017. *Where the Water Goes: Life and Death Along the Colorado River*. New York, NY: Riverhead Books/Penguin Random House, LLC.

Patagonia. 2017. Protect Public Lands. *Patagonia*. http://patagonia.com/protect-public-lands.html.

Pearson, Byron E. 2002. *Still the Wild River Runs: Congress, the Sierra Club, and the Fight to Save Grand Canyon*. Tucson, AZ: University of Arizona Press.

Philippon, Daniel J. 2004. Edward Abbey's Remarks at the Cracking of Glen Canyon Dam. *ISLE: Interdisciplinary Studies in Literature and Environment* 11 (2): 161–166.

Porter, Eliot. 1963. *The Place No One Knew: Glen Canyon on the Colorado*, ed. David Brower. San Francisco, CA: Sierra Club.

Powell, James Lawrence. 2008. *Dead Pool: Lake Powell, Global Warming, and the Future of Water in the West*. Berkeley and Los Angeles, CA: University of California Press.

Powell, John Wesley. 2002 [1895]. *The Exploration of the Colorado River and Its Canyons*. Washington, DC: National Geographic Society.

Prentiss, Sean. 2015. *Finding Abbey: The Search for Edward Abbey and His Hidden Desert Grave*. Albuquerque, NM: University of New Mexico Press.

Proclamation No. 6920, 3 C.F.R. 6920 (1996).

Proclamation No. 9558, 3 C.F.R. 9558 (2016).

Proclamation No. 9681, 3 C.F.R. 9681 (2017).

Proclamation No. 9682, 3 C.F.R. 9682 (2017).

REI. 2017. REI Responds to Administration's Decision on Bear's Ears, Grand Staircase-Escalante National Monuments. *REI*. http://rei.com/blog/news/rei-responds-to-administrations-decision-on-bears-ears-grand-staircase-escalante-national-monuments.

Reisner, Marc. 1993 [1986]. *Cadillac Desert: The American West and Its Disappearing Water*. New York, NY: Penguin Books.

Roberts, David. 2021. *The Bears Ears: A Human History of America's Most Endangered Wilderness*. New York, NY: W.W. Norton & Company, Inc.

Robinson, Rebecca. 2018. *Voices from Bears Ears: Seeking Common Ground on Sacred Land*. Tucson, AZ: University of Arizona Press.

Ross, John F. 2018. *The Promise of the Grand Canyon: John Wesley Powell's Perilous Journey and His Vision for the American West*. New York, NY: Penguin Books.

Schmidt, John C., with Maggi Kraft, Daphnee Tuzlak, and Alex Walker. 2016. *Fill Mead First: A Technical Assessment*. White Paper No. 1. Logan, UT: Center for Colorado River Studies, Utah State University.

Sheehy, Lisa Pike. 2018. It Was Always About Oil, Coal, Gas, and Uranium. *Patagonia*. http://pat.ag/Lies.

Silko, Leslie Marmon. 1991. *Almanac of the Dead: A Novel*. New York, NY: Simon & Schuster, Inc.

Solnit, Rebecca. 2017. *Drowned River: The Death and Rebirth of Glen Canyon on the Colorado*. Santa Fe, NM: Radius Books.

Spurgeon, Sara L. 2009. Miracles in the Desert: Literature, Water, and Public Discourse in the American West. *ISLE: Interdisciplinary Studies in Literature and Environment* 16 (4): 743–759.

Standing, Sarah Ann. 2012. Earth First!'s "Crack the Dam" and the Aesthetics of Ecoactivist Performance. In *Readings in Performance and Ecology*, eds. Wendy Arons and Theresa J. May, 147–155. New York, NY: Palgrave Macmillan.

Stegner, Wallace, ed. 1955. *This Is Dinosaur: Echo Park Country and Its Magic Rivers*. San Francisco, CA: Sierra Club.

Stegner, Wallace. 1992 [1954]. *Beyond the Hundredth Meridian: John Wesley Powell and the Second Opening of the West*. London: Penguin Books.

Stegner, Wallace. 2017 [1969]. *The Sound of Mountain Water: The Changing American West*. New York, NY: Vintage Books.

Strom, Stephen E. 2018. *Bears Ears: Views from a Sacred Land*. Staunton, VA: George F. Thompson Publishing LLC.

Sumner, David Thomas. 2002a. *Testimony*, Landscape, and the West: A Conversation with Stephen Trimble. *Weber Studies* 19 (3): 2–14.

Sumner, David Thomas. 2002b. *Testimony*, *Refuge*, and the Sense of Place: A Conversation with Terry Tempest Williams. *Weber Studies* 19 (3): 15–28.

The North Face. 2017. Homepage. *The North Face*. http://thenorthface.com.

Thoreau, Henry David. 2004 [1854]. *Walden*, 150th Anniversary Edition, ed. J. Lyndon Shanley. Princeton, NJ and Oxford: Princeton University Press.

Trimble, Stephen. 2008. *Bargaining for Eden: The Fight for the Last Open Spaces in America*. Berkeley and Los Angeles, CA and London: University of California Press.

Trimble, Stephen, ed. 2016. *Red Rock Testimony: Three Generations of Writers Speak on Behalf of Utah's Public Lands*. Salt Lake City, UT: Torrey House Press.

Trimble, Stephen, ed. 2017. *Red Rock Stories: Three Generations of Writers Speak on Behalf of Utah's Public Lands*. Salt Lake City, UT: Torrey House Press.

Trimble, Stephen, and Terry Tempest Williams, eds. 1996. *Testimony: Writers of the West Speak on Behalf of Utah Wilderness*. Minneapolis, MN: Milkweed Editions.

Vernon, John. 2002. *The Last Canyon*. New York, NY: Mariner Books/Houghton Mifflin Company.

Waterman, Jonathan. 2010. *Running Dry: A Journey From Source to Sea Down the Colorado River*. Washington, DC: National Geographic.

Wegner, David L. 1991. A Brief History of the Glen Canyon Environmental Studies. In *Colorado River Ecology and Dam Management: Proceedings of a Symposium, May 24–25, 1990, Santa Fe, New Mexico*, ed. Committee to Review the Glen Canyon Environmental Studies, Water Science and Technology Board, and Commission on Geosciences, Environment, and Resources, 226–238. Washington, DC: National Academies Press.

Wegner, David L. 2000. Looking Toward the Future: The Time Has Come to Restore Glen Canyon. *Arizona Law Review* 42 (2): 239–258.

Wehr, Kevin. 2004. *America's Fight Over Water: The Environmental and Political Effects of Large-Scale Water Systems*. New York, NY: Routledge.

Williams, Terry Tempest. 2002. *Red: Passion and Patience in the Desert*. New York, NY: Vintage Books.

Williams, Terry Tempest. 2006. Foreword. In *Postcards from Ed: Dispatches and Salvos from an American Iconoclast*, eds. Edward Abbey and David Petersen, ix-xv. Minneapolis, MN: Milkweed Editions.

Williams, Terry Tempest. 2012. *When Women Were Birds: Fifty-Four Variations on Voice*. New York, NY: Picador.

Worster, Donald. 1985. *Rivers of Empire: Water, Aridity, and the Growth of the American West*. New York, NY: Pantheon Books.

Worster, Donald. 2001. *A River Running West: The Life of John Wesley Powell*. New York, NY: Oxford University Press, Inc.

Wrenched. 2014. Dir. M.L. Lincoln. M.L. Lincoln Films LLC.

Zakin, Susan. 1993. *Coyotes and Town Dogs: Earth First! and the Environmental Movement*. Tucson, AZ: University of Arizona Press.

7

Reflections on Literature, Ecological Restoration, and Activism

7.1 Rewilding America's Places Through Its Pages: The Lasting Environmental Legacy of Thoreau, Muir, Leopold, Douglas, and Abbey

Henry David Thoreau and Walden Pond. John Muir and Yosemite. Aldo Leopold and his sand farm. Marjory Stoneman Douglas and the Everglades. Edward Abbey and Glen Canyon. Each of these entanglements between writer and landscape has secured a place in the history of American conservation. But of greater import for restorationists is how each entanglement has also produced a literary intervention in restoration politics. A literary restoration ethic.

These are entanglements of literature and cultural landscape restoration—and activism. Each landscape is underwritten by ecologies, biographies, histories, stories, ideas, encounters, frictions, and tensions. These literary landscapes are politicized by the writers themselves, and by conservationists, activists, policymakers, and others, in defense of public

L. Smith, *Ecological Restoration and the U.S. Nature and Environmental Writing Tradition*, https://doi.org/10.1007/978-3-030-86148-3_7

lands and the idea of wilderness. Out of these entanglements of page and place also emerges a cacophony of state and federal land protections.

Thoreau enacted one of the earliest restoration projects in Walden Woods, with his 1859 replanting—rewilding—of 400 white pine saplings on his 2.5-acre bean field near his cabin site (recorded in Thoreau 1906 *XII*), although the pines succumbed to wildfire before the end of the century. Moreover, his call later that year to preserve a 'primitive forest' in each town (e.g. Thoreau 1906 *XII*) was realized at Walden 60 years after his death in the 1922 Deed of Gift from the Emerson, Forbes, and Heywood families. The land grant protected the 'Walden of Emerson and Thoreau' (qtd. in Maynard 2004), and the almost 80 acres included in the conveyance lie at the center of the 462-acre Walden Pond State Reservation.

A number of grassroots and nonprofit organizations have emerged to defend—and restore—Walden Woods (both inside and outside the reservation boundary) in the decades since, including the Thoreau Society's Save Walden Committee in the 1950s, Walden Forever Wild and the Thoreau Country Conservation Alliance in the 1980s, the Walden Woods Project in the 1990s, and Friends of Thoreau Country in the 2000s. Walden Pond was designated a National Historic Landmark in 1962. Thoreau's Concord and Walden writings—among them, *Walden* (Thoreau 2004a), his essays (in Thoreau 2001b), and journals (Thoreau 1906)—have helped in restor(y)ing Walden, infiltrating restoration reference models, and restoration programs from Red Cross Beach to shoreline stabilization to the municipal landfill to Brister's Hill. In Maine, the presidential proclamation establishing the Katahdin Woods and Waters National Monument in 2016 briefly noted Thoreau's 1857 float trip on the East Branch of the Penobscot River (Proclamation No. 9476, 2016, cf. Thoreau 2004b), and RESTORE The North Woods' ongoing campaign for a larger Maine Woods National Park and Preserve draws on and honors Thoreau's (2004b) vision for a 'national preserve' in Maine's North Woods.

Muir's twin essays 'The Treasures of Yosemite' (Muir 1890a) and 'Features of the Proposed Yosemite National Park' (Muir 1890b) for *The Century Magazine* helped launch—and fuel—the campaign to add Yosemite to America's fledgling national park system in the late-nineteenth century. But Muir's literary activism would return to Yosemite less than two decades later, as he tried (and ultimately failed) to defend a part of the new national park from a dam and reservoir project in the aftermath of the 1906 San Francisco earthquake—beginning with his 1908 'The Hetch-Hetchy Valley' essay in the *Sierra Club Bulletin* (Muir 1908). Muir's nonprofit Sierra Club, established in 1892 to protect the Sierra Nevada cordillera, brought institutional gravitas to the Hetch Hetchy conservation campaign. Nearly a century later, nonprofit Restore Hetch Hetchy (initially the Sierra Club offshoot, Hetch Hetchy Restoration Task Force) began exploring the water politics of the Hetch Hetchy Valley—and the political intersection of San Francisco, the Tuolumne River, O'Shaughnessy Dam, and Hetch Hetchy Reservoir.

Leopold's restoration ethic was forged in the 1930s and 1940s on the Curtis Prairie at the University of Wisconsin-Madison Arboretum, the world's first ecological restoration project, and on his family's sand farm camp north of Baraboo. At the farmstead—the 'Shack'—Leopold, together with his wife and five children, introduced an annual pine planting program to restore the oak openings, and restored the exhausted and abandoned cornfield to prairie. The cultural landscapes of the sand farm became a theater for ecological restoration, and *A Sand County Almanac* (Leopold 1968), with its seasonal almanac, and essays on conservation and ethics, is Leopold's denouement of the Shack experiment, of his encounters and experiences along the Wisconsin River. His restoration pathology is crystalized, synthesized in his 'land ethic' (in Leopold 1968). The 1,600-acre Leopold Memorial Reserve was established in 1967 by the Sand County Foundation, Aldo Leopold Foundation, and other landowner partners, and the almost 16,000-acre Leopold-Pine Island International Bird Area was designated in 2005. The Leopold farmstead was also protected as the Leopold Shack and Farm National Historic Landmark in 2009. In early-2021, the UW Arboretum was designated a National Historic Landmark, for its place in the annals of American conservation—and restoration ecology—and its commitment to a Leopoldian land ethic.

A pioneer of the U.S. Forest Service's first designated wilderness area, and cofounder of the Wilderness Society, Leopold's wilderness preservation pathology also resounded in the 1964 Wilderness Act. Muir, Leopold, and Douglas all have U.S. wilderness areas named in their honor, in the places that undergird their conservation philosophy and writing (and activism). The John Muir Wilderness extends across Inyo National Forest and Sierra National Forest in California. The Aldo Leopold Wilderness lies in Gila National Forest in New Mexico. The Marjory Stoneman Douglas Wilderness extends across much of Everglades National Park in Florida.

Douglas set up the nonprofit Friends of the Everglades in 1969 to challenge the construction of the Big Cypress Swamp Jetport at the northwestern boundary of Everglades National Park. From its inception, the organization has worked to preserve, protect, and restore the Everglades. Everglades *restoration* is central to the work of Douglas's FOE. Two principal opportunities led Douglas to this activism juncture—her membership of Ernest F. Coe's Tropic Everglades National Park Association and the establishment of Everglades National Park in 1947, and her research for *The Everglades: River of Grass* (Douglas 2017), published six weeks before the national park designation. With the cancellation of the jetport project, FOE turned to tackling other issues in Everglades restoration. 'Everglades restoration' has assumed many guises over the past half-century, culminating in the Comprehensive Everglades Restoration Plan, authorized in 2000. FOE remains a central protagonist in Everglades restoration, and Douglas's legacy as Everglades agitator endures.

It is in the 1975 protest novel, *The Monkey Wrench Gang* (Abbey 2004), and its posthumously published sequel 15 years later, *Hayduke Lives!* (Abbey 1990a), that Abbey presents his most fantastical plan for defending the slickrock plateaus and canyons of the American West—and for restoring Glen Canyon on the Colorado River. Through the conversations, plans, and exploits of Abbey's fictional quartet, Glen Canyon Dam and Lake Powell are presented as a proxy for the widespread industrialization of the Desert Southwest, and the anticipated restoration of the Glen comes to stand for the restoration of wilderness. The radical, anarchic ecotage and vandalism of Abbey's fantasy might be absent from contemporary Colorado River Basin water politics, but Abbey's gentler restoration pathology, centered on decommissioning

Glen Canyon Dam, draining Lake Powell and recovering the Glen's side canyons, and restoring the riparian ecosystems through Glen Canyon and Grand Canyon (in essays in e.g. Abbey 1984, 1990b, 1991), finds political purchase. It echoes and continues in the advocacy work of nonprofit Glen Canyon Institute, and its Fill Mead First proposal and other projects.

A 'restoration ethic' looks quite different across the writings of Thoreau, Muir, Leopold, Douglas, and Abbey. In the philosophy and writings of Thoreau, a restoration ethic is molded by his New England Transcendentalism and the romantic pastoral, and similarly for Muir, it reflects a numinous, spiritual, mystical approach to nature. Leopold's restoration pathology is closely aligned with his conservation or land ethic, and overlaps with Douglas's scaffolding of restoration within bioregionalism narratives. A restoration ethic, in Abbey's canon, is much more anarchic and political. Yet considered together, what emerges is a chronology of the evolution and mutations of a literary restoration ethic—of an ecological restoration rhetoric and refrain in American letters—from the mid-nineteenth century into the late-twentieth century. It is a literary restoration ethic reflective of, and responding to, wider episodes and milestones in the history of American conservation, from the designation of the first U.S. national park in 1872, to the Progressive era political reform of the 1890s–1920s that saw the creation of America's first conservation organizations and introduced some of the earliest conservation laws, through to the birth of the modern environmental movement in the 1960s and legislation including the 1964 Wilderness Act, and beyond. In a literary restoration ethic is a provocation on how literature can variously confront, disrupt, curate, and recast environmental threats, and signal to actions and interventions on and off the page.

7.2 The Story of Public Lands Writing-as-Advocacy in the Trump Administration

Three vignettes epitomize the politicization of writing-as-advocacy in defense of America's public lands across the four years of the Trump administration. And in them, the legacies of Thoreau, Muir, Leopold, Douglas, and Abbey persist, united and entwined, in both subtle and radical ways:

- *The social media activism of the 'rogue ranger'* / Five days into Donald Trump's presidency, on January 24, 2017, the Badlands National Park Twitter account posted a series of statements on climate change.[1] This was a subtle—and defiant—response to the Trump administration's gag orders on staff of the Environmental Protection Agency and the Department of Agriculture earlier that day, and the administration's previous temporary suspension of some of the Department of the Interior's Twitter accounts after the National Park Service (NPS) retweeted photographs comparing inauguration crowd sizes for Obama in 2009 and Trump in 2017. These four climate science tweets were short-lived and were deleted soon after. But Badlands became the touch-paper for an impromptu national parks Twitterstorm on the importance of science communication. Golden Gate National Recreation Area tweeted about record global temperatures.[2] Death Valley National Park shared photographs of Japanese Americans interned during the

[1] @BadlandsNPS. 'The pre-industrial concentration of carbon dioxide in the atmosphere was 280 parts per million (ppm). As of December 2016, 404.93 ppm.' *Twitter*. Twitter, January 24, 2017, 11.40. @BadlandsNPS. 'Today, the amount of carbon dioxide in the atmosphere is higher than at any time in the last 650,000 years.' *Twitter*. Twitter, January 24, 2017, 11.40. @BadlandsNPS. 'Flipside of the atmosphere; ocean acidity has increased 30% since the Industrial Revolution. "Ocean Acidification" #climate #carboncycle' *Twitter*. Twitter, January 24, 2017, 12.25. @BadlandsNPS. 'Burning one gallon of gasoline puts nearly 20lbs of carbon dioxide into our atmosphere. #climate' *Twitter*. Twitter, January 24, 2017, 13.50.

[2] @GoldenGateNPS. '2016 was the hottest year on record for the 3rd year in a row. Check out this @NASA & @NOAA report: https://nasa.gov/press-release/nasa-noaa-data-show-2016-warmest-year-on-record-globally' *Twitter*. Twitter, January 23, 2017, 20.11. twitter.com/GoldenGateNPS/status/823624278230695936.

Second World War, in response to President Trump's Muslim immigration ban.[3] Redwood National and State Parks tweeted about the importance of redwood stands as carbon sinks.[4] The #BadlandsResistance, #resist, and #sciencematters hashtags began trending across social media. A number of alt—or rogue—Badlands accounts also began to appear on Twitter, led by @BadHombreNPS and @BadlandsGonWild, as well as rogue NPS accounts, including @AltNatParkServ (which quickly changed to @NotAltWorld), @ALTNPS, @AltUSNatParkService, and @NatParkUndrgrnd, and several other rogue individual park accounts, all unofficial and unaffiliated, but nonetheless determined to resume—and defend—the conversation on climate science and environmental protection in the U.S.[5] It was the park ranger that became the leading voice of resistance in the parks—celebrated as 'rogue rangers.' The #rogueranger hashtag emerged—and flourished. It is here that Abbey's legacy enters most prominently into the fray, his voice assuming a renewed importance and urgency. The former seasonal park ranger and fire lookout was swiftly adopted as the patriarch of the 'rogue ranger' extended family. And his voice was joined by others—especially that of John Muir and Theodore Roosevelt, two of the founders of the U.S. national parks movement. Indeed, the two rogue Badlands Twitter accounts often turn to the words of Roosevelt and Abbey for solace and inspiration.

[3] Two examples: @DeathValleyNPS. 'During WWII Death Valley hosted 65 endangered internees after the #Manzanar Riot. #JapaneseAmericanInternment' *Twitter*. Twitter, January 25, 2017, 15.09. twitter.com/DeathValleyNPS/status/824273025285705730. @DeathValleyNPS. 'We want the opportunity they have to prove their loyalty. We are asked to accept a denial of that privilege in the name of patriotism.' *Twitter*. Twitter, January 25, 2017, 16.28. twitter.com/DeathValleyNPS/status/824292861625663492.

[4] @RedwoodNPS. 'DYK redwood groves are #1 carbon sink/acre in nature? About 200 tons an acre. More redwoods would mean less #climatechange #climate' *Twitter*. Twitter, January 25, 2017, 17.43. twitter.com/RedwoodNPS/status/824311863362011136.

[5] This social media retaliation was not confined to the national parks, but rapidly spread to other federal government departments and agencies, with rogue accounts emerging for: the Department of Agriculture (@altusda), the Center for Disease Control (@Alt_CDC, @viralCDC), the Environmental Protection Agency (@ActualEPAfacts, @altUSEPA, @ungaggedEPA), the Food and Drug Administration (@alt_fda), the U.S. Fish and Wildlife Service (@AlternateUSFW, @AltUSFWS), the U.S. Forestry Service (@AltForestServ), the Department of Health and Human Services (@AltHHS), NASA (@Alt_NASA, @RogueNASA), the National Institutes of Health (@Alt_NIH), the National Oceanic and Atmospheric Administration (@altNOAA), and the National Weather Service (@AlternativeNWS), among others.

In its Twitter profile, the @BadlandsGonWild account also asks, '#WhatWouldTeddyDo?'

- *Citing a small, furry, mustachioed—and fictional—forest guardian* / In a December 13, 2018 decision filed by the U.S. Court of Appeals for the Fourth Circuit in Richmond, Virginia, revoking Dominion Energy's permit to construct the 604.5-mile Atlantic Coast Pipeline through West Virginia, Virginia, and North Carolina, the final paragraph borrowed from Dr. Seuss's (1971) *The Lorax*, to entrust the USFS to 'speak for the trees, for the trees have no tongues'—and protect the national forests (see Cowpasture River Preservation Ass'n v. USFS 2018: 60, cf. Geisel 2004: 23). A children's book and eco-fable had entered public and political discourse on natural gas pipelines, a conversation already ignited by the highly controversial Keystone XL and Dakota Access pipelines.
- *To write of songbirds* / On August 11, 2020, the U.S. District Court in the Southern District of New York overturned the Department of the Interior's 2017 re-interpretation of 'takings' and 'killings' under the Migratory Bird Treaty Act of 1918, opening its decision with a line adapted from Harper Lee's (1960) *To Kill a Mockingbird*—'It is not only a sin to kill a mockingbird, it is also a crime' (see NRDC v. Interior 2020: 1, cf. Lee 2000: 99). Lee's motif of mockingbird-as-sacred, mockingbird-as-sublime is used to dispute Interior's revised definition, and defend the 1918 law.

There have been other creative mediations on the Trump administration's attacks on public lands. When President Trump's legal team held an impromptu press conference in the parking lot of Four Seasons Total Landscaping in Philadelphia, Pennsylvania in November 2020, after the president erroneously tweeted about a press conference to be held at the Philadelphia Four Seasons hotel as his re-election campaign challenged the result of the presidential election,[6] *context* supplanted *content*. The episode instantly became a pop culture meme (see e.g. Bekiempis

[6] @realDonaldTrump. 'Lawyers Press Conference at Four Seasons, Philadelphia. 11:00 A.M.' *Twitter*. Twitter, November 7, 2020, 06.35. @realDonaldTrump. 'Big press conference today in Philadelphia at Four Seasons Total Landscaping—11:30am!' *Twitter*. Twitter, November 7, 2020, 09.45.

2020, also Fiverr 2021 and its Super Bowl LV commercial). The landscaping company is located on the same block as the Fantasy Island Adult Bookstore and the Delaware Valley Cremation Center, and the landscape gardening, sex, and death motifs synthesized three defining characteristics of Donald Trump's presidency—the battle for public lands, the sexual misconduct and sexual assault allegations filed against the president, and the federal government's response to the COVID-19 pandemic.

There is something distinctive about public lands writing-as-advocacy in the Trump administration. This literary activism still quietly, gently disrupts and unsettles public lands dialogue. But to contend with an administration that favored agitation, shock politics, and disinformation and alt-facts, this literary activism also doubles as a platform to respond to and challenge denialist theories on science, climate change, and more. Environmental writer-activists have turned to humor as well, throwing shade on the administration's actions. And there is no better example than what unfolded with Badlands National Park on Twitter. At the close of the Trump administration, more than 100 environmental policies and rules had been (or were in the process of being) reversed or revoked (Popovich et al. 2020).

Early in the transition period between the U.S. presidential election in November 2020 and the inauguration of Joe Biden as the 46th President of the U.S. in January 2021, conservation discourse and dialogue on U.S. public lands had already begun to shift. One of the biggest public lands controversies of the Trump administration—the 2017 national monument review, and the dramatic downsizing of both Grand Staircase-Escalante and Bears Ears national monuments in southern Utah—reentered conservation politics.

On the campaign trail, the Biden-Harris ticket committed to consolidating and building on the Obama-Biden conservation legacy. Included in the *Biden-Harris Plan for Tribal Nations* (see joebiden.com 2020) was a campaign promise to 'revers[e] Trump's attacks on the Arctic National Wildlife Refuge, Bears Ears, and Grand Staircase-Escalante.' Soon after the election, the recently-established Restoration Project, co-chaired by former NPS director Jonathan B. Jarvis, sent its recommendations for reversing 100 priority Trump-era 'environmental harms' to the Biden-Harris transition team (later released as Restoration Project 2021a; b, c).

#16 on the Restoration Project's list of environmental harms is 'Reduced the size of two national monuments in Utah' (Restoration Project 2021a; b, c).

On his first day in office, President Biden signed Executive Order 13990: Protecting Public Health and the Environment and Restoring Science To Tackle the Climate Crisis. Section 3 is titled, 'Restoring National Monuments' (Exec. Order No. 13990, 2021). E.O. 13990 instructs the Secretary of the Interior to review the 'monument boundaries and conditions' created by President Trump's rollbacks (Proclamation No. 9681, 9682, 2017), and determine if restoration 'would be appropriate.' In an historic vote on March 15, 2021, the Senate confirmed Deb Haaland as Secretary of the Interior, making her the first Indigenous cabinet secretary in U.S. history. As part of the rollback review, Secretary Haaland traveled to Utah in early-April 2021, to meet with tribal leaders and state politicians to discuss restoring protections to Bears Ears and Grand Staircase-Escalante, and tour the national monument. The restorative refrain continued in a BEITC (2021a) press release issued on April 19, titled, 'To Secretary Haaland: Restoration Should Not Wait On Legislation,' and a BEITC (2021b) ad campaign to 'Restore and Expand Bears Ears.' On June 14, Secretary Haaland recommended the restoration of the two Utah monuments (Eilperin and Partlow 2021). Less than four months later, on October 8, President Biden restored environmental protections to three national monuments, reversing the 2017 downsizing actions of the Trump administration. The boundary of Bears Ears National Monument was increased to 1.36 million acres (Proclamation No. 10285, 2021), Grand Staircase-Escalante National Monument was restored to 1.87 million acres (Proclamation No. 10286, 2021), and Northeast Canyons and Seamounts Marine National Monument was also restored (Proclamation No. 10287, 2021). A restoration rhetoric is variously mobilized and politicized across the new Biden administration to speak back to restoring federal commitment to environmental protection and stewardship, reversing environmental policy rollbacks, and more. The *Red Rock Testimony* (Trimble 2016) advocacy chapbook is entering its third administration (and *Red Rock Stories* (Trimble 2017) and *Edge of Morning* (Keeler 2017) their second) in defense of the ancestral, sacred lands of the Bears Ears—and the restoration of the national monument designation. Restorationists and public lands advocates have good reason to be hopeful.

7.3 A Walden Reprise

This book began with a Walden vignette, and it is how it closes, too. The origin story of this book, after all, is entangled with Walden Woods. And with the forest trails on Brister's Hill in particular.

From the first time I read *A Week on the Concord and Merrimack Rivers* (Thoreau 2001a) as a Master's student at Cardiff University, more than 15 years ago, there is one Thoreau passage that has stayed with me more than any other. More so than any of the preservation or wilderness proclamations of *Walden* (Thoreau 2004a) or *The Maine Woods* (Thoreau 2004b). Researching and writing this book over the last three years has only further confirmed and consolidated my love for this passage. In the final chapter of *A Week*, 'Friday,' Thoreau (2001a: 228) reflects, 'When I visit again some haunt of my youth, I am glad to find that nature wears so well. The landscape is indeed something real, and solid, and sincere, and I have not put my foot through it yet.' Distilled into this one statement—for me—is the long game of a literary restoration ethic. When I read again and again the works of Thoreau, or Muir, or Leopold, or Douglas, or Abbey, I am glad to find in them variously a biography of place, an environmental history, a botanical inventory and compendium, an almanac. But more than that, I am glad to find an attendant, entwined, urgent, perennial call to action on conservation—and restoration—*in that place.* A roadmap. A touch-paper. A spark. This environmental literature doubles as a cultural artifact (after Egan and Howell 2001), as a creative template for restoration reference models.

Works Cited

Abbey, Edward. 1984. *Beyond the Wall: Essays from the Outside*. New York, NY: Henry Holt and Company, LLC.

Abbey, Edward. 1990a. *Hayduke Lives!* New York, NY: Back Bay Books/Little, Brown, and Company.

Abbey, Edward. 1990b [1968]. *Desert Solitaire: A Season in the Wilderness*. New York, NY: Touchstone.

Abbey, Edward. 1991 [1982]. *Down the River*. New York, NY: Plume.

Abbey, Edward. 2004 [1975]. *The Monkey Wrench Gang*. London: Penguin Books Ltd.

Bears Ears Inter-Tribal Coalition. 2021a. To Secretary Haaland: Restoration Should Not Wait On Legislation. *BEITC*. https://bearsearscoalition.org/to-madame-secretary-haaland-restoration-of-bears-ears-should-not-wait-for-legislation/.

Bears Ears Inter-Tribal Coalition. 2021b. Restore and Expand Bears Ears. Ad Campaign. *BEITC*. https://bearsearscoalition.org/restore-bears-ears-ad-campaign/.

Bekiempis, Victoria. 2020. 'Make America Rake Again:' Four Seasons Total Landscaping Cashes in on Trump fiasco. *The Guardian*. https://www.theguardian.com/us-news/2020/nov/09/four-seasons-total-landscaping-trump-gardening-merchandise.

Cowpasture River Preservation Ass'n v. USFS. USCA4 Appeal: 18–1144. Doc: 104 (4th Cir. 2018).

Douglas, Marjory Stoneman. 2017 [1947]. *The Everglades: River of Grass*, 70th Anniversary Edition. Sarasota, FL: Pineapple Press, Inc.

Egan, Dave, and Evelyn A. Howell, eds. 2001. *The Historical Ecology Handbook: A Restorationist's Guide to Reference Ecosystems*. Washington, DC: Island Press.

Eilperin, Juliet, and Joshua Partlow. 2021. Haaland Urges Biden to Fully Protect Three National Monuments Weakened by Donald Trump. *The Washington Post*. https://www.washingtonpost.com/climate-environment/2021/06/14/haaland-biden-national-monuments/.

Exec. Order No. 13990, 86 Fed. Reg. 7037 (January 20, 2021).

Fiverr. 2021. Opportunity Knocks | Fiverr Big Game Commercial Extended Cut (2021). *YouTube*. https://youtu.be/XelsNvpibpQ.

Geisel, Theodor Seuss (Dr. Seuss). 2004 [1971]. *The Lorax*. London: Harper Collins Children's Books.

joebiden.com. 2020. Biden-Harris Plan for Tribal Nations. *joebiden.com*. https://joebiden.com/tribalnations/.

Keeler, Jacqueline, ed. 2017. *Edge of Morning: Native Voices Speak for the Bears Ears*. Salt Lake City, UT: Torrey House Press.

Lee, Harper. 2000 [1960]. *To Kill a Mockingbird*. London: Vintage Classics.

Leopold, Aldo. 1968 [1949]. *A Sand County Almanac, and Sketches Here and There*. New York, NY: Oxford University Press.

Maynard, W. Barksdale. 2004. *Walden Pond: A History*. Oxford and New York, NY: Oxford University Press.

Muir, John. 1890a. The Treasures of Yosemite. *The Century Magazine* XL (4): 483–500.

Muir, John. 1890b. Features of the Proposed Yosemite National Park. *The Century Magazine* XL (5): 656–667.

Muir, John. 1908. The Hetch-Hetchy Valley. *Sierra Club Bulletin* VI (4): 212–220.

NRDC v. Interior. Case: 1:18-cv-04596-VEC. Doc: 89 (S.D.N.Y. 2020).

Popovich, Nadja, Livia Albeck-Ripka, and Kendra Pierre-Louis. 2020. The Trump Administration Is Reversing More Than 100 Environmental Rules. Here's the Full List. *The New York Times*. https://www.nytimes.com/interactive/2020/climate/trump-environment-rollbacks-list.html.

Proclamation No. 9476, 3 C.F.R. 9476 (2016).

Proclamation No. 9681, 3 C.F.R. 9681 (2017).

Proclamation No. 9682, 3 C.F.R. 9682 (2017).

Proclamation No. 10285, 3 C.F.R. 10285 (2021).

Proclamation No. 10286, 3 C.F.R. 10286 (2021).

Proclamation No. 10287, 3 C.F.R. 10287 (2021).

Restoration Project. 2021a. Reversing the Environmental Harms of the Trump Administration: Part I. *The Restoration Project*. https://rproject.world/public/RestorationProject-PartOne.pdf.

Restoration Project. 2021b. Reversing the Environmental Harms of the Trump Administration: Part II. *The Restoration Project*. https://rproject.world/public/RestorationProject-PartTwo.pdf.

Restoration Project. 2021c. Part III: The 100 Environmental Harms in Priority Order with Updates. *The Restoration Project*. https://rproject.world/part-iii.

Thoreau, Henry David. 1906. *The Writings of Henry David Thoreau, Journals*, 14 vols., eds. Bradford Torrey and Francis H. Allen. Boston, MA and New York, NY: Houghton Mifflin and Company.

Thoreau, Henry David. 2001a [1849]. *A Week on the Concord and Merrimack Rivers*. Mineola, NY: Dover Publications, Inc.

Thoreau, Henry David. 2001b. *Collected Essays and Poems*, ed. Elizabeth Hall Witherell. New York, NY: The Library of America.

Thoreau, Henry David. 2004a [1854]. *Walden*, 150th Anniversary Edition, ed. J. Lyndon Shanley. Princeton, NJ and Oxford: Princeton University Press.

Thoreau, Henry David. 2004b [1864]. *The Maine Woods*, ed. Joseph J. Moldenhauer. Princeton, NJ and Oxford: Princeton University Press.

Trimble, Stephen, ed. 2016. *Red Rock Testimony: Three Generations of Writers Speak on Behalf of Utah's Public Lands*. Salt Lake City, UT: Torrey House Press.

Trimble, Stephen, ed. 2017. *Red Rock Stories: Three Generations of Writers Speak on Behalf of Utah's Public Lands*. Salt Lake City, UT: Torrey House Press.

Index

L. Smith, *Ecological Restoration and the U.S. Nature and Environmental Writing Tradition*, https://doi.org/10.1007/978-3-030-86148-3

The manufacturer's authorised representative in the EU is Springer Nature Customer Service Centre GmbH, Europaplatz 3, 69115 Heidelberg, Germany. If you have any concerns regarding our products, please contact ProductSafety@springernature.com

Printed and bound by CPI Group (UK) Ltd, Croydon, CR0 4YY

30/09/2026

02225703-0002